Introduction to Particle Physics

Introduction to Particle Physics

Roland Omnès
Faculté des Sciences d'Orsay

Translation:
G. Barton
University of Sussex

WILEY–INTERSCIENCE
a division of John Wiley & Sons Ltd
LONDON · NEW YORK · SYDNEY · TORONTO

First published under Roland Omnès
Introduction à l'étude des particules élémentaires by Ediscience S.A. 7 rue Buffon Paris 5e.

Dépôt légal: 1er trimestre 1970.

Library of Congress catalog card No. 75-172471

ISBN 0 471 65372 1

Set by Monophoto in Northern Ireland by The Universities Press, Belfast and printed by lithography in Great Britain by The Pitman Press, Bath.

Preface

That part of natural philosophy which we call the physics of high energies or of elementary particles is both very extensive and difficult to approach. In the last twenty years thousands of experimenters and theorists have, unceasingly, multiplied data, facts, and ideas. So many particles are now known that because of their very abundance one is increasingly reluctant to call them elementary. Though their properties are beginning to fall into some kind of order, it is not yet known whether they will soon be deducible from some unified theory, or whether we must be resigned to a long prospect of classifying the regularities without being able to explain them.

It is a difficult task to impart knowledge acquired in a field which changes so quickly and which is always growing. One should provide reliable foundations well confirmed by experiment, and sketch also the outlines of current research. Given an inclination to rigour, one could try to present well-constructed theories, with a risk of the student becoming sophisticated to a point where he can no longer confront the raw facts and the phenomenology. Contrariwise, one could remain on the level of experimental data, at the risk of drowning in them. Already, there is an increasingly noticeable tendency for particle physics to split into narrower styles, schools, and specialities. We must ensure that this fragmentation is not reflected from the outset in our teaching; accordingly, as soon as we can we must give an overall view of the problems and of the results that have been obtained so far.

The present book tries to answer both these contradictory requirements. In the first part, we introduce some basic concepts of experiment and theory, which allow one to define what one means by a particle, and how the essential characteristics of particles are determined. In some sense this part explains grammatical rules which are reliable and necessary, and here we try to be precise as well as elementary, insofar as these two qualities are compatible. The second part is a far less classical enterprise. It makes no attempt at a step-by-step explanation of the various disciplines constituting particle physics: electrodynamics, strong interactions, weak interactions, field theory, high energy phenomena, invariance groups, all of them topics which are currently treated in special and often bulky monographs. On the contrary, we have tried to make them accessible at the level one reaches by mastering the first part of this book. At this level it is unusual to offer so superficial an introduction; and we feel that we must offer a justification, if only to forestall the accusation that we are presenting ideas in a way that is incomplete and useless for research. In fact we believe it good for the student to be offered a preliminary panoramic survey of what he will encounter later on, calculated to arouse his interest and his curiosity. In other words, as a sequel to the grammar in part I we offer a guide to the literature. The drawbacks of this approach are complementary to those of the more usual specialised and systematic treatments, which concentrate primarily on certain selected portions of the material.

The book is intended primarily for first-year postgraduate students. The prerequisite is a postgraduate course in quantum mechanics.

Because of the way it is written, the second part aims, also, to fill a frequently lamented gap by giving physicists specialising in other fields an overall survey of the most important ideas and researches that are current in particle physics.

I am grateful to Georges Valladas and to the Information Service of the Lawrence Radiation Laboratory for their help in collecting the illustrations.

ROLAND OMNÈS
Orsay, May 1, 1969

Preface to the English Edition

The present book is a first introduction to elementary particle physics. Its level is elementary and it can be used as a textbook after a one-semester course in quantum mechanics.

It is divided into two parts: The first one is devoted to the basic notions and the second one is a survey of elementary particle physics which may be found useful to physicists working in other fields.

In the first part, some of the material is standard. I have made an effort towards pedagogy in some chapters, such as an introduction to group theory and to the theory of collisions which may be found to be an easy first introduction. There are two chapters which differ markedly from previous treatments. One contains a discussion of relativistic kinematics using the Poincaré group. This approach, which is due to Wigner, is by far the most illuminating and, in my opinion, simpler than conventional treatments of relativistic particles with spin. I tried to show this simplicity in chapter 4. In chapter 10, the Dirac equation is derived from this point of view: although this discussion will probably be found instructive by a serious student, it can be skipped in a first reading.

The aim of the second part is to provide a first glimpse of the whole field of particle physics. The subject is becoming increasingly wide and specialized and I thought that such a general survey would help a beginner to understand what is already known and what problems are being investigated. The price to pay for the broader scope of this second part is a corresponding disregard for rigour, but references are given to more technical sources.

Another uncommon feature of the book is a short historical notice indicating the main discoveries in the field. I feel that an apology is needed here: I found that such a notice was a useful pedagogical aid in giving more life and reality to the subject and this is why it was included. On the other hand, it is not at all the outcome of a serious historical analysis and it suffers obviously from a personal bias.

The book was written recently in French and not enough time has elapsed since its completion to justify a significant revision of the text for the English edition except of course for the correction of some errors.

I am most grateful to Dr. G. Barton who not only translated the book excellently, but also made very useful suggestions.

ROLAND OMNÈS
Paris, April 1971

Contents

Part I

CHAPTER 1

THE ELEMENTARY PARTICLES: A PRELIMINARY SURVEY

In this chapter we suggest a preliminary definition of particles as entities which obey the laws of quantum mechanics for point objects endowed with sharply defined mass and charge. We show how particles are created and observed, and are thus led to an outline account of accelerators and detectors (counters, bubble chambers etc...). Since several detection methods depend on the ionisation of matter by charged particles, a simplified theory of ionisation is given in passing.

Finally we give a rapid sketch of the experimental facts and show that interactions (and, through them, the particles), can be classified into three general categories: the strong, the electromagnetic, and the weak interactions.

1. Preliminary definition of a particle

In physics, the word *particle* is generally applied to an object whose spatial extension is not one of its essential properties, and which in first approximation can be considered as a point. By thus denoting say the electrons issuing from a Crooke's tube, one implies that their motion through an electric or a magnetic field is well described by the laws of mechanics for a point object. If one measures their mass or their electric charge, one finds that these quantities are sharply defined.

The ionised atoms observed in a mass spectrograph have a similarly well determined electric charge and mass. In fact we know much more than this about them: we know that in many situations they appear to all intents and purposes to consist of a nucleus surrounded by electrons whose motion obeys the laws of quantum mechanics. The attraction between the nucleus and the electrons is due to the Coulomb force with which we are well acquainted and which, surprisingly enough, remains applicable in these submicroscopic systems. In this way we can explain the spectra of atoms, their chemical properties, and the structure of molecules, more or less quantitatively, and sometimes with great precision. Closer to our everyday macroscopic scale, we can describe and often calculate the characteristics of solids and of gases, such as their transport properties (electric and thermal conductivities), or their magnetic, dielectric, optical, elastic, and thermal properties; so much so that we can claim to have a clear mental picture of atoms, and an accurate description of them.

What do we learn from this description? First of all, that the atom does in fact have an extension in space. This extension is not sharply defined because it is governed by quantum mechanics. However, we know perfectly well in what sense one can legitimately say that the hydrogen atom has a radius of the order of 10^{-8} cm, and that its nucleus, the proton, is much smaller.

How does the electron look to us by comparison with the atom? We know neither its radius, even supposing it has one, nor its structure, supposing this expression is meaningful. Nevertheless we accelerate and observe the movements of electrons and of ionised atoms by techniques which in all essential respects are very alike. Moreover we always accept the measured values of its mass and of its charge as the signature of an individual atom, for in practice this is the only way we have of recognising atoms one by one. All other methods are statistical and serve only to indicate the presence of a greater or lesser number of atoms of a given kind, without specifying the characteristics of any one particular atom; examples being the analysis of spectral lines and methods of chemical separation.

Summarising, atoms and electrons share the feature that in the apparatus used for producing and detecting them, i.e. practically speaking in electric or magnetic fields, they move essentially like point objects of sharply defined mass and charge. Thus we would be well advised to use this property as the definition of a particle, especially since it is a conveniently operational one and provides us with a simple rule for identifying particles. It has the disadvantage of not stressing that the atom actually consists of a nucleus and of electrons, but who knows whether this really is always a disadvantage? This we can decide only in the course of future work and we now make the following definition:

Definition:

A particle is an object which in a macroscopic electric or magnetic field, or more generally in any experimental layout, behaves like a point having well-defined mass and charge.

We shall avoid calling particles "elementary" or "fundamental", because to do so would imply that we can distinguish those which are not elementary from others that are more so. Since this is not the case, we shall avoid burdening our language with adjectives that are useless and possibly insiduous.

Our definition gains in significance from the fact that the observed masses and charges are discrete: for instance, the electron has a mass of 0.510976 MeV, and we never see particles with masses 1.414 or 0.513 times that of the electron; (or if we do, we know how to repair our apparatus). As for atomic charges, they are always an exact multiple of that of the electron, to a very high accuracy (see Problem 1). In other words, our definition identifies the particles well in practice, at least as regards atomic particles. One should note that strictly speaking all the excited states of an atom have different masses, and are therefore different particles by virtue of our definition. This consequence we accept; what we consider important is that these masses should be well defined. In conclusion, we consider the charges and the masses of particles as their signature, and we shall assign different names or symbols to particles whose masses or charges differ.

This definition of a particle is meaningful only if we specify the laws of mechanics according to which it behaves "like" a point object. In many kinds of measuring apparatus it is enough to appeal to classical mechanics. But often one needs quantum mechanics, as for instance in the Stern–Gerlach experiment which features the atomic spin. Since quantum mechanics includes classical mechanics, we adopt the following hypothesis, which though natural is essential:

Hypothesis:

We shall assume that the properties of particles can be studied according to the rules of quantum mechanics.

2. Production of particles

Our definition of particles appeals explicitly to the experimental methods of observation whereby we can study them. Hence it will be useful to consider more closely how they can be produced and observed in practice.

The principal sources of particles are the following:

Radioactive sources;
cosmic rays;
accelerators;
secondary beams;
collisions between particles;
the decay of other particles.

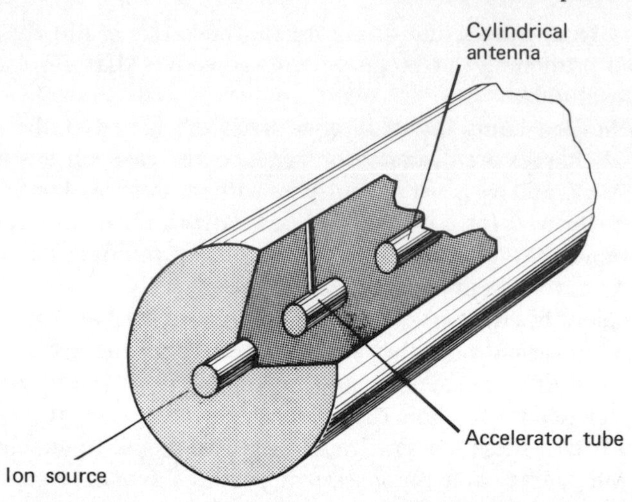

FIG. 1.2.1. Sketch plan of a linear accelerator. Particles emerging from an ion source pass through a series of tubes which form cavities containing an oscillating electric field. The particles accelerate if they enter a tube at a time when the electric field is parallel to their velocity. The cylindrical antennae are for focussing; they set up a field which prevents too great a spreading of the particles. The entire system is evacuated. There are definite phase relations between the antennae and the tubes.

(a) *Radioactive sources*

Radioactive sources produce mainly photons, electrons, positrons, protons, neutrons and α particles (which are helium nuclei). These particles are always produced with low energies (a few MeV at most).*

(b) *Cosmic rays*

Basically these are protons of high energy (up to 10^9 MeV). Some, of relatively lower energy, come from the sun. The most energetic ones probably originate in the galaxy.

(c) *Accelerators*

These are by far the most important sources of particles, and we shall therefore review the principles behind the most important types.

(c1) *Linear accelerators*

In their simplest form one is speaking here of machines which accelerate a charged particle by an electric field according to the law

$$\frac{d\mathbf{p}}{dt} = e\mathbf{E} \tag{2.1}$$

where e is the charge of the particle, $\mathbf{p}$ its momentum, and $\mathbf{E}$ the electric field.

* Recall that one MeV is the energy of a charge equal to an electronic charge at an electrostatic potential of one million volts. A GeV equals one thousand MeV.

In the most modern versions, a system of magnetic and electric lenses confines the electrons to a straight line trajectory while they are accelerated by an alternating electric field (figure 1.2.1). At present electrons can be accelerated up to 20 GeV, as in the accelerator at Stanford. At these energies the electron is ultra-relativistic, momentum and speed being related by

$$\mathbf{p} = m\mathbf{v}\left(1 - \frac{v^2}{c^2}\right)^{-1/2} \tag{2.2}$$

where m is the particle mass and c the speed of light.

It is noteworthy that the correct functioning of these machines provides the most detailed known verification of the dynamical laws of special relativity.

(c2) *Cyclotron*

In the cyclotron, the particles being accelerated do not follow a linear trajectory, but are confined by means of a magnetic field (figure 1.2.2).

Consider the motion of a charged particle in a constant magnetic field **B**. One has the equation of motion

$$\frac{d\mathbf{p}}{dt} = e\mathbf{v} \wedge \mathbf{B} \tag{2.3}$$

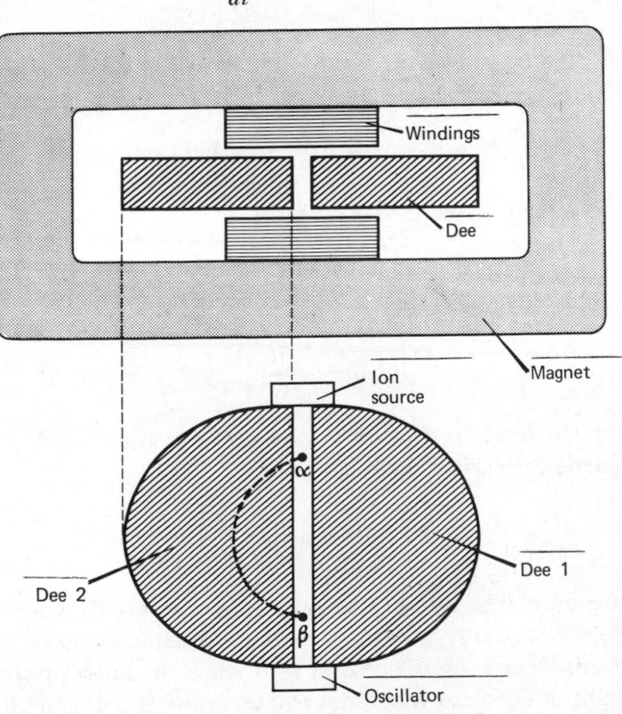

FIG, 1.2.2. Sketch plan of a cyclotron.

where the right hand side is the Lorentz force acting on a particle moving with the velocity $\mathbf{v}$. Since $\mathbf{p}$ and $\mathbf{v}$ are collinear, we can take the scalar product of this equation first with $\mathbf{p}$ and then with $\mathbf{B}$, obtaining

$$\mathbf{p} \cdot \frac{d\mathbf{p}}{dt} = \frac{1}{2} \frac{d\mathbf{p}^2}{dt} = 0 \tag{2.4}$$

$$\frac{d\mathbf{p}}{dt} \cdot \mathbf{B} = 0. \tag{2.5}$$

In the light of (2.2), the first of these equations shows that the squared velocity $\mathbf{v}^2$ is constant.

Equation (2.5) shows that, if the particle is injected either with zero velocity, or perpendicularly to the field, then one always has $\mathbf{p} \cdot \mathbf{B} = 0$. Hence the particle always remains in the same plane perpendicular to $\mathbf{B}$. In this plane we introduce two fixed orthonormal vectors $\mathbf{e}_1$ and $\mathbf{e}_2$, and write

$$\mathbf{v} = v_1 \mathbf{e}_1 + v_2 \mathbf{e}_2 . \tag{2.6}$$

It is convenient to use a complex notation and, in order to integrate equation (2.3), to introduce the complex number

$$v = v_1 + i v_2 . \tag{2.7}$$

The vector product $\mathbf{v} \wedge \mathbf{B}$ is normal to $\mathbf{v}$, so that in this notation equation (3) becomes

$$m\gamma \frac{dv}{dt} = ievB \qquad \text{with} \qquad \gamma = (1 - | v^2 | / c^2)^{-1/2}. \tag{2.8}$$

The solution of equation (2.8) is given by

$$v = v_0 e^{i\omega t} \tag{2.9}$$

where we have introduced a frequency ω:

$$\omega = \frac{eB}{m\gamma} . \tag{2.10}$$

Integrating the relation $dr/dt = v$, one gets $r = (r_0/\omega)\, ie^{i\omega t}$. This means that the particle describes a circle of radius

$$R = \frac{m v_0 \gamma}{eB} \tag{2.11}$$

with frequency ω. The sign of the radius of curvature depends on that of the charge. If the charge is known, then in principle a measurement of R determines the speed. Note however that when the latter approaches the speed of light, R becomes very large and therefore it is difficult to measure accurately.

The principle on which the cyclotron works depends on these properties. It is, to confine the particle to a limited region of space by means of a magnetic field, accelerating it meanwhile by an electric field. In practice the accelerator consists of two flat half-cylinders separated by a small distance δ.

A magnetic field **B** is applied parallel to the axis of the machine. An alternating potential difference $2Ve^{i\omega t}$ is applied between the two half-cylinders, in such a way that the half-cylinder (1) is at a potential $-Ve^{i\omega t}$ when the other half (2) is at a potential $Ve^{i\omega t}$. If at this moment a particle passes through the point α shown in the figure, it experiences an electric field $2Ve^{i\omega t}\delta^{-1}$ during a period of time $t = \delta/v$. According to equation (2.1), its momentum is therefore increased by

$$\Delta p = eE\Delta t = 2eVv^{-1} \tag{2.12}$$

Then it enters the half-cylinder (1), inside of which it is not exposed to any electric field, and describes a circle. Because the frequency of the alternating voltage equals the rotational frequency of the particle in the magnetic field, the electric field has changed sign by the time the particle arrives at the point β between the cylinders, and again accelerates it by an amount Δp. The particle increases its momentum by a total of $2\Delta p$ in each revolution.

As the speed grows, γ grows; according to equation (2.10) ω diminishes and the radius of the orbit grows, so that the particle describes a spiral. In order to continue the acceleration, the frequency of the applied potential must be varied if γ differs appreciably from 1. In that case the machine is called a *synchrocyclotron*.

In the synchrocyclotron one uses a series of magnets whose field B forces the particles into a circular orbit which is *fixed*. The particles are accelerated by an electric field which oscillates at the same frequency with which they rotate. This field is produced in resonant cavities interspaced between the magnets. The magnetic field as well as the frequency must be varied synchronously in order to keep constant the radius of the orbit. (Figure 1.2.3).

At present, the most powerful machines of this kind accelerate protons, and are located at CERN (Centre Européen de Recherches Nucléaires, European Centre for Nuclear Research) at Geneva, and at Brookhaven (U.S.A.). They produce protons whose energy can reach 25 GeV. A more powerful machine designed to reach 70 GeV is under construction at Serpukhov (U.S.S.R.). Still more powerful machines are under study.*

* Translator's note: (1970): the Serpukhov machine is now in operation; a 300 GeV machine is under construction near Chicago, and it appears possible that it will actually reach 500 GeV.

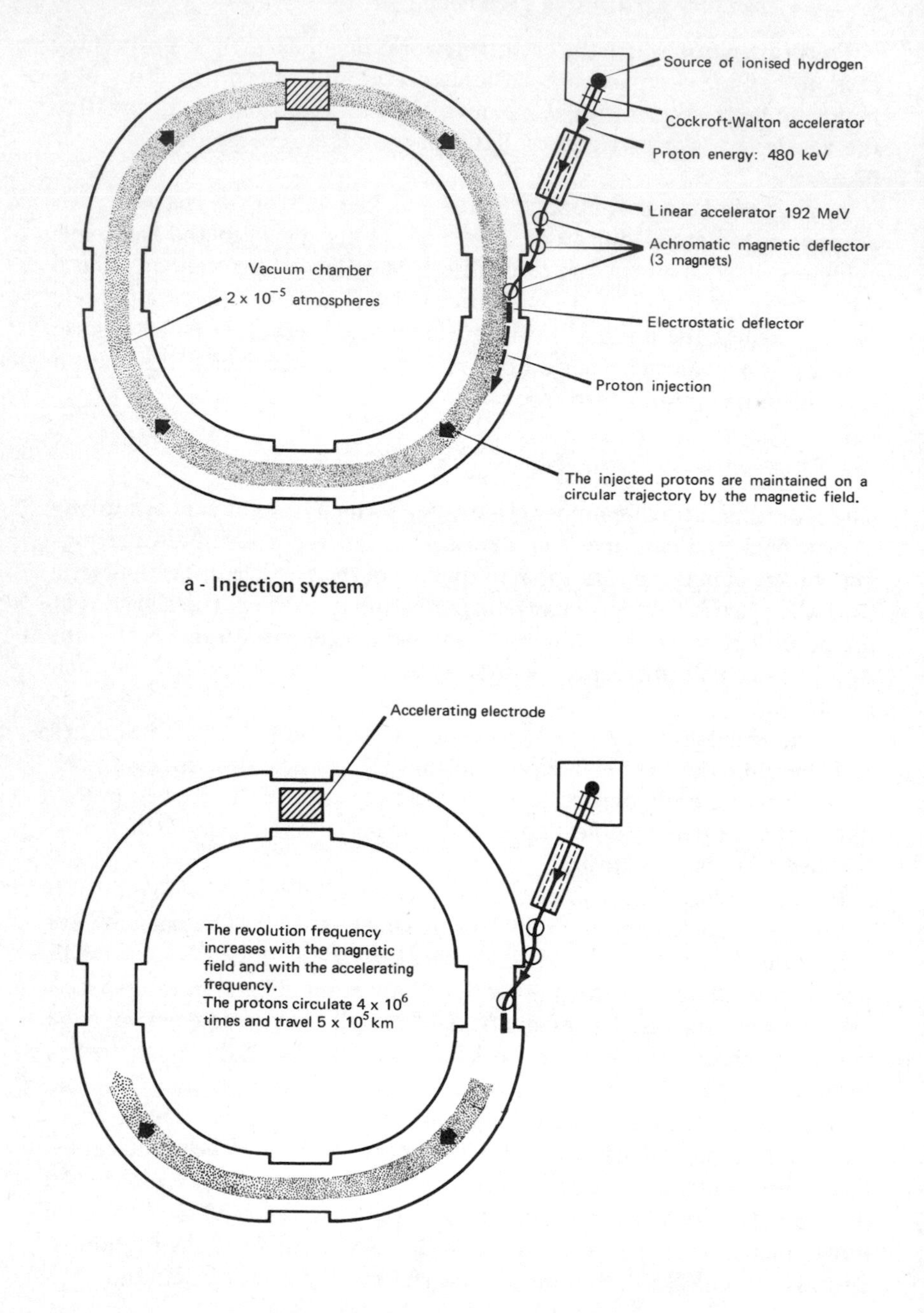

a - Injection system

b - Acceleration

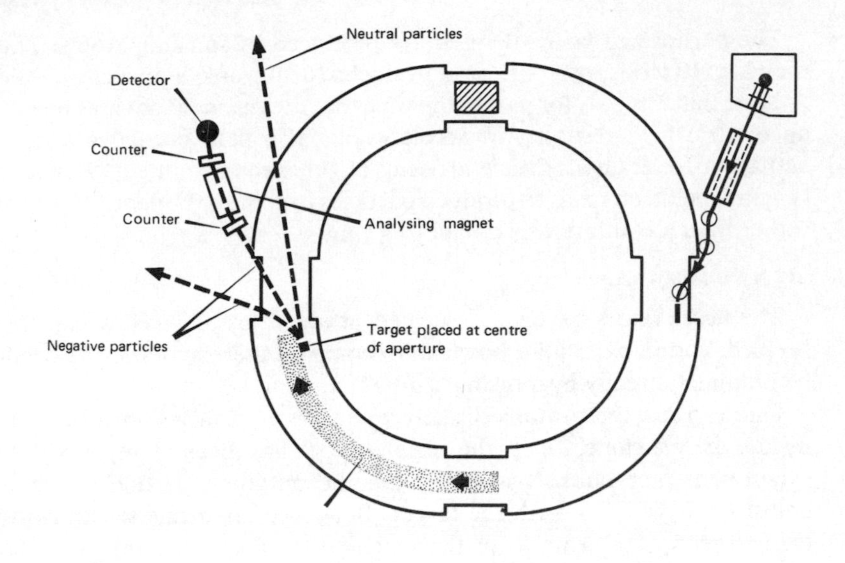

c - Experiment with internal beam

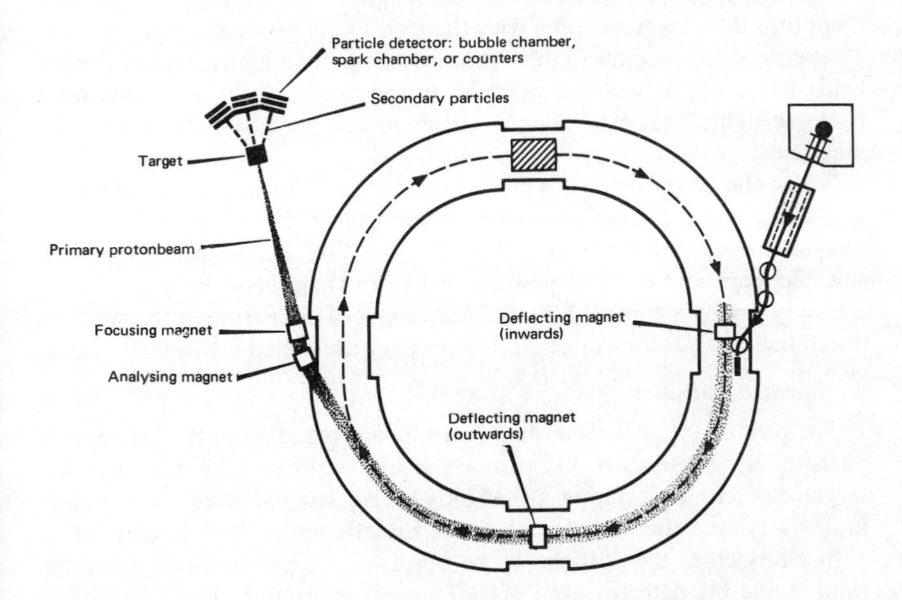

d - Experiment with external beam

FIG. 1.2.3. Sketch plan of a synchroton (the Bevatron at Berkeley).

The particles to be accelerated are produced by ionising atoms. Thus, they are electrons, protons, or light nuclei (deuterons, α particles).

Note that it is only for part of the time that the resonant cavities produce an electric field resulting in acceleration. The field oscillates and can actually decelerate a particle arriving at the wrong moment. Hence the synchrocyclotron tends to produce particles in groups: bunches of protons rather than a continuously circulating ring.

(d) *Secondary beams*

The accelerators we have described produce protons. How can these be used, and in particular, how can one make other particles that cannot be obtained directly by ionising atoms?

One can use the protons directly by inserting a target into the beam inside the machine, once the acceleration has been completed. The experiment then consists in observing the outgoing particles from the collision of the beam with the target. But more often this set-up is used for extracting the beam from the machine, or for producing secondary beams.

To extract the beam, one relies on the fact that the particles are slowed down in the target by ionisation. Afterwards they are no longer in phase with the accelerating cavities, and the magnetic field ceases to constrain them into their former orbits, with the result that they leave the machine. They can be focussed into a new beam by the aid of magnets and of electric fields which act on charged particles in the same way as lenses act on a beam of light. This new beam is easier to use because it is outside the machine.

When the beam collides with a target placed inside the accelerator, ionisation is not the only process taking place; there occur also many reactions which sometimes result in the creation of new particles. With suitable beam-optics these new particles can themselves be focussed. In this way one can make beams of π mesons, of K mesons, etc., with an initial beam of protons or of electrons. They are called secondary beams.

(e) *Particle collisions*

The production of a secondary beam is one example of the creation of particles in collisions of the primary beam particles with nuclei in the target. Sometimes it happens that only a very few particles can be made in this way, and that they are observed directly on leaving the target.*

In many cases the particles to be observed are produced in collisions right inside the detector itself. This happens in bubble chambers and in the spark chambers that we shall consider later on.

* In practice, a target is a piece of metal or of polyethylene (plastic) or liquid hydrogen contained in a reservoir, etc. We shall see later that from the point of view which interests us it is essentially an assemblage of nuclei.

(f) *Particle decays*

The π^+ mesons (positive pions) in a secondary beam decay in 10^{-8} seconds, producing leptons called μ^+ (positive muons). Through this decay a beam of π^+ becomes a beam of μ^+ which can then be stripped of its remaining pions by sending it through a thickness of material, since pions interact with matter much more strongly than do muons, and are absorbed much more quickly. This is but one example of a procedure that is often used, particularly for making beams of muons or of neutrinos.

3. Detection of particles: ionisation

Most detectors depend on the ionisation of matter by charged particles in motion. This is a fundamental phenomenon, and in order to exhibit its most salient features we shall given an extremely elementary theory for it.

Consider a charged particle moving through matter with velocity v, having charge q, and a mass M which we assume much larger than the electron mass m. It gives rise to an electric field $\mathbf{E}$ capable of accelerating an electron attached to an atom at a distance b from the trajectory; (b is commonly known as the impact parameter). In fact this field is appreciable only for a period of time of the order of b/v during which the particle is not too far from the electron. If this collision time b/v is much smaller than the revolution period of the electron in the atom, then one can neglect the fact that the electron is bound and treat it as a free particle. Under these conditions the effect of the field $\mathbf{E}$ is to impart to the electron a momentum $\mathbf{p}$ which we aim to calculate for the case where the particle moves nonrelativistically.

Note first of all that the momentum imparted to the electron is small, and that the electron moves little during the collision time. Hence by neglecting this displacement one can write

$$\mathbf{p} = \int_{-\infty}^{+\infty} e\mathbf{E}(t)\, dt \tag{3.1}$$

where $\mathbf{E}(t)$ is evaluated at the initial position of the electron, and where e is its charge. It is clear from symmetry that the component of $\mathbf{E}(t)$ parallel to the line of motion of the particle will not contribute to the integral (3.1). If we denote by $\mathbf{E}_\perp(t)$ the components of $\mathbf{E}(t)$ normal to the trajectory, then our task is to calculate

$$\mathbf{p} = \int_{-\infty}^{+\infty} \mathbf{E}_\perp(t)\, dt \tag{3.2}$$

The variation of $\mathbf{E}(t)$ with time is the same as one would observe if the particle remained at rest and if one measured the field at a point moving with velocity v along a line parallel to the (actual) trajectory. Hence it is convenient to consider a cylinder of radius b with its axis along the trajectory. Along this cylinder one has

$$\int_{-\infty}^{+\infty} \mathbf{E}_{\perp}(t)\, dt = \int_{-\infty}^{+\infty} E_{\perp}'(x)\, \frac{dx}{v} = \frac{1}{v} \int_{-\infty}^{+\infty} E'(x)\, dx \tag{3.3}$$

where $E_{\perp}'(x)$ is the component normal to the cylinder of the field due to a charge q fixed at a point on the axis. The integral (3.3) can be evaluated easily by appeal to Gauss' theorem, and yields

$$4\pi q = 2\pi b \int_{-\infty}^{+\infty} E_{\perp}'(x)\, dx. \tag{3.4}$$

From equations (3.2), (3.3), and (3.4) we obtain

$$p = \frac{2qe}{vb}. \tag{3.5}$$

Hence the energy acquired by the electron is given by

$$\frac{p^2}{2m} = \frac{2q^2e^2}{mv^2b^2}. \tag{3.6}$$

The electron evidently acquires this energy at the expense of the particle. While the latter moves unit distance through a medium with N electrons per unit volume, it loses an amount of energy

$$dE(b) = 2\pi b\, db N \frac{2q^2e^2}{mv^2b^2} = \frac{4\pi q^2e^2N}{mv^2b}\, db \tag{3.7}$$

to electrons having impact parameters between b and $b + db$. Hence the total energy loss of the particle over a distance dx is given by

$$\frac{dE}{dx} = -\frac{4\pi q^2e^2N}{mv^2} \operatorname{Log} \frac{b_{\max}}{b_{\min}} \tag{3.8}$$

where $b_{\max}$ and $b_{\min}$ are the limits on the impact parameter beyond which the above arguments cease to apply.

The limit $b_{\max}$ is reached when the collision time becomes comparable with the revolution period of the electron in the atom. When the Coulomb field varies slowly, the disturbance of the atom is effectively adiabatic and transfers no energy. Thus one has

$$\frac{b_{\max}}{v} \approx \frac{1}{\bar{\nu}} \tag{3.9}$$

$\bar{\nu}$ being an average atomic frequency. At this point one can introduce relativistic corrections. If v is relativistic, $E_{\perp}$ must be multiplied by $\gamma = (1 - v^2)^{-1/2}$ while t is divided by γ. The overall integral (3.2) remains unchanged. But the collision time occurring in equation (3.9) will be divided by γ, and one gets

$$b_{\text{max}} = \frac{\gamma v}{\bar{\nu}}. \tag{3.10}$$

Similarly, the theory ceases to apply when b becomes comparable to the de Broglie wavelength of the electron in the rest frame of the particle (which is practically the same as the centre of mass frame). In this frame the electron has velocity v, momentum $mv\gamma$, and wavelength $1/mv\gamma$, whence

$$b_{\text{min}} = \frac{1}{mv\gamma} \tag{3.11}$$

so that equation (3.8) becomes

$$\frac{dE}{dx} = -\frac{4\pi q^2 e^2 N}{mv^2} \operatorname{Log} \frac{mv^2\gamma^2}{\bar{\nu}}. \tag{3.12}$$

The form of this function is shown in figure (1.3.1).

At low energies (along BDC) one can neglect the variation of the logarithm, and gets essentially a $1/v^2$ law, v tending to c along the CD portion of the curve. At energies that are too low (along AB) the speed of the particle is comparable to that of the electrons, and there is hardly any ionisation. The portion CE of the curve shows the effect of the logarithm at ultra-relativistic energies. Note the existence of an energy where the ionisation has a minimum (at the point C).

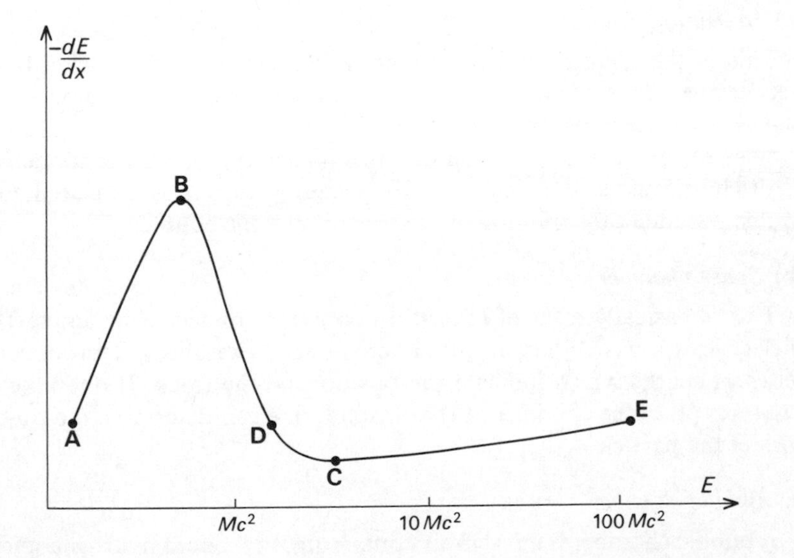

FIG. 1.3.1. Charged particle energy loss through ionisation as a function of the particle energy E.

Because of this energy loss, the particle slows down and travels only a distance

$$D = \int_0^{E_0} \frac{dE}{f(E)} \tag{3.13}$$

where

$$f(E) = \frac{mv^2}{4\pi q^2 e^2 N} \frac{1}{\log \dfrac{mv^2\gamma^2}{\bar{\nu}}} \tag{3.14}$$

which it is easy to integrate numerically. It follows that the range D of a given particle in matter is a well-defined function of its initial energy E_0. Hence a measurement of D yields the energy E_0. Similarly, if one knows E_0 and D one can determine the mass of the particle. With appropriate calibration, this property is the starting point of many techniques of measurement in particle physics.

It is found empirically that the number of ionised atoms is approximately proportional to dE/dx.

4. Detectors

We shall review some particle detectors which exploit the ionisation of matter by charged particles. There exist other types of detectors (Cerenkov counters, scintillation counters, photomultipliers) which we shall not consider.

(a) *Ionisation counter*

This is the Geiger-Müller counter. It consists of a dielectric placed between condenser plates whose form varies for different counters. A charged particle crossing the dielectric ionises it; if the potential difference is close to the breakdown potential, this ionisation allows the condenser to discharge across the dielectric. The resulting current is measured, and in this way one can count the particles crossing the counter.

(b) *Spark chamber*

This is basically a set of ionisation counters connected in series. The dielectric is a gas. When a particle crosses the chambers, it produces a series of sparks which indicate the position of the track. If one takes a photograph of the chamber at that instant, one can determine the trajectory of the particle.

(c) *Bubble chamber*

A bubble chamber is a reservoir containing, say, liquid hydrogen under pressure. On sudden decompression the liquid hydrogen attains a metastable state because it cannot boil if the chamber is very clean. (Figure 1.4.1).

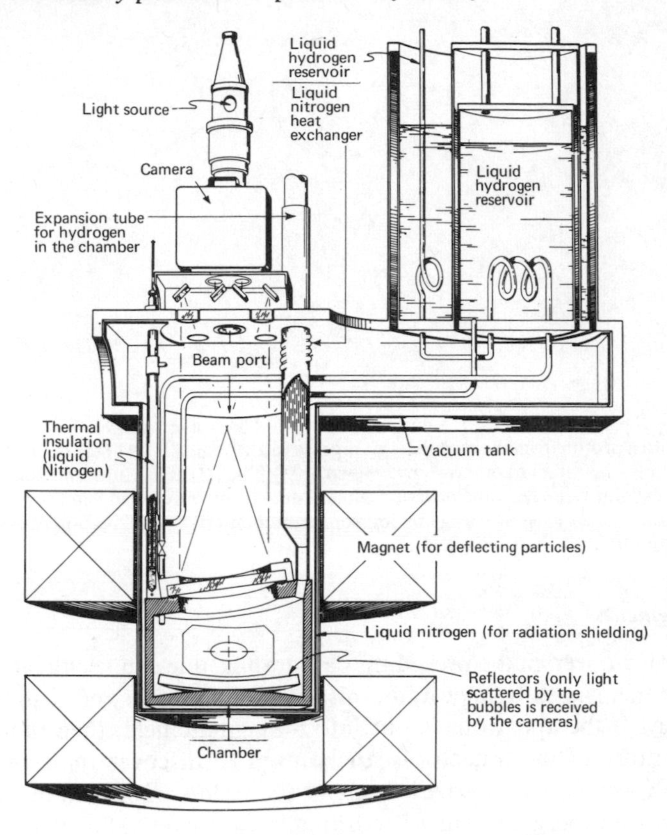

FIG. 1.4.1. Sketch plan of a bubble chamber.

Suppose a charged particle crosses the chamber at this instant; it ionises some of the atoms along its path. These ionised atoms then catalyse the formation of bubbles of gaseous hydrogen. If one photographs the chamber at that moment, one sees a line of bubbles which determine the particle trajectory very accurately. (Fig. 1.4.2.)

(d) *Wilson cloud chamber*

This works on the same principle as the bubble chamber, but uses alcohol vapour which has been suddenly compressed. The particle then catalyses the formation of liquid droplets along its trajectory.

(e) *Photographic emulsion*

Ionisation of atoms in an emulsion catalyses the usual photochemical reactions. One needs only to develop the picture in order to see the trajectory of the particle.

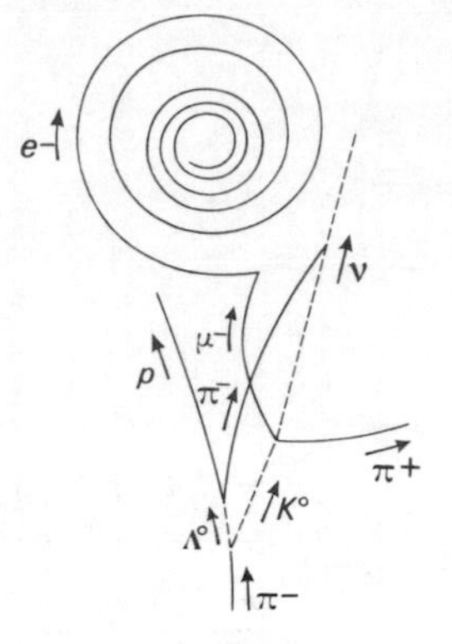

FIG. 1.4.2. What one sees in a bubble chamber. Here, a π^- arrives from below, and collides with a proton from the chamber, giving rise to a Λ^0 and a K^0. (These neutral particles are not visible; their tracks are shown by broken lines). The Λ^0 decays into a proton and a π^-. The K^0 decays into a π^+, μ^-, and neutrino; the μ^- decays into an electron and two neutrinos. The electron quickly loses its energy by ionisation, and describes a succession of circles with decreasing radii in the magnetic field.

Measurements

As we have seen in the preceding paragraphs, one can use the measurements of the particle's range and ionisation to identify it and to determine its energy. If the apparatus is put into a magnetic field, then the sign of the curvature of the trajectory can be used to discover the sign of the charge. In principle, a measurement of the radius of curvature can also determine the energy, because according to formula (3.1) one has

$$R = \frac{vE}{eB} \tag{4.1}$$

It is possible to measure accurately the angles between the trajectories of different particles.

Note finally that the energy of an initial particle is often known quite accurately (to better than 1 %) simply through the fact that it has passed through a particular arrangement of magnetic lenses.

Such measurements furnish the basic data about interparticle reactions occurring, say, in a bubble chamber.

Detection of neutral particles

The procedures outlined so far enable one to detect and to measure the momenta of charged particles. To detect and to make measurements on neutral particles is a more delicate task. Let us sketch some of the methods in current use.

(a) *Photons*

A very energetic photon can be detected through pair production, i.e. through the reaction

$$\gamma \rightarrow e^+ + e^- \tag{4.2}$$

where e^- is an electron and e^+ a positron. In actual fact the reaction as written in (4.2) is forbidden by energy and momentum conservation. However, if the photon passes through the electric field of a nucleus N, then one can have

$$\gamma + N \rightarrow e^+ + e^- + N'. \tag{4.3}$$

In chapter 10 we shall see that the probability for reaction (4.3) varies as Z^2, the squared charge of the nucleus. Hence the distance which the photon can travel without initiating the reaction (4.3) is inversely proportional to Z^2 (this is what one calls the *radiation length:* it signifies only the *average* distance travelled by a photon). Hence the photon detection efficiency will be higher the higher the Z value of the working material (bubble chambers containing iodine, freon, etc.).

Similarly one can deduce the energy distribution of a photon sample from a measurement of electron and positron energies, by exploiting in detail the known theory for reaction (4.3).

(b) *Energy-momentum balance*

When observing a reaction with only a single neutral participant, as for instance in

$$p + p \rightarrow p + p + \pi^0$$

where p is a proton and π^0 a certain neutral particle, the 'neutral pion', it is possible to measure the energies and momenta of all the charged particles, and to deduce that of the neutral particle which must balance them. Let E and $\mathbf{p}$ be the energy and momentum thus measured. Then the mass of the particle can be found from the relation

$$m^2 = E^2 - \mathbf{p}^2. \tag{4.4}$$

(c) *Decay*

If a neutral particle decays into charged ones, as happens for instance to the Λ^0 and K^0 particles:

$$\Lambda^0 \rightarrow p + \pi^-$$

$$K^0 \rightarrow \pi^+ + \pi^-$$

then one can measure the energies and momenta of the charged particles and deduce from them those of the neutral particle, to which they add up. Again, the mass is given by the formula (4.4).

It may be necessary to elaborate such methods: thus, π^0's are sometimes detected through their decay

$$\pi^0 \rightarrow \gamma + \gamma$$

the photons being detected in their turn through pair creation.

5. Experimental data

Let us now look at the information which these experimental methods provide.

(a) *Particles*

Since we know how to measure the mass and charge of any particle, we can draw up a catalogue listing every observed particle. One assigns a different name or symbol to particles differing in mass or charge.

This list is given in table (5.1). It includes only those particles which are neither atoms nor nuclei (except for the proton). It is these that by convention are called *elementary particles*.

(b) *Decays*

Experiment shows that some particles decay. For instance

$$\Lambda^0 \rightarrow p + \pi^-. \tag{5.1}$$

Since this is a quantum phenomenon with a well-defined probability for it to occur per unit time, the individual lifetimes in a sample of Λ^0's will be distributed exponentially. If we start at zero time with a sample of N Λ^0 particles, then $Ne^{-t/\tau}$ will still survive at time t; τ is called the mean life of the Λ^0. One determines τ from one's knowledge of the Λ^0 momenta by measuring the distance they travel before decaying, from which their lifetime follows.

It often happens that a particle can decay in several different ways. Thus a Λ^0 can equally well decay by the mode

$$\Lambda^0 \rightarrow n + \pi^0. \tag{5.2}$$

The frequency of a particular mode relative to the sum for all decay modes is called the branching ratio of the mode.

Table (5.1) gives also the particles' mean lives, decay modes, and branching ratios.

(c) *Reactions*

Observation shows that a collision between two particles can create other particles, as in the examples

$$p + p \rightarrow p + n + \pi^+$$

$$\pi^- + p \rightarrow \Lambda^0 + K^0$$

$$\pi^+ + p \rightarrow \Sigma^+ + K^+, \text{ etc...}$$

It is equally interesting to note that certain reactions never occur, for instance

$$p + p \rightarrow p + n$$

$$\pi^- + p \rightarrow \Sigma^+ + \pi^-$$

$$p + p \rightarrow p + \pi^+$$

$$\pi^- + p \rightarrow e^- + p, \text{ etc...}$$

It would take too long to list all reactions that do or do not occur. But later on we shall mention some of them which help one to discover some of the laws which they obey. This problem is rather similar to that met in chemistry when one wishes to establish the rules governing chemical reactions and combinations. Later on we shall encounter the equivalent for elementary particles to the concept of chemical valency.

(d) *Characteristics of particles*

We have assumed that a particle is characterised by its mass and its electric charge. Is this working hypothesis confirmed experimentally?

There has been discovered a particle, the antiproton ($\bar{p}$), which can be made in reactions of the type

$$p + p \rightarrow \bar{p} + p + p + p \tag{5.3}$$

by very high energy collisions between two protons. Within experimental error the antiproton has exactly the same mass as the proton, but its charge is negative and therefore different from that of the proton, in accordance with our definition of a new particle.

Proton and antiproton have very different properties. Thus, when two low energy protons collide, all that happens is that their momenta change. In other words, the only reaction possible under these conditions is

$$p + p \rightarrow p + p$$

By contrast, when a low energy antiproton meets a proton, they can 'annihilate' in reactions like

$$\bar{p} + p \rightarrow \pi^+ + \pi^- + \pi^0 + \pi^+ + \pi^-$$

$$\bar{p} + p \rightarrow \pi^+ + \pi^- + \pi^0 + \pi^0 \text{ etc...}$$

Another interesting reaction that can ensue from collisions between protons and antiprotons gives two neutral particles *each having the mass of*

TABLE 5.1

MASSES AND CHARGES OF PARTICLES

Name of the particle	Symbol	Mass (in MeV)	Charge (in units of the electron charge)	Mean life (seconds)	Observed decays	Branching ratio (%)
photon	γ	0	0	infinite		
neutron	ν	0	0	infinite		
electron	e (or e^-)	0,510976	-1	infinite		
positron	e (or e^+)	0,510976	$+1$	infinite		
negative muon	μ^-	105,65	-1	$2{,}2 \times 10^{-6}$	$e^- \nu\nu$	100
{positive muon	μ^+	105,65	$+1$	$2{,}2 \times 10^{-6}$	$e^+ \nu\nu$	100
{π^+ meson (or pion)	π^+	139,58	$+1$	$2{,}5 \times 10^{-8}$	$\mu^+ \nu$	$\sim$100
					$e^+ \nu$	10^{-2}
π^- meson	π^-	139,58	-1	$2{,}5 \times 10^{-8}$	$\mu^- \nu$	$\sim$100
					$e^- \nu$	10^{-2}
π^0 meson	π^0	134,97	0	10^{-16}	$\gamma\gamma$	98,8
					$e^+ e^- \gamma$	1,2
K^0 meson	K^0	497,9	0	two mean lives		
				$0{,}86 \times 10^{-10}$ }	$\pi^+ \pi^-$	66,5
					$\pi^0 \pi^0$	33,5
				$5{,}4 \times 10^{-8}$ }	$\pi^+ \pi^- \pi^0$	8,7
					$\pi^0 \pi^0 \pi^0$	3,8
					$\pi^+ e^- \nu$	} 28,3
					$\pi^- e^+ \nu$	
					$\pi^+ \mu^- \nu$	} 25,0
					$\pi^- \mu^+ \nu$	
K^+ meson	K^+	493,7	1	$1{,}2 \times 10^{-8}$	$\mu^+ \nu$	64,2
					$\pi^+ \pi^0$	18,6
					$\mu^+ \pi^0 \nu$	4,8

					$e^+ \pi^0 \nu$	5,0
					$\pi^+ \pi^+ \pi^-$	5,7
					$\pi^+ \pi^0 \pi^0$	1,7
K^- meson	K^-	493,7	−1	$1{,}2 \times 10^{-8}$	$\mu^- \nu$	64,2
					$\pi^- \pi^0$	18,6
					$\mu^- \pi^0 \nu$	4,8
					$e^- \pi^0 \nu$	5,0
					$\pi^- \pi^- \pi^+$	5,7
					$\pi^- \pi^0 \pi^0$	1,7
proton	p	938,21	1	infinite		
neutron	n	939,50	0	1000	$p\, e\, \nu$	100
antiproton	$\bar{p}$	938	−1	infinite		
antineutron	$\bar{n}$	939	0	?	not observed	
lambda zero	Λ^0	1115,4	0	$2{,}6 \times 10^{-10}$	$p\pi^-$	66
					$n\, \pi^0$	34
antilambda	$\bar{\Lambda}^0$	1115	0	?	$\bar{p}\, \pi^+$	
					$\bar{n}\, \pi^0$	
sigma +	Σ^+	1189,3	1	$0{,}78 \times 10^{-10}$	$p\, \pi^0$	50,7
					$n\, \pi^+$	49,3
sigma 0	Σ^0	1193,2	0	?	$\Lambda\gamma$	~100
sigma −	Σ^-	1197,6	−1	$1{,}59 \times 10^{-10}$	$n\, \pi^-$	~100
antisigma +	$\overline{\Sigma^+}$	1189	−1		$\bar{p}\, \pi^0$	
					$\bar{n}\, \pi^+$	
antisigma zero	$\overline{\Sigma^0}$	1193	0		$\bar{\Lambda}^0\, \gamma$	
antisigma −	$\overline{\Sigma^-}$	1189	+1		$\bar{n}\, \pi^+$	
xi zero	Ξ^0	1315	0	$2{,}8 \times 10^{-10}$	$\Lambda\, \pi^0$	~100
xi −	Ξ^-	1321,2	−1	$1{,}75 \times 10^{-10}$	$\Lambda\, \pi^-$	~100
antixi zero	$\overline{\Xi^0}$	1315	0		$\bar{\Lambda}\, \pi^0$	
antixi −	$\overline{\Xi^-}$	1321	1		$\bar{\Lambda}\, \pi^+$	
omega −	Ω^-	1680	−1		ΛK^-	
					$\Xi^0\pi^-$	

the neutron. One is then led to ask whether one has seen the reaction

$$\bar{p} + p \rightarrow n + n$$

Now this is certainly not the case, because if one of these two particles is an ordinary neutron, then the other behaves rather like an antiproton. Accordingly, if we call it an antineutron and denote it by $\bar{n}$, then we can observe reactions like

$$\bar{n} + p \rightarrow \pi^+ + \pi^- + \pi^+ + \pi^0 \text{ etc...}$$

so that the reaction we saw initially was in fact

$$\bar{p} + p \rightarrow \bar{n} + n.$$

The neutron and the antineutron have the same mass and the same charge (zero) but are nevertheless quite distinct, and observation reveals several other like cases. This will evidently lead us to look for characteristics other than mass and charge which we can attribute to a particle in order to identify it unambiguously.

(e) *Cross sections*

More than the mere fact of its occurrence can be learnt from a reaction like

$$\pi^- + p \rightarrow \Sigma^+ + K^- \tag{5.4}$$

$$\pi^+ + p \rightarrow \pi^+ + p \tag{5.5}$$

$$p + p \rightarrow p + n + \pi^+. \tag{5.6}$$

By observing a sufficiently large number of reactions of a given kind, one can also measure the probability that it should occur and that the particles in the final state should have certain definite momenta.

In order to exploit as thoroughly as possible the information yielded by such measurements, it is desirable to display these probabilities in the most easily interpretable form. In practice, the experiment consists in directing onto a target a beam of particles of known energy E and velocity v. The number of particles in the beam is specified by the flux Φ, i.e. by the number of particles per second crossing an area of one cm^2 normal to the beam direction. On the other hand, the target contains N_0 nuclei capable of participating in the reaction under study (N_0 protons in reactions (5.4), (5.5), (5.6)). We observe, per second, n events of a predetermined kind (for example, events of the kind (5.4)).

Evidently n is proportional to N_0 and to Φ, and the probability of observing one event of the given kind when there is one proton in the target is

$$P = \frac{n}{N_0}. \tag{5.7}$$

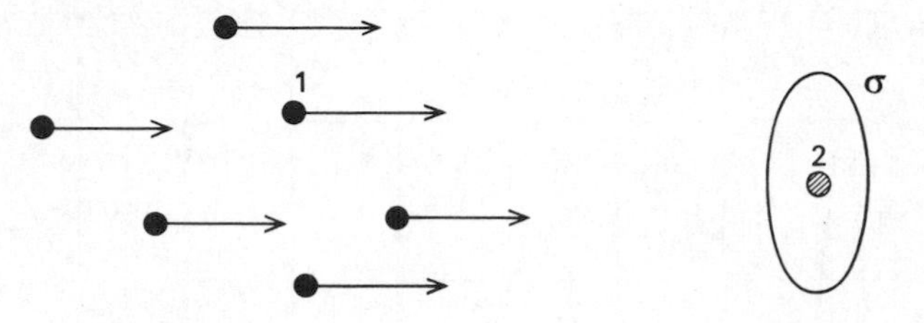

FIG. 1.5.1. The cross-section presented by particle 2 to the particle 1 arriving in the beam can be represented as a surface element of area σ, normal to the beam direction.

This probability is itself proportional to Φ. A parameter independent of the beam intensity is defined by the ratio

$$\sigma = \frac{P}{\Phi} \tag{5.8}$$

The parameter σ depends solely on the individual collision between one beam particle and one target nucleus. It has the physical dimensions of an area, and is called a cross-section.

A cross-section can be pictured very simply: imagine that each target proton presents to the beam a shield of area σ. Then the number of events is equal to the number of beam particles impinging on the shields. (Figure 1.5.1).

In general, the results of an experiment are given in the form of cross-sections as functions of the energy of the beam particles.

The unit for measuring cross-sections is the barn (10^{-24} cm), with its subunits the millibarn (*mb*) and the microbarn (*μb*).

Such cross-section curves can look very remarkable, as in the example of figure (1.5.2). This shows the *total cross-section* for $\pi^+ p$ collisions, i.e. the sum of the cross-sections for all processes initiated by the collision of a π^+ and a proton at fixed energy. One notices peaks and dips which are evidently quite remarkable dynamical phenomena that need to be understood and interpreted.

Experimental results can be displayed in greater detail through *differential cross-sections*. Consider for instance the reaction

$$\pi^- + p \rightarrow \pi^- + p. \tag{5.9}$$

There exists a well defined Galilean reference frame, called the centre of mass, frame, where the momenta of the initial π^- and proton are equal and opposite, say $\mathbf{q}$ for the π^- and $-\mathbf{q}$ for the proton. In this frame their energies are

$$E_\pi \equiv \omega = (\mu^2 + \mathbf{q}^2)^{1/2} \qquad E_p \equiv E = (m^2 + \mathbf{q}^2)^{1/2} \tag{5.10}$$

where μ and m are the π^- and the proton masses respectively (Figure 1.5.3).

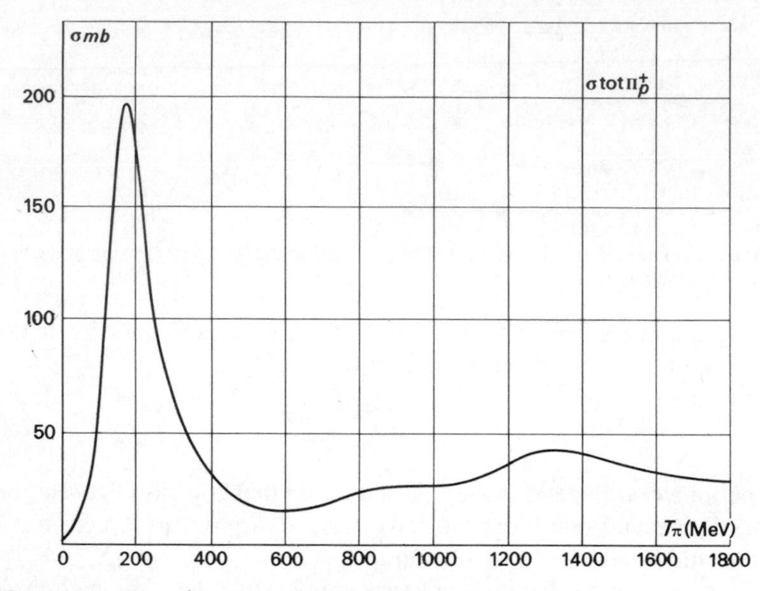

FIG. 1.5.2. Total π^+p cross-section in millibarns, shown as a function of the energy in the laboratory frame.

If momentum is conserved in the reaction, then in the centre of mass frame the total momentum of the π^-p system must still be zero after the collision. In other words the momenta of the π^- and the proton will be equal and opposite, say $\mathbf{q}'$ for the π^- and $-\mathbf{q}'$ for the proton. Their energies are

$$\omega' = (\mu^2 + \mathbf{q}'^2)^{1/2} \qquad E' = (m^2 + \mathbf{q}'^2)^{1/2} \tag{5.11}$$

If energy is conserved, one has

$$\omega + E = \omega' + E' \tag{5.12}$$

whence

$$|\,\mathbf{q}\,| = |\,\mathbf{q}'\,|. \tag{5.13}$$

It follows that $\mathbf{q}$ can change only in direction but not in magnitude. Hence at a fixed energy the process can be specified completely by giving the number of π^-'s which, in the centre of mass frame, emerge from the collision into a solid angle $d\Omega$ about a given direction (θ, φ); (θ is the angle between $\mathbf{q}$ and $\mathbf{q}'$, and φ an azimuthal angle). With dn the number of such events, we write

$$d\sigma = \frac{dn}{N_0\Phi}, \tag{5.14}$$

Evidently $d\sigma$ is proportional to $d\Omega$, and the individual collision process is fully characterised by $d\sigma/d\Omega$, which is called the differential cross-section. (It is measured in barn per steradian). For reaction (5.9) it is shown in

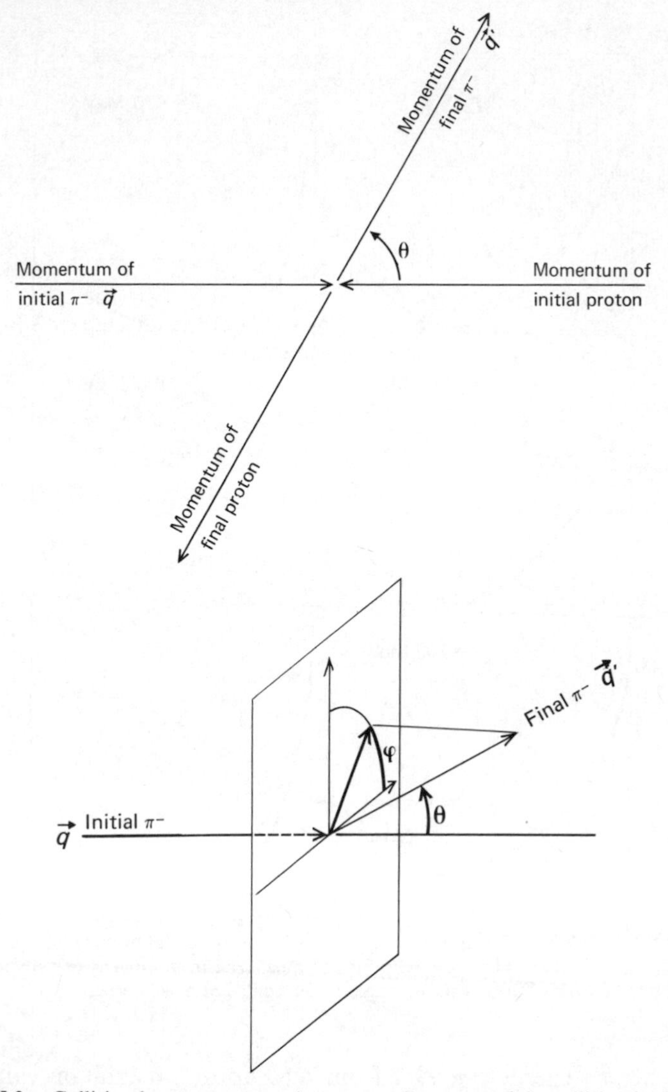

FIG. 1.5.3. Collision between a π^- and a proton. Geometry of the collision in the centre of mass frame.

figure (1.5.4) for a π^- energy of the order of 1 GeV. (It is found that in general the cross-section is independent of φ, due as we shall see later to invariance under rotations).

In reactions like (5.6) where there are three particles in the final state, their individual momenta in the centre of mass frame are not uniquely determined. Thus one can give the number of events per unit neutron kinetic energy, as for instance in figure (1.5.5).

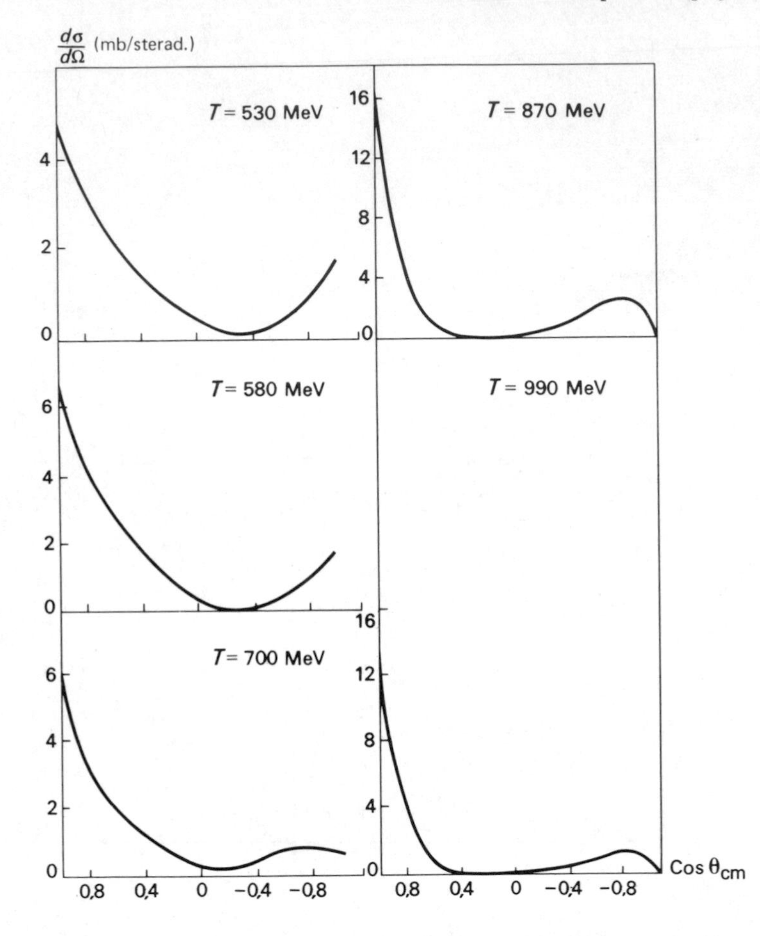

FIG. 1.5.4. Differential $\pi^- p$ scattering cross-section for π^- laboratory energies between 530 MeV and 990 MeV. The cross-sections are measured in millibarns per steradian. The abscissa is the cosine of the scattering angle in the centre of mass frame.

This kinetic energy can vary from zero up to a maximum value E_{max}. Such a curve is called an energy spectrum of the neutron (by analogy with a frequency spectrum in optics, frequency corresponding to energy).

Conclusions

The study of elementary particles subdivides into two extensive fields of activity:

1. An experimental search resulting in a census of particles and of reactions that can happen, and in a quantitative description of their

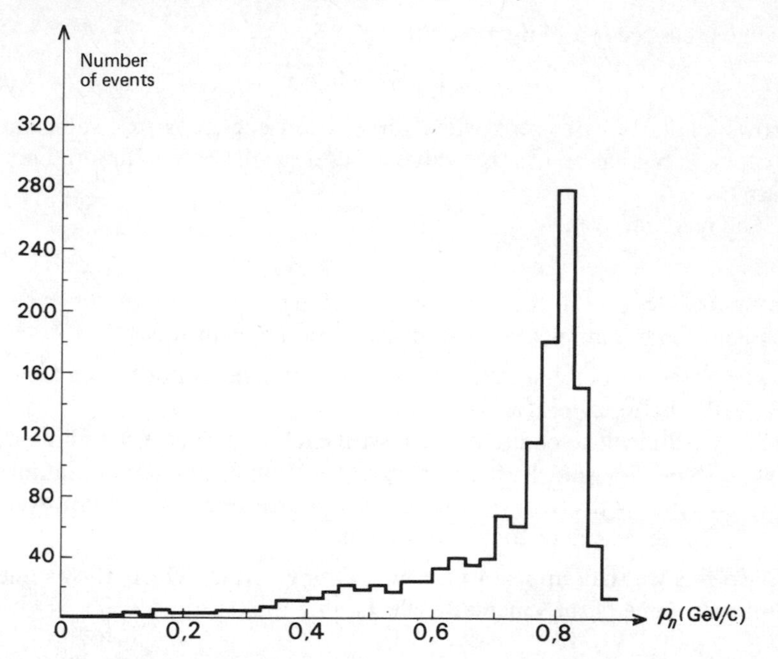

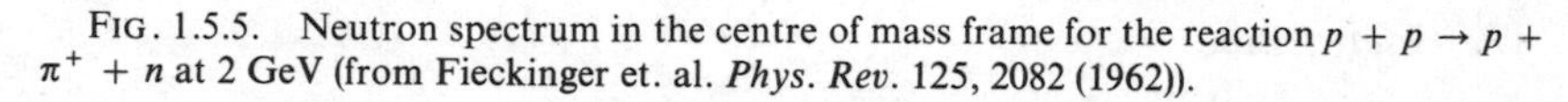
FIG. 1.5.5. Neutron spectrum in the centre of mass frame for the reaction $p + p \to p + \pi^+ + n$ at 2 GeV (from Fieckinger et. al. *Phys. Rev.* 125, 2082 (1962)).

characteristics: mean lives, branching ratios, cross-sections, and spectra.

2. A theoretical interpretation of these results.

The theoretical approach can be subdivided into two parts:

(a) Study of qualitative regularities (e.g. of reactions that do or do not happen).

(b) Interpretation of quantitative data on the basis of hypotheses about the nature of the interactions between particles.

6. Classes of interactions

A first orientation in a new branch of physics consists in acquiring a good feeling for orders of magnitude.

Consider the orders of magnitude of cross-sections. The cross-section of the reaction

$$\pi^+ + p \to \pi^+ + p \tag{6.1}$$

grows from practically zero for small pion energies, reaches a maximum of the order of 75 mb, and then falls and stays for a long time around a few tens of millibarns. (Figure 5.2)

The cross-section of the reaction

$$\gamma + p \rightarrow \pi^+ + n \tag{6.2}$$

grows similarly from zero when the photon energy is just sufficient to create a π, but has an average value around the order of a hundred micro-barns.

The reaction

$$\nu + p \rightarrow e^+ + n \tag{6.3}$$

has a cross-section of the order of 10^{-14} barns.

From these examples one can draw certain conclusions:

1. Cross-sections can vary by many orders of magnitude (by a factor of 10^{13} between (6.1) and (6.3)).
2. It is difficult to compare them with each other because of kinematic effects (they vanish at low energies, or, sometimes, tend to infinity as the energy decreases). It does seem worth one's while, however, to analyse factors of 10^{13} in more detail.

To do this we shall appeal to Fermi's Golden Rule, which allows one to eliminate some of the kinematic effects.

The Golden Rule

In quantum mechanics one has the following formula for the transition probability per unit time from an initial state $|i\rangle$, (e.g. a state with one π^+ and one proton), to a final state $|f\rangle$, (e.g. a π^+p state with different momenta):

$$P_{fi} = 2\pi\, |\langle f \mid H \mid i\rangle|^2\, \rho_f(E) \tag{6.4}$$

This formula involves the interaction Hamiltonian H. Strictly speaking it applies only when H is small and when the effect under consideration can be treated by first order perturbation theory. But the formula (6.4) will give us an idea of the order of magnitude of P even when perturbation theory is not applicable, as long as we are not afraid of being mistaken by a factor of 10. Recall that the factor $\rho_f(E)$ is the density of final states at total energy E, E being the initial energy of the system.

A general method for calculating $\rho_f(E)$ is the following: Let $\mathbf{P}$ be the total momentum of the initial state and suppose there are n particles in the final state. Let $\mathbf{p}_1, \mathbf{p}_2, \ldots \mathbf{p}_n$ be the momenta of these particles, and $E_1, E_2, \ldots, E_n$ their energies. Then,

$$\rho_f(E) = \int \delta[E - (E_1 + E_2 + \cdots + E_n)]\, \delta[(\mathbf{p}_1 + \mathbf{p}_2 + \cdots + \mathbf{p}_n) - \mathbf{P}] \times \frac{d^3\mathbf{p}_1\, d^3\mathbf{p}_2 \cdots d^3\mathbf{p}_n}{(2\pi)^{3n}} \tag{6.5}$$

These expressions presuppose that the states are normalised to one particle per unit volume, corresponding to the normalisation

$$\langle \mathbf{p}' \mid \mathbf{p} \rangle = (2\pi)^3 \, \delta(\mathbf{p} - \mathbf{p}') \tag{6.6}$$

whence the Schroedinger (configuration space) representative of the state $|\mathbf{p}\rangle$ is the wavefunction $e^{i\mathbf{p}\cdot\mathbf{x}}$.

If the initial state $|i\rangle$ is a single-particle state, then P gives immediately the *mean life* of this unstable particle:

$$\tau^{-1} = \sum_f P_{fi} \, . \tag{6.7}$$

The sum runs over all decay modes and over all the variables $\mathbf{p}_1, ..., \mathbf{p}_n$ that are not uniquely determined by the δ functions in (6.5).

If the initial state is a two-particle state, and if again one normalises the states of the initial particles, individually, to one particle per unit volume, then the flux Φ of these particles through unit area per unit time is equal to their speed v. Hence the cross-section for the process $i \to f$ is given by

$$\sigma_{fi} = \frac{P_{fi}}{\Phi} = \frac{2\pi}{v} |\langle f \mid H \mid i \rangle|^2 \, \rho_f(E). \tag{6.8}$$

We see that Fermi's Golden Rule displays explicitly the kinematic factors v and ρ_f, and allows one to deduce the matrix element $\langle f | \, H \, | i \rangle$ of the Hamiltonian, at least as to order of magnitude even if perturbation theory is not applicable. Further, by appeal to the formula (6.7) it allows one to exploit mean lives on the same footing as cross-sections for information about the Hamiltonian.

A note about dimensions: in the natural system of units, one has $\hbar = 1 = c$, and the only physical dimension is that of energy; the units of mass and momentum are the same as that of energy E, while those of length and time are E^{-1}. In these units, equation (6.4) shows that $\langle f | \, H \, | i \rangle$ has the dimension E^{-2} when $|f\rangle$ contains two particles.

Results

Some calculations of $\langle f | \, H \, | i \rangle$ are set below as problems. If one confines oneself to reactions with only two particles in the final state, then one notes the following facts:

1. Most observed reactions correspond to values of $\langle f | \, H \, | i \rangle$ of the order of 10^{-4} MeV^{-2}. Actually this is only an average of values that are spread rather widely.
2. Reactions with a photon in the initial or final state have matrix elements smaller by a factor of ten. In fact, when one can compare

two reactions with the same final state such as

$$\pi^- + p \rightarrow \pi^- + p \tag{6.8a}$$

$$\gamma + n \rightarrow \pi^- + p \tag{6.8b}$$

then discounting kinematic factors the cross-section for the second reaction is always smaller than that for the first, by a factor of about 100.

Electrodynamics helps us to understand this fact. If we assume that effectively the photon always couples to a charge, then this charge will be of the order of the electron charge, which is that carried by all elementary particles that are not neutral. Thus the observed factor of 100 is simply the coupling constant $e^2/\hbar c \approx \frac{1}{137}$. This explanation is strongly supported by the mean life of the π^0. The π^0 decays into 2γ and thereby involves the photon coupling twice, giving a factor of e^2 in the matrix element and of e^4 in the decay probability. The mean life of 10^{-16} seconds does indeed correspond to a value of $|\langle f|\,H\,|i\rangle|$ of the order of 10^{-8} MeV^{-2}.

3. Reactions induced by neutrinos have matrix elements of the order of 10^{-11} MeV^{-2}. The same order of magnitude is found on calculating the matrix elements for the decays

$$n \rightarrow p + e + \nu$$

$$\mu \rightarrow e + \nu + \nu$$

$$K \rightarrow \pi + \mu + \nu$$

(Here we have picked decays into three particles in order to have matrix elements with the same physical dimensions as those considered above.)

Classes of interactions

In the light of these observations we shall accept that there exist three kinds of basic interaction, which can be distinguished by the orders of magnitude of their matrix elements:

1. strong interactions;
2. electromagnetic interactions;
3. weak interactions.

Most reactions observed in the laboratory are due to the strong interaction.

Electromagnetic interactions have been studied in detail with the help of quantum electrodynamics, and are by far the best known. In particular, it is only possible to understand the agreement between the measured values of the gyromagnetic ratios (ratio of magnetic moment to spin) of the electron and the μ, and their theoretical values as calculated by quantum electrodynamics, if one assumes that the electron and the μ have only electromagnetic but no strong interactions.

The weak interactions are responsible for the decay of almost all particles with long mean lives, i.e. longer than 10^{-21} seconds. In fact, the only stable particles known are the electron, the positron, the neutrino, the photon, and the proton (and most likely the antiproton). Some particles decay by electromagnetic interactions, like the π^0 and the Σ^0:

$$\pi^0 \to \gamma + \gamma$$

$$\Sigma^0 \to \Lambda^0 + \gamma.$$

Most particles (including the neutron) are unstable, and decay through weak interactions.

We admit that such a classification relying on orders of magnitude needs to be interpreted with some reticence. However, we shall see later on that there exist qualitative differences between the three classes of interactions, and it is through these that the distinction between them attains its full weight.

Classification of particles

Particles which sometimes take part in strong interactions are called *hadrons*.

Leptons are particles which never take part in strong interactions but do sometimes take part in weak interactions; (the only leptons are electrons, positrons, μ^+, μ^-, and neutrinos).

There remains the photon which has only electromagnetic interactions.

The stability of unstable particles

Most particles are unstable. Does this imply that, in studying the electromagnetic or the strong interactions, stable and unstable particles must be treated on an altogether different footing? For instance, must one employ different mathematical methods to study the proton and the neutron?

Very fortunately, the answer is no. We saw that matrix elements of the strong interactions are effectively of the order of $(100 \text{ MeV})^{-2}$. Hence they would be of order unity if we took 100 MeV for our unit of energy. In fact, with $\hbar = 1 = c$, this unit of energy corresponds to a unit of length of the order of 10^{-13} cm, i.e. to a unit of cross-section of $10^{-26} \text{ cm}^2 =$ 10 millibarns, which is indeed the correct order of magnitude for strong interactions. Then the unit of time is of the order of $10^{-13} \text{ cm}/c \approx 10^{-23}$ seconds. This is the time scale characteristic of strong interactions. Now, even the electromagnetically decaying π^0 has a mean life of 10^{-16} seconds, which it is safe to consider as infinitely long in comparison to 10^{-23} seconds.

This remark greatly simplifies our study of strong interactions.

7. Supplementary remarks

We have considered the determination of particle masses through ionisation. This is not the method used for precise mass measurements. For this topic we refer to G. Källen: Elementary particles.

The theory of accelerators can be found in the book on general electronics by Blanc-Lapierre, Fortet and Lapostolle.

Experimental methods are treated in many books; we mention only 'Methods of nuclear physics', edited by E. Segré.

Problems

1. Suppose that the charge of the proton is $(-1 + \epsilon)$ times that of the electron. For what value of ϵ would the electrostatic repulsion exceed the gravitational attraction between earth and moon; earth and sun; sun and galaxy? Make reasonable assumptions (and justify them) about the proton and electron contents of these astronomical bodies; you will be surprised at the result.
2. To what energy must a proton be accelerated in order that, on colliding with a target proton, it should give rise to (a) a π meson; (b) an antiproton? The relevant reactions are

$$p + p \rightarrow p + n + \pi^+$$

$$p + p \rightarrow p + p + \bar{p} + p \qquad (\bar{p} = \text{antiproton})$$

3. What energy must a photon have in order that the following photoproduction reaction can taken place on a target proton:

$$\gamma + p \rightarrow n + \pi^+$$

4. One has available electromagnets giving a maximum field of 100,000 gauss. What is the minimum radius of a synchrocyclotron for accelerating protons, (a) to 100 MeV; (b) to 100 GeV? Similarly for electrons.
5. Show that energy-momentum conservation forbids the reaction

$$\gamma \rightarrow e^+ + e^-$$

6. The following formula gives the cross-section for pair production from a nucleus of charge Z, when the photon energy ω is such that

$\omega - 2m \ll m$, where m is the electron mass:

$$\sigma(\omega) = \frac{\pi}{12} Z^2 r_0^2 \left(\frac{\omega}{m} - 2\right)^3$$

In this expression r_0 is the classical electron radius:

$$r_0 = e^2/4\pi m = 2.8 \times 10^{-13} \text{ cm.}$$

Calculate the radiation length as a function of ω in hydrogen and in lead.

7. What must be the laboratory energy of a proton in order that on colliding with a proton it should produce an antiproton by the following reaction

$$p + p \rightarrow \bar{p} + p + p + p.$$

8. Consider the reaction

$$\pi^- + p \rightarrow n + \pi^+ + \pi^-$$

where the π^- has an energy of 1 GeV in the laboratory. What is the maximum neutron energy in the centre of mass frame?

9. Evaluate the volume of phase space

$$\int \delta(E - \Sigma E_i)\, \delta(\mathbf{P} - \Sigma \mathbf{p}_i) \frac{d^3\mathbf{p}_1 \cdots d^3\mathbf{p}_n}{(2\pi)^{3n}}$$

for the following cases

1. Two nonrelativistic particles in their centre of mass frame ($E_i = \mathbf{p}_i^2/2m_i$)
2. Two nonrelativistic particles in the laboratory frame.
3. Two relativistic particles in their centre of mass frame (for the case where one mass is zero).
4. Two relativistic particles in the laboratory frame (for the case where one mass is zero).
5. Three particles in the centre of mass frame, one of the three being nonrelativistic (i.e. very heavy).

10. Calculate $|\langle f|\, H\, |i\rangle|$ for each of the following processes, given that
    (a) The cross section for $\pi^+ p \rightarrow \pi^+ p$ is 75 mb when the π^+ laboratory energy is 280 MeV.
    (b) The cross-section for $\nu + p \rightarrow \mu + n$ is 10^{-34} cm^2 when the neutrino laboratory energy is 1 GeV.
    (c) The mean life of the π^+, decaying mainly into $\mu^+ + \nu$, is 10^{-8} seconds.

11. Given that the deuteron has spin 1 and positive parity, explain why the neutron in the deuteron is stable. More generally explain why neutrons can be stable inside nuclei.

12. Calculate the speed of the initial proton at the threshold of the reaction

$$p + p \rightarrow p + p + \pi^0.$$

13. By using the Golden Rule, show that for an 'exothermic reaction' like

$$\Sigma^- + p \rightarrow \Lambda^0 + n$$

where the sum of the masses of the initial particles exceeds that of the final particles, the cross-section is such that

$$v\sigma \rightarrow \text{constant}$$

when the relative speed v of the initial particles tends to zero; (σ tends to infinity but the probability P remains finite).

CHAPTER 2

CONSERVATION LAWS: ADDITIVE QUANTUM NUMBERS

In this chapter we examine the conservation of electric charge, and the conservation of additive quantities analogous to charge: baryon number, lepton number, electron and muon numbers, and strangeness.

1. Charge conservation

We have seen that it is particularly easy to determine the charge of a particle from the ionisation it produces, and from the direction in which its trajectory curves in a magnetic field.

However, it is possible for charged particles to enter a reaction without being detected directly: this is the case for instance with the protons in a target. Nevertheless there are cases where one can be sure that a particle involved in a collision was a proton, for example if the target consists of liquid hydrogen, and where in consequence its charge is known.

Hence one can readily ascertain the charges of all the particles taking part in a reaction, and thus establish

The law of conservation of charge

The sum of the charges of the initial particles entering a reaction is equal to the sum of the charges of the final particles.

Examples can be found in strong interactions:

$$p + p \to p + p + \pi^0 \qquad p + p \to p + n + \pi^+ \qquad \pi^+ + p \to \pi^+ + p$$

$$\pi^- + p \to \pi^- + p \quad \pi^- + p \to n + \pi^0 \quad \pi^- + p \to p + \pi^- + \pi^+ + \pi^- \text{ etc,}$$

in electromagnetic interactions:

$$\gamma + e \to \gamma + e \qquad e + p \to e + p \qquad \pi^0 \to \gamma + \gamma$$

and in weak interactions:

$$\pi^- \to \mu^- + \nu \qquad \pi^+ \to \mu^+ + \nu \qquad \mu^- \to e^- + \nu + \nu \qquad n \to p + e^- + \nu.$$

This rule is very useful from a practical point of view, because it allows us to rule out from consideration as impossible a large number of reactions, as for instance

$$\pi^- + p \rightarrow \pi^+ + p \qquad K^- + p \rightarrow K^0 + p + \pi^+$$

When it is used in this way, a conservation law is called a *selection rule*.

2. Baryon number

We know that the neutron is unstable, but how does it come about that the proton is stable? It has enough energy to decay in many ways, such as

$$\begin{gathered} p \rightarrow e^+ + \gamma \qquad p \rightarrow \pi^+ + \pi^0 \\ p \rightarrow \pi^+ + \gamma \cdots p \rightarrow \pi^+ + \pi^- + \pi^0 + \pi^+ + \pi^- + \pi^0 \end{gathered} \tag{2.1}$$

Along lines made venerable by tradition, we explain this by saying: the proton has a certain inherent property which is conserved. More precisely, we assume that there exists a selection rule which forbids the reactions (2.1).

Is this selection rule analogous to those which follow from charge conservation? To answer this we must consider a set of reactions which will allow us to assign a 'protonic charge' to every particle. Having done this, we must then verify whether this protonic charge is conserved in all observed reactions.

The method is rather lengthy and we shall have to apply it several times, so that we give the procedure in detail, once and for all, in the special case of the proton.

(a) *Assigning charges*

Let us assign to the proton a protonic charge of 1. Then we determine the protonic charges of other particles. Note to begin with that photon emission by an accelerating proton: $p \rightarrow p + \gamma$, and the reaction

$$p + p \rightarrow p + p + \pi^0 \tag{2.2}$$

force one to assign a protonic charge 0 to the γ and the π^0.

At this point we are held up by the fact that there exist few particles other than the γ and the π^0 which are created in reactions involving nothing but protons. For instance, it is impossible to determine the protonic charge of the neutron, because charge conservation stops it from taking part in such a reaction. But in view of the similarities between the properties of proton and neutron, we nevertheless assign to the neutron a protonic charge of 1. This choice is convenient but not necessary. Once it is made, one can determine the protonic charges of the charged pions, by appeal to

$$p + p \to p + n + \pi^+ \qquad \pi^- + p \to n + \pi^0$$

Thus, denoting protonic charge by B:

$$B(p) = B(n) = 1 \qquad B(\pi^0) = B(\pi^+) = B(\pi^-) = 0. \tag{2.3}$$

The decays

$$K^+ \to \pi^+ + \pi^0 \quad K^0 \to \pi^+ + \pi^- \quad K^- \to \pi^- + \pi^0 \quad K^0 \to \pi^+ + \pi^- + \pi^0$$

give

$$B(K^+) = B(K^-) = B(K^0) = 0. \tag{2.4}$$

The reactions

$$\pi^- + p \to \Lambda^0 + K^0 \qquad \pi^+ + p \to \Sigma^+ + K^+$$

$$\pi^- + p \to \Sigma^- + K^+ \qquad \pi^- + p \to \Sigma^0 + K^0$$

$$K^- + p \to \Xi^- + K^+ \quad K^- + p \to \Xi^0 + K^0 \quad K^- + p \to \Omega^- + K^+ + K^0$$

give

$$B(\Lambda^0) = B(\Sigma^-) = B(\Sigma^0) = B(\Sigma^+) = B(\Xi^-) = B(\Xi^0) = B(\Omega^-) = 1. \tag{2.5}$$

Leptons are produced only in pairs by the particles whose protonic charges we have determined. Hence it is once again impossible to determine their protonic charges unambiguously. We make the choice

$$B(e^-) = 0.$$

which is convenient but not necessary. Then the reactions

$$\gamma \to e^+ + e^- \quad n \to p + e + \nu \quad \mu \to e + \nu + \nu \quad \pi^+ \to \mu^+ + \nu \quad \pi^- \to \mu^- + \nu$$

give

$$B(e^+) = B(\mu^+) = B(\mu^-) = B(\nu) = 0. \tag{2.6}$$

We see that the following particles have non-zero protonic charge: $(p, n, \Lambda^0, \Sigma^+, \Sigma^0, \Sigma^-, \Xi^-, \Xi^0, \Omega^-)$. All these are manifestly heavy particles. For this reason they are denoted collectively as baryons (a greek word meaning heavy), and, instead of calling it the protonic charge, B is called the baryonic charge or *baryon number*.

(b) *Antiparticles*

We have seen already that antiprotons can be produced in proton collisions at high energy:

$$p + p \to p + p + p + \bar{p}$$

whence

$$B(\bar{p}) = -1. \tag{2.7}$$

The following reactions are observed:

$$\bar{p}+p \to \bar{n}+n \qquad \bar{p}+p \to \bar{\Lambda}^0+\Lambda^0 \qquad \bar{p}+p \to \bar{\Sigma}^0+\Sigma^0$$

$$\bar{p}+p \to \overline{\Sigma^-}+\Sigma^- \qquad \bar{p}+p \to \overline{\Sigma^+}+\Sigma^+ \qquad \bar{p}+p \to \overline{\Xi^-}+\Xi^-$$

where each antiparticle has the same mass as, and a charge opposite to, the corresponding particle. These reactions show that

$$B(\bar{\Lambda}^0) = B(\bar{n}) = B(\overline{\Sigma^0}) = B(\overline{\Sigma^+}) = B(\overline{\Sigma^-}) = B(\overline{\Xi^-}) = -1. \tag{2.8}$$

(c) *The law of conservation of baryon number*

With the baryon numbers assigned as above, experiment shows that *The sum of baryon numbers for initial particles is equal to the sum for final particles in all reactions and in all decays that have been observed.*

3. Lepton number

We can play the same game with the electron as with the proton. Let us attribute to the electron an 'electronic charge' $\ell(e) = 1$. Because of baryon number conservation it is possible to assign arbitrarily the 'electronic' charge of the proton, and for convenience we take it as 0. Because of electric charge conservation, the choice for the neutron is also arbitrary, and we shall assign it, too, an 'electronic' charge of zero.

Some 'electronic' charges can be determined from the reactions

$$p+e \to p+e+\gamma \qquad \gamma \to e^+ + e^- \qquad \pi^0 \to 2\gamma \qquad p+\pi^- \to n+\pi$$

$$p+\pi^0 \to n+\pi^+$$

whence one deduces

$$\ell(e^+) = -1 \qquad \ell(\gamma) = \ell(\pi^0) = \ell(\pi^-) = \ell(\pi^+) = 0 \tag{3.1}$$

By appeal to the reactions already studied when considering baryon number, it is easy to show that the 'electronic' charge is zero for all particles other than leptons.

Let us now study the following decays, which, though rare, have been observed:

$$\pi^- \to e^- + \nu \tag{3.2}$$

$$\pi^+ \to e^+ + \nu \tag{3.3}$$

From (3.2) one deduces $\ell(\nu) = -1$; but by (3.3), $\ell(\nu) = +1$. Hence there are two possibilities: either, 'electronic' charge does not actually exist, or there are two neutrinos with opposite 'electronic' charges. Let us call them ν and $\bar{\nu}$, with $\ell(\nu) = +1, \ell(\bar{\nu}) = -1$.

Let us explore the second hypothesis. From the reactions

$$\pi^- \to \mu^- + \nu \qquad \pi^+ \to \mu^+ + \nu \tag{3.4}$$

we can conclude that $\ell(\mu) = \pm 1$. There are no reactions which would allow us to fix the value of $\ell(\mu)$ without further assumptions as to the number of distinct kinds of neutrinos. By convention, one assigns

$$\ell(\mu^-) = +1 \qquad \ell(\mu^+) = -1 \tag{3.5}$$

whence the reactions (3.4) must be written as

$$\pi^- \to \mu^- + \bar{\nu} \qquad \pi^+ \to \mu^+ + \nu \tag{3.6}$$

in order that 'electronic' charge should be conserved.

This whole construction rests on the assumption of two different neutrinos, ν and $\bar{\nu}$. Does there exist a selection rule which embodies the conservation of 'electronic' charge and which can be verified experimentally? Suppose we make a π^- beam which is then allowed to decay according to (3.6), (the reaction (3.2) is less frequent by $\times 10^{-4}$ and can be neglected), and that the μ^- are removed from the beam by a magnetic field. One can make a very clean neutrino beam, by exploiting the fact that neutrinos are the only particles with exclusively weak interactions. By passing the ν beam through a great thickness of concrete, all other 'parasitic' particles are eliminated. The interactions of the neutrinos can be observed in a bubble or in a spark chamber. It is at this point that one needs an extremely pure neutrino beam, because neutrino interactions are very rare, and any contamination whatever by other particles would horribly complicate the analysis of the data.

When a neutrino interacts with a proton or a neutron, it can lead to

$$\bar{\nu} + p \to n + \mu^+ \tag{3.7}$$

$$\bar{\nu} + n \not\to p + \mu^- \tag{3.8}$$

Reaction (3.8) is forbidden if 'electronic' charge is conserved. In fact, reaction (3.7) is observed while reaction (3.8) is not.

Since ℓ is non-zero only for leptons, it is called leptonic charge or *lepton number*. A survey of all reactions confirms

The law of lepton conservation:

The total lepton number is conserved in all reactions and in all decays.

4. Electron number and muon number

Returning to the neutrino experiments described in the last paragraph, one notes the very remarkable fact that the following reaction does not take place:

$$\bar{\nu} + p \nrightarrow n + e^+ \tag{4.1}$$

This can be understood if there exist distinct electron and muon numbers E and M, with

$$E(e) = -E(\bar{e}) = 1 \qquad E(\mu^-) = -E(\mu^+) = 0$$
$$M(e) = M(\bar{e}) = 0 \qquad M(\mu^-) = -M(\mu^+) = 1$$

and if one assumes that there exist two neutrinos ν_e and ν_μ as well as two antineutrinos $\bar{\nu}_e$ and $\bar{\nu}_\mu$, with

$$E(\nu_e) = -E(\bar{\nu}_e) = 1 \qquad E(\bar{\nu}_\mu) = E(\bar{\nu}_\mu) = 0$$
$$M(\nu_\mu) = -M(\bar{\nu}\ ) = 1 \qquad M(\nu_e) = M(\bar{\nu}_e) = 0.$$

Thus, in the decay $\mu \rightarrow e + \nu + \nu$, one actually has

$$\mu^- \rightarrow e^- + \bar{\nu}_e + \nu_\mu$$
$$\mu^+ \rightarrow e^+ + \nu_e + \bar{\nu}_\mu$$

This assumption is compatible with the absence of the decay

$$\mu^- \rightarrow e^- + \gamma \qquad \mu^- \rightarrow e^- + e^+ + e^-$$

Note that lepton number is the sum of electron and muon numbers.

5. Strangeness

We have seen that the absence of certain reactions can lead to the discovery of conservation laws. In this respect it is interesting to note the striking absence of reactions like

$$p + n \nrightarrow p + \Lambda \tag{5.1}$$
$$p + p \nrightarrow p + n + K^+ \tag{5.2}$$
$$p + n \nrightarrow p + p + K^- \tag{5.3}$$

which are allowed by all the conservation laws we have met hitherto. By contrast, the reactions

$$p + p \rightarrow p + \Lambda + K^+ \tag{5.4}$$
$$K^- + p \rightarrow \Lambda + \pi^0 \tag{5.5}$$

have cross-sections which are quite large (several millibarns). It is clear that by introducing a new charge S such that

$$S(K^+) = 1 \quad S(\Lambda) = -1 \quad S(p) = 0 \quad S(n) = 0 \quad S(K^-) = -1 \tag{5.6}$$

one can explain simultaneously why the reactions which do happen are allowed and why those which do not happen are forbidden.

The charges S of other particles can be determined by the method we have used several times already. Thus we show, as for the 'electronic' charge, that

$$S(\pi^+) = S(\pi^-) = S(\pi^0) = S(\gamma) = 0. \tag{5.7}$$

Then the occurrence of the reaction

$$\bar{p} + p \rightarrow \pi^+ + \pi^- + \pi^+ + \pi^-$$

shows that

$$S(\bar{p}) = 0 \tag{5.8}$$

At this point one meets a difficulty. Both the following reactions are observed

$$\bar{p} + p \rightarrow K^+ + K^0 + \pi^- \tag{5.9}$$

$$\bar{p} + p \rightarrow K^- + K^0 + \pi^+ \tag{5.10}$$

and they entail, respectively, $S(K^0) = -1$ and $S(K^0) = +1$. If there really is a charge S, then we must assume that there exist two different K^0 which we shall call K^0 and $\bar{K}^0$, with the convention

$$S(K^0) = +1 \qquad S(\bar{K}^0) = -1. \tag{5.11}$$

The occurrence of the reactions

$$K^- + p \rightarrow \Sigma^- + \pi^+ \quad K^- + p \rightarrow \Sigma^0 + \pi^0 \quad K^- + p \rightarrow \Sigma^+ + \pi^- \tag{5.12}$$

$$K^- + p \rightarrow \Xi^- + K^+ \qquad K^- + p \rightarrow \Xi^0 + K^+ + \pi^- \tag{5.13}$$

$$K^- + p \rightarrow \Omega^- + K^0 + K^+ \tag{5.14}$$

leads one to assign

$$S(\Sigma^-) = S(\Sigma^0) = S(\Sigma^+) = -1 \tag{5.15}$$

$$S(\Xi^-) = S(\Xi^0) = -2. \tag{5.16}$$

In view of the reaction (5.14), the charge S of the Ω^- depends on that of the K^0. If this actually is a K^0, then $S(\Omega^-) = -3$; if it is a $\bar{K}^0$, then $S(\Omega^-) = -1$. The second possibility is ruled out by the absence of the reaction

$$K^- + p \nrightarrow \Omega^- + \pi^+$$

whence

$$S(\Omega^-) = -3. \tag{5.17}$$

All the interactions considered so far have been strong interactions. The weak interactions present us with a problem, for the observed decays

$$K^- \to \pi^- + \pi^0 \qquad \Lambda \to p + \pi^- \qquad \Sigma^+ \to p + \pi^0 \qquad \Xi^- \to \Lambda + \pi^- \tag{5.18}$$

are incompatible with the conservation of S. As regards the electromagnetic interactions, they are seen to conserve S by noting the absence of the decays

$$K^- \nrightarrow \pi^- + \gamma \qquad \Sigma^+ \nrightarrow p + \gamma \tag{5.19}$$

and the occurrence of

$$\Sigma^0 \to \Lambda^0 + \gamma \tag{5.20}$$

S is called *strangeness*. A systematic survey of all interactions confirms the following law:

The law of conservation of strangeness

The strong and the electromagnetic interactions conserve strangeness.

We see that this is a conservation law of an altogether new kind, in that it applies only to some of the interactions.

Comments: 1. Because of the large mass of the K, high energies are needed to produce strange particles from non-strange ones. This explains why strange particles were discovered rather late, being then dubbed 'strange', i.e. unusual, whence their name.

2. Only the weak interactions involve strange particles and leptons simultaneously. Since these interactions do not conserve strangeness, there is no compelling reason to assign strangeness to leptons. However, one often adopts the convention that their strangeness is zero.

3. We have seen that the strong and the electromagnetic interactions conserve three 'charges': electric charge, baryon number, and strangeness. One can ask whether there are others. A systematic survey of all observed reactions leads to the following conclusion:

There exist no 'charges' other than those we have discussed, provided that there exist no long-lived particles beyond those that are already known, and provided that none of the latter has a multiplicity higher than 1 (for instance, if there do not exist two different Ξ^-).

6. Nomenclature

We mention, or recall, some of the names commonly applied to families of particles, chosen in view of the conservation laws.

1. Particles with non-zero lepton number are *leptons*. The neutrinos have only weak interactions. The other leptons, (e, $\bar{e}$, μ^-, μ^+), have both weak and electromagnetic interactions.

2. The photon has only electromagnetic interactions. All its 'charges' are zero.
3. The other particles, which have strong interactions, are *hadrons*. Amongst them, one distinguishes between
4. Particles with baryon number ± 1, called baryons (for $+1$) or anti-baryons (for -1).
5. Particles with baryon number 0, called mesons. They include π^+, π^-, π^0, K^+, K^-, K^0, $\bar{K}^0$; (it is only by accident that the muon has long been called the μ meson. This name must disappear).
6. Particles with non-zero strangeness, called *strange particles*.
7. The strange baryons are called *hyperons* (Λ^0, Σ^-, Σ^0, Σ^+, Ξ^-, Ξ^0, Ω^-).

7. Antiparticles

We have seen that particles with at least one 'charge' different from zero occur in particle-antiparticle pairs, whose members have equal masses and equal and opposite charges (this applies to all their 'charges'). We list all these pairs:

particle	antiparticle	
ν_e	$\bar{\nu}_e$	
ν_μ	$\bar{\nu}_\mu$	
e^-	e^+	
μ^-	μ^+	
γ		which is its own antiparticle (all 'charges' zero)
π^0		ditto
Ξ^-	$\overline{\Xi^-}$	
Ξ^0	$\overline{\Xi^0}$	
Ω^-	$\bar{\Omega}^-$	not observed
π^-	π^+	
p	$\bar{p}$	
n	$\bar{n}$	
K^0	$\bar{K}_0$	
K^+	K^-	
Λ^0	$\bar{\Lambda}^0$	
Σ^+	$\overline{\Sigma^+}$	
Σ^0	$\overline{\Sigma^0}$	
Σ^-	$\overline{\Sigma^-}$	

Problems

Indicate whether the following reactions are allowed or forbidden

$$\pi^- + p \to \Sigma^- + K^0 + \pi^0 \qquad e + p \to n + e + \pi^-$$

$$e + p \to e + p + \pi^0 + \gamma \qquad e + \mu^- \to \nu_e + \bar{\nu}_\mu$$

$$e^+ + \mu^- \to \bar{\nu}_e + \nu_\mu \qquad \mu^+ \to e^+ + \bar{\nu}_e + \bar{\nu}_\mu$$

$$\mu^- \to e^- + \gamma \qquad \mu^- \to e^- + e^+ + e^-$$

$$K^- + p \to \Sigma^- + K^+ \qquad \pi^- + p \to K^- + \Sigma^- + \pi^0$$

$$\bar{p} + p \to \overline{\Sigma^+} + \Lambda^0 + \pi^+ \qquad \bar{p} + p \to \bar{K}^0 + K^+ + \pi^- + \pi^0$$

$$\bar{p} + p \to \bar{\Lambda}^0 + \Lambda^0 \qquad \bar{p} + \pi \to \bar{\Lambda} + K^0$$

CHAPTER 3

ELEMENTARY GROUP THEORY: ISOTOPIC SPIN

In this chapter we give a brief review of group theory, mostly through particular examples, and with no pretensions to rigour. Next, we show how the existence of multiplets of particles leads to the concept of isotopic spin, and what kinds of experiments confirm its underlying assumptions.

1. Multiplets

On considering the set of all hadrons, one notices that they are grouped into families. Thus, the mesons π^+, π^-, and π^0 all have the same strangeness (0), the same baryon number (0), and differ only by their electric charge. The difference in mass between the charged π's and the π^0 is 4.6 MeV, while the mass difference between π^+ and K^+ is 354 MeV, in spite of the fact that the K^+ is the particle immediately following the π^+ in an order of increasing masses (Figure 3.1.1). By convention one assigns a common generic name to the members of each family; thus the π^+, π^-, and π^0 are all called π's or pions, regardless of their charge. Such families are

$$
\begin{array}{llll}
\pi: & \pi^-, \pi^+, \pi^0 & S=0 & B=0 \\
K: & K^+, K^0 & S=+1 & B=0 \\
\bar{K}: & K^-, \bar{K}^0 & S=-1 & B=0 \\
N(\text{nucleons}) & p, n & S=0 & B=1 \\
\Lambda: & \Lambda^0 & S=-1 & B=1 \\
\Sigma: & \Sigma^+, \Sigma^0, \Sigma^- & S=-1 & B=1 \\
\Xi: & \Xi^-, \Xi^0 & S=-2 & B=1 \\
\Omega: & \Omega^- & S=-3 & B=1
\end{array}
\tag{1.1}
$$

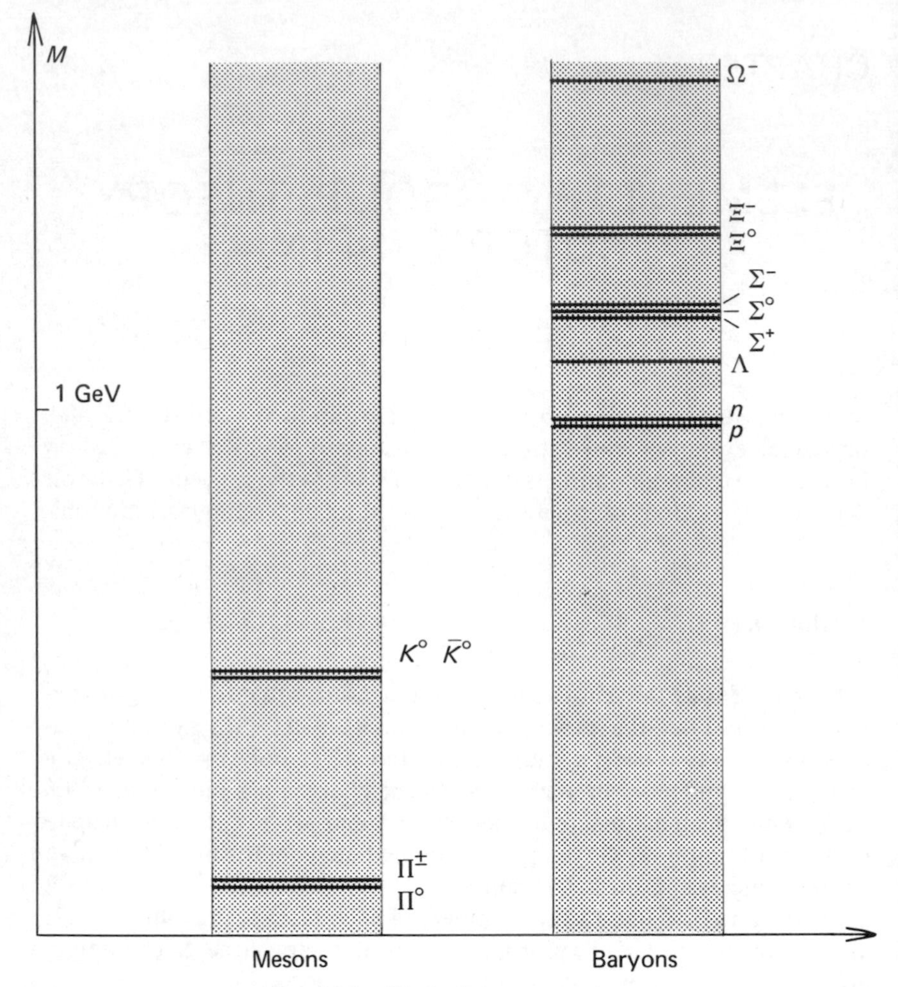

FIG. 3.1.1. The hadron mass spectrum.

This situation is most remarkable and calls for an interpretation. It is analogous to that encountered in the splitting of atomic energy levels in a weak magnetic field: we know that the energy levels of an atom in field-free space can be classified by the angular momentum ℓ and by magnetic quantum number $m = \ell, \ell - 1, \ell - 2, ..., -\ell$. When no magnetic field is applied, states corresponding to different values of m all have the same energy, and are said to be degenerate; the degeneracy being the number of levels with a common value of the energy, $2\ell + 1$ in this case. In a sufficiently weak magnetic field, the levels corresponding to different values of m acquire different energies, and one observes the appearance of families of levels (called *multiplets*) differing but little from each other in energy, so that the energies separating two different multiplets are in general much larger than the energy differences within one particular multiplet (Figure 3.1.2).

Quantum mechanics shows that the basic reason for the degeneracy of atomic levels is the invariance of the atomic Hamiltonian under rotations, itself a consequence simply of the isotropy of space. Application of a magnetic field destroys this isotropy by singling out a preferred direction, that of the field. One can formulate this by saying that there exists a group of operations (in this case the group of rotations) which commute with the Hamiltonian in the absence of a magnetic field, but which do not commute with the interaction Hamiltonian coupling the atom to the magnetic field.

Proceeding by analogy as regards particle multiplets, we are led naturally to ask: does there exist a group G of operations commuting with the strong interaction Hamiltonian, with strangeness, and with baryon number, and such that the elementary particles can be classified in a manner analogous to atomic levels? The small mass differences within a multiplet could then be attributed to an interaction that is less strong, and whose Hamiltonian does not commute with the operations of the group G. Since the particles of a multiplet differ in charge, it would be natural to identify this second interaction with the electromagnetic interaction, which evidently does depend on the charge.

Actually it is easy to put forward such a group: if for instance G is the group of rotations in a three-dimensional space, then we know that its multiplets are specified by a quantum number j, the multiplicity being given by $2j + 1$. Calling this quantum number I to distinguish it from the

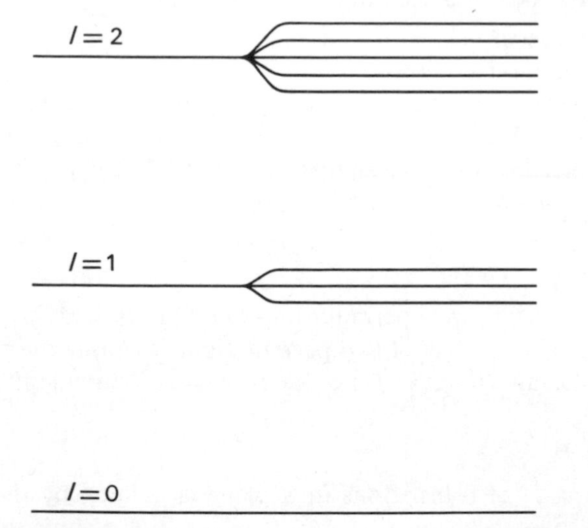

FIG. 3.1.2. Illustration of the lifting of rotational degeneracy between atomic energy levels by the Zeeman effect.

ordinary angular momentum, we see that the observed multiplets could all be catered for by the following choices of I:

$$I_\pi = 1 \quad I_K = \tfrac{1}{2} \quad I_{\bar{K}} = \tfrac{1}{2} \quad I_N = \tfrac{1}{2} \quad I_\Lambda = 0 \quad I_\Sigma = 1 \quad I_\Xi = \tfrac{1}{2} \tag{1.2}$$

Two important questions arise at this point:

1. Is the group R_3 of rotations in a three-dimensional space the only possible choice, or are there others?
2. Once we have assigned the quantum numbers I, what consequences follow from the invariance of the strong interactions under G? We need to answer this question if we wish to judge the appositeness of our assumption by comparison with experiment.

To analyse the first question one needs a deep knowledge of group theory. The answer turns out to be in the affirmative, and the rotation group R_3 is indeed the only one possible; hence the second question could be tackled at once by exploiting our knowledge of this particular group. Nevertheless, in view of the importance of group theory in other applications that will be needed, we shall devote some effort to its treatment in a more general form.

2. The concept of a group

By definition, a group G is a set of elements $a, b, c, \ldots$ for which a multiplication rule is defined such that

1. the product ab belongs to G if a and b belong to G; this is expressed by $[a \in G, b \in G] \Rightarrow ab \in G$.
2. multiplication is associative: $a(bc) = (ab)c$ for all a, b, c.
3. there exists a unit element $e \in G$ such that $ae = ea = a$ for all $a \in G$.
4. every element $a \in G$ has an inverse a^{-1} also belonging to G, such that $aa^{-1} = a^{-1}a = e$.

Example 1: S_n

S_n denotes the group of permutations of n objects. This group contains $n!$ elements. The product of two permutations is simply the permutation producing the same result as its two 'factors' carried out one after the other.

Example 2: T_2

The group T_2 of translations in a plane is defined by the geometric transformations

$$T_{\mathbf{a}} : \mathbf{x}' = \mathbf{x} + \mathbf{a} \tag{2.1}$$

where $\mathbf{x}$, $\mathbf{x}'$, and $\mathbf{a}$ are two-dimensional vectors. The group elements are translations. In the light of (2.1), the multiplication rule is given by

$$T_{\mathbf{a}+\mathbf{a}'} = T_{\mathbf{a}} + T_{\mathbf{a}'} \,. \tag{2.2}$$

The unit element is the null translation $T_0 : \mathbf{x}' = \mathbf{x}$. The inverse of $T_{\mathbf{a}}$ is $T_{-\mathbf{a}}$.

Note a fundamental difference between the groups under 1 and 2: $T_{\mathbf{a}}$ can be varied continuously by varying $\mathbf{a}$, while S_n consists of discrete operations. S_n is said to be a *discrete* group, while the translation group is *continuous*. In the following we shall confine ourselves exclusively to continuous groups. There are good reasons for this which we do not go into. We mention only that if the symmetry group of the strong interactions were discrete, then electric charge, baryon number, and strangeness would be conserved not absolutely, but only modulo some prime number n.

Example 3: R_3

Consider the group R_3 of rotations in a three-dimensional space. A rotation is a linear operation of the form

$$x_i' = R_{ij}x_j \qquad (i = 1, 2, 3) \qquad (j = 1, 2, 3) \tag{2.3}$$

It conserves distance, i.e. R_{ij} is a proper orthogonal matrix. In other words, one has

$$\det R = 1 \quad \text{and} \quad R_{ij}R_{kj} = \delta_{ik} \,. \tag{2.4}$$

In writing equations (2.3) and (2.4) we have adopted Einstein's convention by which repeated suffices are to be summed over; δ_{jk} is the Kronecker symbol,

$$\begin{cases} \delta_{jk} = 1 & \text{for} \quad j = k \\ \delta_{jk} = 0 & \text{for} \quad j \neq k. \end{cases}$$

The multiplication rule is the composition law for rotations, which can be written as a matrix product: if r_1 and r_2 are rotations having the matrices (R_{ij}^1) and (R_{ij}^2), then the matrix of the product rotation r_2r_1 is the matrix product (R^2R^1). The unit element is the rotation whose matrix is the unit matrix. The rotations r_1 and r_2 are the inverses of each other if their matrices are inverse matrices. This correspondence between, on the one hand, geometrical transformations interpretable as matrix operations, and the corresponding matrices on the other hand, is of general validity; in the following, whenever we speak of groups of matrix operations, the ideas of multiplication, of identity, and of the inverse, will always correspond to the same ideas applied to the matrices themselves.

A rotation can always be regarded as rotation through an angle θ about an axis with given orientation. Therefore it depends on three parameters (two direction cosines for fixing the axis, plus the angle θ).

Comments: 1. It follows that T_2 and R_3 depend on 2 and 3 parameters respectively. The number of such parameters is called the order of the continuous group.

2. Two translations commute, but this is not generally true for two rotations. A group all of whose elements commute is called *Abelian*.

Example 4: $SU(2)$

The group $SU(2)$ is the group of unitary and unimodular 2×2 matrices (two rows and two columns), i.e. the group of 2×2 matrices U satisfying the conditions

$$UU^{+} = I \quad \text{and} \quad \det U = 1. \tag{2.5}$$

It is of order 3.

Example 5: $SL(2C)$

The group $SL(2C)$ is the group of complex unimodular 2×2 matrices A, obeying

$$\det A = 1 \tag{2.6}$$

It is of order 6.

Example 6: $SU(3)$

The group $SU(3)$ is the group of unitary and unimodular 3×3 matrices, i.e. of 3×3 matrices such that

$$UU^{+} = I \quad \text{and} \quad \det U = 1. \tag{2.7}$$

It is of order 8.

Comment. All the continuous groups which we have introduced are matrix groups, with the exception only of T_2. Actually it is possible to write a translation as a matrix transformation, if one introduces homogeneous coordinates (y_1, y_2, y_3), where $x_1 = y_1/y_3$ and $x_2 = y_2/y_3$. Then equation (2.1) can be written as

$$y_1' = y_1 + a_1 y_3 \qquad y_2' = y_2 + a_2 y_3 \qquad y_3' = y_3 \tag{2.8}$$

which can be rewritten immediately in matrix form.

In the following, we shall consider only groups that are defined as groups of matrices from the start. In this way we lose some generality, but avoid being unnecessarily abstract.

3. Infinitesimal operations

By definition, a matrix A belonging to a matrix group has an inverse. Hence its determinant is non-zero, and there exists a matrix α such that

$$A = e^{\alpha} = I + \alpha + \frac{\alpha^2}{2!} + \cdots \tag{3.1}$$

Evidently the matrix α is the logarithm of A. In group theory, the set of all matrices α whose exponentials belong to G is also called the *Lie algebra* of G. The matrix α is connected with A by the well known relation

$$A = \lim_{n\to\infty} \left(I + \frac{\alpha}{n}\right)^n \tag{3.2}$$

which entails

$$\alpha = \lim_{n\to\infty} n(A^{1/n} - I). \tag{3.3}$$

This means that for n large enough, the matrix $1 + \alpha/n$ is an operator of the group, and that it yields A by iteration. (It is not selfevident that $A^{1/n} \approx 1 + \alpha/n$ is an operator of the group. But it can be verified easily in all the cases we shall meet). In this sense, $1 + \alpha/n$ is called an *infinitesimal operation*, since for large enough n it differs infinitesimally from the unit element.

Example 1: T_2

Return to the expression (2.1) for $T_\mathbf{a}$. It shows immediately that $T_{\mathbf{a}/n} = (T_\mathbf{a})^{1/n}$, whence by eq. (3.3):

$$\alpha_\mathbf{a} = \lim_{n\to\infty} [T_{\mathbf{a}/n} - I]\ \mathrm{n}.$$

In the matrix representation (2.8), $T_{\mathbf{a}/n}$ is represented by the matrix

$$T_{\mathbf{a}/n} : \begin{bmatrix} 1 & 0 & a_1/n \\ 0 & 1 & a_2/n \\ 0 & 0 & 1 \end{bmatrix} \tag{3.4}$$

whence we obtain the element of the Lie algebra

$$\alpha_\mathbf{a} = \begin{bmatrix} 0 & 0 & a_1 \\ 0 & 0 & a_2 \\ 0 & 0 & 0 \end{bmatrix} \tag{3.5}$$

Example 2: $SU(2)$

Here we write

$$U = e^{ih} \tag{3.6}$$

in other words, $\alpha = ih$. The unitarity condition gives

$$UU^+ = e^{ih}e^{-ih^+} = I, \tag{3.7}$$

whence we have $U^+U = (UU^+)^+ = I = UU^+$. Hence U and U^+ commute; it follows that their logarithms h and h^+ also commute. When h and h^+ commute, one can write

$$e^{ih}e^{-ih^+} = e^{i(h-h^+)} = 1 = e^0$$

whence

$$h - h^+ = 0. \tag{3.8}$$

This shows that h is a 2×2 Hermitean matrix. Reversing the argument, we see that e^{ih} is unitary if h is Hermitean.

Since the matrix h is Hermitean, it can be diagonalised, and its eigenvalues are real. This means that one can find a unitary operation V such that the matrix

$$h_\Delta = VhV^{-1} \tag{3.9}$$

is diagonal. Equation (3.6) then shows that

$$U_\Delta = VUV^{-1} \tag{3.10}$$

is also diagonal. Denoting the eigenvalues of h by λ_1 and λ_2, one has

$$h_\Delta = \begin{pmatrix} \lambda_1 & 0 \\ 0 & \lambda_2 \end{pmatrix} \qquad \mathscr{U}_\Delta = \begin{pmatrix} e^{i\lambda_1} & 0 \\ 0 & e^{i\lambda_2} \end{pmatrix} \tag{3.11}$$

whence

$$\det \mathscr{U}_\Delta = e^{i\lambda_1}e^{i\lambda_2} = e^{i(\lambda_1+\lambda_2)} = e^{i\ \mathrm{Trace}\ h_\Delta} \tag{3.12}$$

so that $\det \mathscr{U}_\Delta = \det \mathscr{U}$ and $\mathrm{Trace}\ h_\Delta = \mathrm{Trace}\ h$. Therefore the unimodularity condition (2.5) on $\mathscr{U}$ is equivalent to

$$\mathrm{Trace}\ h = 0 \tag{3.13}$$

Thus the elements of the Lie algebra of $SU(2)$ are the *Hermitean matrices with zero trace*. One can readily check that they depend on three parameters, just like those in $SU(2)$.

Example 3: $SU(3)$

The treatment is exactly similar to that of $SU(2)$: the elements of the Lie algebra of $SU(3)$ are the Hermitean 3×3 matrices with zero trace.

Example 4: R_3

Note that the conditions (2.3) and (2.4) defining the matrices of R_3 are precisely the same as the conditions (2.7) defining the matrices of $SU(3)$, if one adds the extra condition that the matrices be real. It follows that R_3 is a *subgroup* of $SU(3)$. Consequently a matrix $R \in R_3$ must have the form

$$R = e^{ih}$$

h being Hermitean with zero trace. Since R must be real, it follows that ih is an antisymmetric matrix. (It is unnecessary to add 'and with zero trace', since this condition is obeyed automatically by any antisymmetric matrix.)

Hence the Lie algebra of the group of 3 dimensional rotations consists of the real antisymmetric 3×3 matrices which, as expected, do indeed depend on three parameters.

Example 5: $SL(2C)$

One can show (c.f. Problem 2) that the Lie algebra of $SL(2C)$ consists of the 2×2 matrices with zero trace.

In all three examples we note that the matrices of the Lie algebra form a linear vector space. They depend on n real parameters (n being the order of the group) $\lambda_1, \lambda_2, \ldots, \lambda_n$ through the formula

$$\alpha = \lambda_1 g_1 + \lambda_2 g_2 + \cdots + \lambda_n g_n \tag{3.14}$$

where α is an arbitrary matrix of the Lie algebra, and $g_1 \cdots g_n$ are n linearly independent matrices belonging to this algebra. The matrices $g_1 \cdots g_n$ are said to form a basis of the Lie algebra. They can be chosen arbitrarily as long as they are linearly independent.

Example 1: T_2

There are two generators, and we can take them as

$$g_1 = \begin{pmatrix} 0 & 0 & 1 \\ 0 & 0 & 0 \\ 0 & 0 & 0 \end{pmatrix} \quad g_2 = \begin{pmatrix} 0 & 0 & 0 \\ 0 & 0 & 1 \\ 0 & 0 & 0 \end{pmatrix} \tag{3.15}$$

so that

$$\alpha_a = a_1 g_1 + a_2 g_2 . \tag{3.16}$$

Example 2: $SU(2)$

Introduce the Pauli matrices:

$$\sigma_1 = \begin{pmatrix} 0 & 1 \\ 1 & 0 \end{pmatrix} \quad \sigma_2 = \begin{pmatrix} 0 & -i \\ i & 0 \end{pmatrix} \quad \sigma_3 = \begin{pmatrix} 1 & 0 \\ 0 & -1 \end{pmatrix} \tag{3.17}$$

Any Hermitean matrix with zero trace can be written as

$$h = \lambda_1 \sigma_1 + \lambda_2 \sigma_2 + \lambda_3 \sigma_3 . \tag{3.18}$$

Example 3: $SU(3)$

One can introduce 8 Hermitean matrices with zero trace:

$$\begin{aligned}
&g_1 = \begin{pmatrix} 0 & 1 & 0 \\ 1 & 0 & 0 \\ 0 & 0 & 0 \end{pmatrix} \quad g_2 = \begin{pmatrix} 0 & 0 & 1 \\ 0 & 0 & 0 \\ 1 & 0 & 0 \end{pmatrix} \quad g_3 = \begin{pmatrix} 0 & 0 & 0 \\ 0 & 0 & 1 \\ 0 & 1 & 0 \end{pmatrix} \\
&g_4 = \begin{pmatrix} 0 & -i & 0 \\ i & 0 & 0 \\ 0 & 0 & 0 \end{pmatrix} \quad g_5 = \begin{pmatrix} 0 & 0 & i \\ 0 & 0 & 0 \\ -i & 0 & 0 \end{pmatrix} \quad g_6 = \begin{pmatrix} 0 & 0 & 0 \\ 0 & 0 & -i \\ 0 & i & 0 \end{pmatrix} \\
&g_7 = \begin{pmatrix} 1 & 0 & 0 \\ 0 & -1 & 0 \\ 0 & 0 & 0 \end{pmatrix} \quad g_8 = \begin{pmatrix} 1 & 0 & 0 \\ 0 & 1 & 0 \\ 0 & 0 & -2 \end{pmatrix}
\end{aligned} \tag{3.19}$$

Then any Hermitean matrix with zero trace can be written as

$$h = \sum_{i=1}^{8} \lambda_i g_i$$

Example 4: R_3

$$J_1 = \begin{pmatrix} 0 & 0 & 0 \\ 0 & 0 & 1 \\ 0 & -1 & 0 \end{pmatrix} \qquad J_2 = \begin{pmatrix} 0 & 0 & -1 \\ 0 & 0 & 0 \\ 1 & 0 & 0 \end{pmatrix} \qquad J_3 = \begin{pmatrix} 0 & 1 & 0 \\ -1 & 0 & 0 \\ 0 & 0 & 0 \end{pmatrix} \tag{3.20}$$

From a geometric point of view it is clear from (3.2) that J_1, J_2, and J_3 generate rotations about the axes 1, 2, and 3 respectively.

Any antisymmetric 3 × 3 matrix with zero trace can be written as a linear combination of the matrixes (3.20).

Example 5: $SL(2c)$

One can introduce the 6 matrices

$$g_1 = \sigma_1 \quad g_2 = \sigma_2 \quad g_3 = \sigma_3 \quad g_4 = i\sigma_1 \quad g_5 = i\sigma_2 \quad g_6 = i\sigma_3 \,. \tag{3.21}$$

Any 2 × 2 matrix with zero trace is a linear combination of these matrices.

Infinitesimal generators:

We shall accept without proof that for any continuous group, an element of the Lie algebra can be expressed by the linear relation (3.14). Once $g_1 \cdots g_n$ are known, we see that all the matrices α of the Lie algebra can be constructed by the formula (3.14) and all elements of the group by the formula (3.1). In this sense these n matrices generate the group; they are called the *infinitesimal generators* of the group.

4. Commutation relations

If the group is Abelian, then for any two of its elements A and B one has $AB = BA$. If the group is not Abelian, then in general AB is different from BA.

The product $C = AB(BA)^{-1} = ABA^{-1}B^{-1}$ is an element of the group. It is easy to find the corresponding element of the Lie algebra:

$$\gamma = \log C = \lim_{n\to\infty} n[C^{1/n} - I]. \tag{4.1}$$

To this end, introduce the elements α and β of the Lie algebra which corresponds to A and B:

$$A = I + \alpha + \alpha^2/2! + \cdots \qquad B = I + \beta + \beta^2/2! + \cdots \tag{4.2}$$

A simple calculation shows that to second order in α and β one has

$$C = ABA^{-1}B^{-1} = I + (\alpha\beta - \beta\alpha) + \cdots \tag{4.3}$$

whence, by (1):

$$\gamma = \alpha\beta - \beta\alpha. \tag{4.4}$$

Thus the commutator of two elements of the Lie algebra is itself an element of the Lie algebra. In particular, if for α and β we chose the infinitesimal generators g_i and g_j, then by the formula (3.14) their commutator must be a linear combination of $g_1 \cdots g_N$; in other words

$$[g_i, g_j] = c_{ij}^k g_k. \tag{4.5}$$

There are $n(n - 1)/2$ relations of this kind; they are called the *commutation relations* of the group, and the $n^2(n - 1)/2$ coefficients c_{ijk} are called the *structure constants* of the group.

Example 1: T_2

This is an Abelian group: its generators commute.

Example 2: $SU(2)$

One checks immediately that

$$[\sigma_1, \sigma_2] = 2i\sigma_3 \qquad [\sigma_2, \sigma_3] = 2i\sigma_i \qquad [\sigma_3, \sigma_1] = 2i\sigma_2 \tag{4.6}$$

Example 3: R_3

A direct calculation yields

$$[J_1, J_2] = iJ_3 \qquad [J_2, J_3] = iJ_1 \qquad [J_3, J_1] = iJ_2. \tag{4.7}$$

The connection between $SU(2)$ *and* R_3

One notes that $SU(2)$ and R_3 have the same commutation relations, in the sense that the matrices J_i and $\frac{1}{2}\sigma_i$ have the same commutation relations.

This similarity can be pushed further. To this end it is convenient to associate a 2×2 matrix X with each real three-dimensional vector $\mathbf{x}$ having components (x_1, x_2, x_3) by means of

$$X = x_1\sigma_1 + x_2\sigma_2 + x_3\sigma_3 = \mathbf{x} \cdot \boldsymbol{\sigma} = \begin{pmatrix} x_3 & x_1 - ix_2 \\ x_1 + ix_2 & -x_3 \end{pmatrix} \tag{4.8}$$

The following relations are readily verified:

$$X = X^+ \qquad (X \text{ is Hermitean}) \tag{4.9}$$

$$\text{Trace } X = 0 \tag{4.10}$$

$$\det X = -\mathbf{x}^2 \tag{4.11}$$

The inverse relations

$$\text{Trace } \sigma_i\sigma_j = 2\delta_{ij} \tag{4.12}$$

allow one to find the components of the vector $\mathbf{x}$ from the matrix X, by means of the equation

$$x_i = \tfrac{1}{2}\,\mathrm{Trace}(\sigma_i X). \tag{4.13}$$

Let A be a matrix belonging to $SU(2)$, so that one has by definition

$$AA^+ = I \qquad \det A = 1. \tag{4.14}$$

If we write

$$X' = AXA^+ \tag{4.15}$$

then we can check immediately that X', like X, is a Hermitean matrix with zero trace, because of

$$X'^+ = AX^+A^+ = AXA^+ = X'$$

$$\mathrm{Trace}\; X' = \mathrm{Trace}\; AXA^+ = \mathrm{Trace}\; A^+AX = \mathrm{Trace}\; X = 0.$$

Hence one can write

$$X' = \mathbf{x}' \cdot \boldsymbol{\sigma}'. \tag{4.16}$$

From the fact that A is unimodular we deduce

$$-\mathbf{x}'^2 = \det X' = \det AXA^+ = \det A \det X \det A^+ = \det X = -\mathbf{x}^2 \tag{4.17}$$

The relationship between $\mathbf{x}'$ and $\mathbf{x}$ is linear, as is shown by the equation

$$x_i' = \tfrac{1}{2}\,\mathrm{Trace}(\sigma_j A \sigma_j x_j A^+) \tag{4.18}$$

Moreover this relationship is norm-preserving. Hence it is a rotation, either with or without inversion of the axes. We can see that in fact $\mathbf{x}'$ and $\mathbf{x}$ are related by a pure rotation (without inversion of axes), because (4.15) can be constructed from a series of consecutive transformations

$$X_1 = A^{1/n}X(A^+)^{1/n} \quad X_2 = A^{1/n}X_1(A^+)^{1/n} \cdots X_n = A^{1/n}X_{n-1}(A^+)^{1/n} = X'$$

where $A^{1/n}$ differs only little from the identity, and where $\mathbf{x}'$ consequently differs only little from $\mathbf{x}$.

Thus, with every matrix A of $SU(2)$ there is associated an element R of the group R_3. From the fact that the relationship (4.15) is quadratic in A we see that one and the same rotation R corresponds to both the matrices $+A$ and $-A$.

If A is taken to be an infinitesimal matrix

$$\alpha_A = \tfrac{1}{2} i \boldsymbol{\lambda} \cdot \boldsymbol{\sigma} \tag{4.19}$$

then it is easy to show by direct calculation that it corresponds to

$$\alpha_R = i\boldsymbol{\lambda} \cdot \mathbf{J}. \tag{4.20}$$

For instance, for $\boldsymbol{\lambda} = (0, 0, \lambda)$ one has

$$A = e^{i\frac{1}{2}\lambda\sigma_3} = \cos\frac{\lambda}{2} + i \sin\frac{\lambda}{2}\,\sigma_3$$

$$R = e^{i\lambda J_3}. \tag{4.21}$$

The fact that both the matrices $\pm A$ correspond to the same matrix R is apparent from equation (21) if one puts $\lambda = 2\pi$. Then one has $A = -I$, while $R = I$ because the eigenvalues of J_3 are $-1, 0, 1$. This means that a rotation by 2π, which is the identity, corresponds in $SU(2)$ to the matrix $-I$. (There is some topological subtlety here, which can be shown to derive from the fact that R_3 is doubly-connected while $SU(2)$ is simply connected.)

Apart from this topological subtlety, $SU(2)$ and R_3 are essentially the same group if one is interested only in the algebraic properties of the group operations, and if one is not concerned with the difference between their matrix representations. One says that $SU(2)$ and R_3 are two different representations of the same abstract group.

This is a general result, and one can show that two matrix groups which have the same commutation relations are always representations of the same abstract group, apart from certain topological subtleties.

5. Transformations of states

In order to discover the consequences of the invariance of interactions under a group G, we must investigate how the operations of the group act on a state.

We know that the states of a physical system correspond to the vectors of a Hilbert space, and that two vectors $|\lambda\rangle$ and $e^{i\varphi}|\lambda\rangle$, which differ only by a phase factor, correspond to the same state.

Information about a state can only be obtained by measurements. Apparatus designed to measure a quantity A (e.g. one component of the momentum) will react differently to different eigenstates of the operator $\hat{A}$ corresponding to A. Let us denote by $|\alpha\rangle$ the eigenstates of $\hat{A}$ and by α the corresponding eigenvalues, so that

$$\hat{A}\,|\,\alpha\rangle = \alpha\,|\,\alpha\rangle \tag{5.1}$$

Then there is a certain probability of finding the value α in a measurement of the quantity A. According to the axioms of quantum mechanics this probability is given by

$$P_\alpha = |\langle\alpha\,|\,\lambda\rangle|^2 \tag{5.2}$$

Suppose that we now apply a transformation to the system, for instance a rotation. This rotation could be envisaged as an *active* one: for instance, we rotate the accelerator which produces the particle. Alternatively it could be a *passive* one: for instance, we rotate the coordinate system to which we refer both the position of the accelerator and also the position of the detecting apparatus (Figure 3.5.1). It is easiest to adopt the passive point of view.

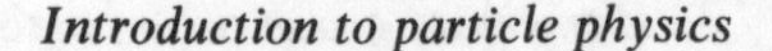

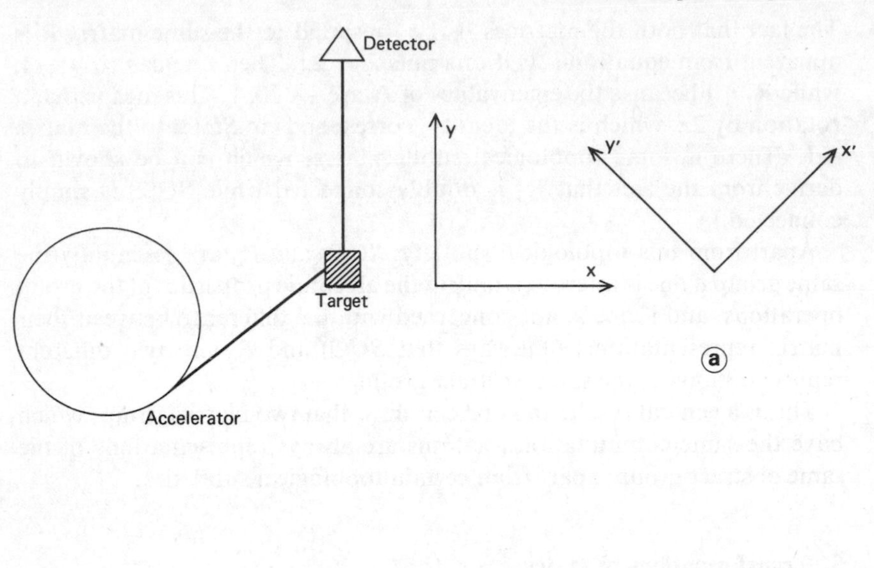

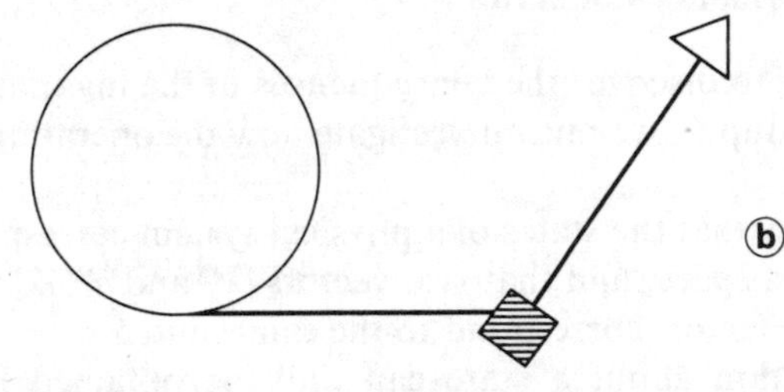

FIG. 3.5.1. Effect of a rotation on a measurement.
(a) Passive transformation: rotation of the coordinate axes;
(b) Active transformation: rotation of the apparatus.

Performing the same experiment after this change of reference frame, the detectors will read differently, and one must introduce new eigenvalues α_R which obey

$$\hat{A}_R \mid \alpha_R\rangle = \alpha_R \mid \alpha_R\rangle \tag{5.3}$$

where A_R is now the quantity measured by the apparatus but referred to the second coordinate system, and $\hat{A}_R$ is the corresponding operator.

The state of the system is now described by a new vector $|R\lambda\rangle$. The correspondence between $|\lambda\rangle$ and $|R\lambda\rangle$ must clearly be such that observable probabilities are left unchanged. In the light of equation (5.2) we must therefore have

$$|\langle\alpha_R \mid R\lambda\rangle|^2 = |\langle\alpha \mid \lambda\rangle|^2 \tag{5.4}$$

while the correspondence between $\langle\alpha_R|$ and $\langle\alpha|$ must maintain the superposition principle. Under these conditions it is possible to prove the following important theorem (Wigner):

Theorem

The correspondence between $|\lambda\rangle$ *and* $|R\lambda\rangle$ *is either unitary or antiunitary.*

Comment. By saying that the correspondence $|\lambda\rangle$ and $|R\lambda\rangle$ is unitary, we mean that there exists a linear relation

$$| R\lambda\rangle = \mathscr{U}_R \mid \lambda\rangle$$

such that

$$\mathscr{U}_R\mathscr{U}_R^+ = I.$$

By saying that the correspondence is antiunitary we mean that there exists an *antilinear* relation

$$| R\lambda\rangle = V_R K \mid \lambda\rangle$$

where K is the complex conjugation operator defined by

$$K(\alpha_1 \mid \lambda_1\rangle + \alpha_2 \mid \lambda_2\rangle) = \alpha_1{}^*K \mid \lambda_1\rangle + \alpha_2{}^*K \mid \lambda_2\rangle$$

and where V_R is unitary. Such a transformation clearly conserves the moduli of scalar products. In fact

$$\langle R\alpha \mid R\lambda\rangle = \langle\alpha \mid \overleftarrow{K} V_R{}^+ V_R \overrightarrow{K} \mid \lambda\rangle = \langle\alpha \mid \overleftrightarrow{KK} \mid \lambda\rangle = \langle\alpha \mid \lambda\rangle^*$$

where the arrow shows in which direction the complex conjugation is put into effect.

In cases where the transformations R form a continuous group, it seems convenient to preserve the continuity of the transformations R by assigning state vectors differing only infinitesimally to the result of transformations which differ only infinitesimally. This can be achieved only if the operations are *unitary*. Indeed, suppose that two transformations R_1 and R_2 both very close to the identity are represented by antiunitary operators, and that the group multiplication law is preserved; then one would have

$$| R_1R_2\rangle = V_1KV_2K \mid \lambda\rangle = V_1V_2{}^* \mid \lambda\rangle,$$

in other words, the transformation R_1R_2, which is also very close to the identity, would be represented by a unitary operator.

One must investigate separately transformations which cannot be realised by continuous variation starting from the identity, as for instance space reflection or time reversal; and antiunitary operations may indeed be needed to represent them.

6. Representations of continuous groups

We have seen that an operation R_1, belonging to a continuous group, and capable of being generated by continuous variation starting from the identity, is represented by a unitary transformation $\mathscr{U}_{R_1}$ acting on the state vectors. If one acts on a state $|\lambda\rangle$ with the two operations R_1 and R_2 successively, one obtains a new state vector

$$\mathscr{U}_{R_2}\mathscr{U}_{R_1} \mid \lambda\rangle. \tag{6.1}$$

On the other hand, the product of the two operations R_1 and R_2 is simply another operation of the group, $R = R_2R_1$, so that the state which results can also be written as

$$\mathscr{U}_R \mid \lambda\rangle = \mathscr{U}_{R_2R_1} \mid \lambda\rangle. \tag{6.2}$$

Hence (6.1) and (6.2) represent the same state. It is unfortunate that because of the indistinguishability of two vectors differing only by a phase factor, one cannot conclude that the two vectors (6.1) and (6.2) are identical, but only that they are related by

$$\mathscr{U}_{R_2R_1} \mid \lambda\rangle = e^{i\varphi(R_1,R_2)}\mathscr{U}_{R_2}\mathscr{U}_{R_1} \mid \lambda\rangle. \tag{6.3}$$

It is very delicate task to analyse the phase factor $\varphi(R_1, R_2)$. Nevertheless it can be shown that the factor $e^{i\varphi(R_1,R_2)}$ can be set equal to 1 in all the cases that we shall have to consider, and in particular for the rotation group. This result is far from self-evident, for there are cases of physical interest in which no such simplification is possible. It depends essentially on the structure of the group under consideration. (For instance, one can show that the phase factor must differ from unity when considering the transformations of Galilean reference frames with no changes in time, under which non-relativistic quantum mechanics is invariant.) To elucidate this point we formulate the following proposition:

Proposition:

For all the continuous groups G which we shall have occasion to consider, and for all group operations R capable of being generated by continuous variation starting from the identity, the action of R on a state vector $|\lambda\rangle$ is represented by a unitary operation $\mathscr{U}_R$ such that the transformed state vector is $\mathscr{U}_R|\lambda\rangle$. This transformation enjoys the following group properties:

a) $$U_{R_1}U_{R_2} = U_{R_1R_2} \tag{6.4}$$

b) $$U_e = I \tag{6.5}$$

e being the unit element of G;

c) $$U_RU_{R^{-1}} = I \tag{6.6}$$

by virtue of (6.4) and (6.5). In fact because of $U_R U_R^+ = I$, one has

$$U_{R^{-1}} = U_R^+ = (U_R)^{-1} \tag{6.7}$$

(d) *Continuity:* as R tends to a fixed group element R_0, one has $U_R |\lambda\rangle \rightarrow U_{R_0} |\lambda\rangle$. More precisely, we mean by this limit that for any vector $|\alpha\rangle$

$$\langle\alpha \mid U_R \mid \lambda\rangle \rightarrow \langle\alpha \mid U_{R_0} \mid \lambda\rangle.$$

The U_R are said to constitute a *unitary representation* of the group G.

Representation of infinitesimal operations

Recall that R is defined as a function of the element $\alpha = \log R$ of the Lie algebra of G by

$$R = e^\alpha = I + \alpha + \alpha^2/2! + \cdots \tag{6.8}$$

It is always possible to define uniquely the logarithm iH of a unitary operator $\mathscr{U}_R$ so that

$$\mathscr{U}_R = e^{iH}. \tag{6.9}$$

As we have seen already, the unitarity condition $UU^+ = I$ ensures that H is a Hermitean operator. It follows that the elements of the Lie algebra of a group are represented by anti-Hermitean operators (i.e. by Hermitean operators multiplied by i).

Given a basis of infinitesimal generators $g_1 \cdots g_n$ of the Lie algebra, such that every element of the algebra can be written as

$$\alpha = \lambda_1 g_1 + \cdots + \lambda_n g_n \,, \tag{6.10}$$

and if the operators $iG_1, iG_2, \ldots, iG_n$ represent $g_1 \cdots g_n$ while iH represents α, then one can show

$$H = \lambda_1 G_1 + \cdots + \lambda_n G_n \tag{6.11}$$

The G_i obey the same commutation relations as do the g_i, up to factors of i, so that

$$[G_i G_j] = -ic_{ij}^k G_k \tag{6.12}$$

(c.f. problem 3).

7. Representations of the rotation group

We shall investigate in greater detail the case where G is the group of rotations in three dimensions. Actually the unitary representations of the rotation group are well known. Since we shall need them repeatedly, we recall the most important results.

1. *Finite representations*

There exist infinitely many finite-dimensional unitary representations; they include infinitely many irreducible representations. This means that every such representation operates in a finite dimensional Hilbert space E, and that E has no subspace which is globally invariant under all the operations in the group. More specifically, one can say that it is impossible to find a basis in E such that the matrices representing the rotations all have block-diagonal form (Figure 3.7.1).

The total angular momentum $\mathbf{J}^2 = J_1{}^2 + J_2{}^2 + J_3{}^2$ is sharply defined in each irreducible representation: every vector of E is an eigenvector of $\mathbf{J}^2$ belonging to the eigenvalue $j(j+1)$, where j is a non-negative integer or half integer, $j = 0, \frac{1}{2}, 1, \frac{3}{2}, \ldots$. An irreducible representation can therefore be specified by $\{j\}$.

In the space E one can define a basis which we shall denote by E_j, by means of the eigenvectors of J_3. In E_j the eigenvalues m of J_3 assume the $(2j+1)$ distinct values $(-j, -j+1, \ldots, j-1, j)$. They are non-degenerate, so that the dimensionality of E_j is $(2j+1)$, and the basis vectors can be denoted by $|jm\rangle$. One has

$$\mathbf{J}^2 \,|\, jm\rangle = j(j+1) |\, jm\rangle \tag{7.1}$$

$$J_3 \,|\, jm\rangle = m \,|\, jm\rangle. \tag{7.2}$$

The representation of the infinitesimal generators is uniquely determined if one writes

$$J^+ = J_1 + iJ_2 \qquad J^- = J_1 - iJ_2 \tag{7.3}$$

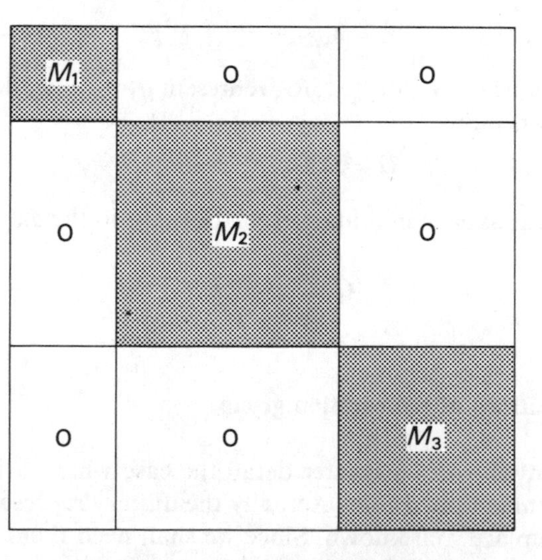

FIG. 3.7.1. Block-diagonal matrix.

and if one choses the relative phases of the basis vectors so that

$$J^+ \mid jm\rangle = \sqrt{j(j+1) - m(m+1)} \mid j, m+1\rangle \tag{7.4}$$

$$J^- \mid jm\rangle = \sqrt{j(j+1) - m(m-1)} \mid j, m-1\rangle. \tag{7.5}$$

We see that J^+ increases m by 1, while J^- diminishes m by 1. (For this reason Hermann Weyl called them the sharp and flat operators.)

2. *Special cases*

For $j = \frac{1}{2}$ the representation $\{\frac{1}{2}\}$ is simply $SU(2)$, and the infinitesimal generators are represented by

$$\mathbf{J} = \tfrac{1}{2}\boldsymbol{\sigma} \tag{7.6}$$

For $j = 1$ the representation $\{1\}$ coincides with the group of concrete three-dimensional rotations defined as above.

3. *Euler angles*

It is convenient to specify a rotation by its Euler angles (α, β, γ) defined as follows: consider a co-ordinate frame $(Oxyz)$, and another frame $(Ox'y'z')$ got from the first by a rotation R. This rotation can be obtained by performing the following operations successively:

(a) A rotation about the axis Oz through an angle α, which transforms $(Oxyz)$ into (Ox_1y_1z). Call this rotation R_1.
(b) A rotation about the axis Oy_1 through an angle β, which transforms (Ox_1y_1z) into (Ox_2y_1z'). Call this R_2.
(c) A rotation about the axis Oz' through an angle γ, which transforms (Ox_2y_1z') into $(Ox'y'z')$. Call this R_3.

Hence (Figure 3.7.2)

$$R = R_3R_2R_1\,.$$

Note that α and β are simply the polar angles of Oz' in spherical polar co-ordinates.

4. *Rotation matrices*

It can be shown that R is given by

$$R = e^{-i\alpha J_3}e^{-i\beta J_2}e^{-i\gamma J_3} \tag{7.8}$$

(see, for instance, M. E. Rose, *Elementary theory of angular momentum*, (1957), Wiley). Hence the elements of the rotation matrix R in E_j, which by

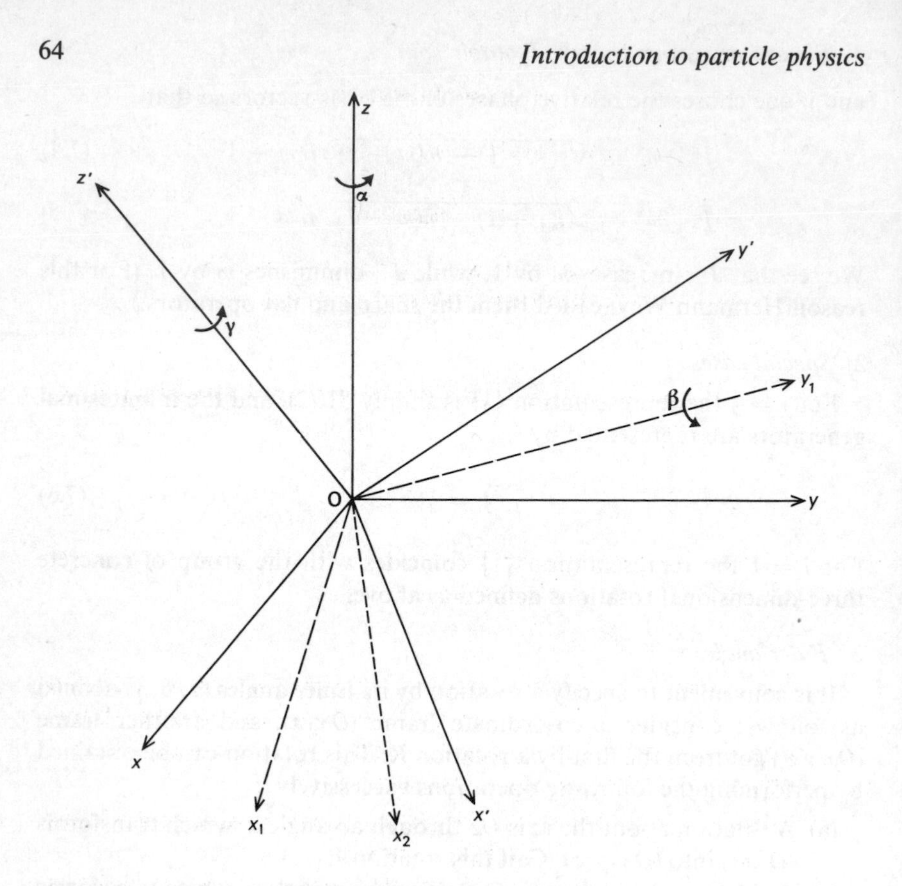

FIG. 3.7.2. The Euler angles.

definition are given according to (7.8) by

$$\begin{aligned}\mathscr{D}^j_{m'm}(\alpha, \beta, \gamma) &= \langle jm' \,|\, e^{-i\alpha J_3} e^{-i\beta J_2} e^{-i\gamma J_3} \,|\, jm\rangle \\ &= e^{-im'\alpha} \langle jm' \,|\, e^{-i\beta J_2} \,|\, jm\rangle \, e^{-im\gamma}\end{aligned} \tag{7.9}$$

If a new matrix $d^j_{m'm}(\beta)$ is defined by

$$\langle jm' \,|\, e^{-i\beta J_2} \,|\, jm\rangle = d^j_{m'm}(\beta) \tag{7.10}$$

one has therefore

$$\mathscr{D}^j_{m'm}(\alpha, \beta, \gamma) = e^{-im'\alpha} \, d^j_{m'm}(\beta) \, e^{-im\gamma} \tag{7.11}$$

$d^j_{m'm}(\beta)$ can be calculated explicitly from the matrix elements of **J** and the definition (7.10). This gives, in particular,

$$d^{(1/2)}(\beta) = \begin{array}{c|cc|} & m = 1/2 & m = -1/2 \\ \hline m' = 1/2 & \cos\beta/2 & -\sin\beta/2 \\ m' = -1/2 & \sin\beta/2 & \cos\beta/2 \end{array} \tag{7.12}$$

$$d^{(1)}(\beta) = \begin{array}{c} \\ m' = 1 \\ m' = 0 \\ m' = -1 \end{array} \begin{array}{c} \begin{array}{ccc} m = 1 & m = 0 & m = -1 \end{array} \\ \begin{bmatrix} 1/2(1+\cos\beta) & -\frac{1}{\sqrt{2}}\sin\beta & \frac{1}{2}(1-\cos\beta) \\ \frac{1}{\sqrt{2}}\sin\beta & \cos\beta & -\frac{1}{\sqrt{2}}\sin\beta \\ \frac{1}{2}(1-\cos\beta) & \frac{1}{\sqrt{2}}\sin\beta & \frac{1}{2}(1+\cos\beta) \end{bmatrix} \end{array} \tag{7.13}$$

5. *Symmetry properties of the rotation matrices*

By inspecting equations (7.4) and (7.5) one can establish certain symmetry properties of the $d^j_{m'm}(\beta)$, namely

$$d^j_{m'm}(-\beta) = d^j_{mm'}(\beta) \tag{7.14}$$

$$d^j_{m'm}(\pi) = (-1)^{j+m'}\,\delta_{m',-m} \tag{7.15}$$

$$d^j_{m'm}(-\pi) = (-1)^{j-m'}\,\delta_{m',-m} \tag{7.16}$$

$$d^j_{m'm}(\beta+\pi) = (-1)^{j+m'}\,d^j_{-m'm}(\beta) \tag{7.17}$$

$$d^j_{m'm}(\pi-\beta) = (-1)^{j+m'}\,d^j_{m-m'}(\beta) \tag{7.18}$$

$$d^j_{m'm}(\beta) = (-1)^{m'-m}\,d^j_{-m'-m}(\beta) \tag{7.19}$$

$$d^j_{m'm}(\beta) = (-1)^{m'-m}\,d^j_{mm'}(\beta) \tag{7.20}$$

6. *Invariant measure*

It is possible to introduce a measure on the rotations, (i.e. a volume element for integration), which is independent of our choice of reference frame. The simplest way to calculate this measure is to evaluate the a priori probability, assuming space to be isotropic, that a randomly chosen direction in space is specified by Euler angles which fall into the range $d\alpha\, d\beta\, d\gamma$ about (α, β, γ).

The probability of thus chosing the axis Oz' is evidently measured by the area $\sin\beta\, d\beta\, d\alpha$ on the unit sphere. Once the axis Oz' is fixed, there is a probability $d\gamma$ that the axis Ox' falls into its indicated plane. Therefore the invariant measure on the rotations is $\sin\beta\, d\alpha\, d\beta\, d\gamma$, and one has

$$\int_0^{2\pi} d\alpha \int_0^{\pi} \sin\beta\, d\beta \int_0^{2\pi} d\gamma = 8\pi^2. \tag{7.21}$$

7. *Orthogonality properties of the rotation matrices*

For the rotation matrices one can prove the following orthogonality relation involving the invariant measure:

$$\frac{1}{8\pi^2}\int_0^{2\pi} d\alpha \int_0^{\pi} \sin\beta\, d\beta \int_0^{2\pi} d\gamma\, \mathscr{D}^{j_1 *}_{m_1' m_1}(\alpha,\beta,\gamma)\, \mathscr{D}^{j_2}_{m_2' m_2}(\alpha,\beta,\gamma)$$

$$= \frac{1}{2j_1+1}\, \delta_{m_1' m_2'}\, \delta_{m_1 m_2}\, \delta_{j_1 j_2} \tag{7.22}$$

This formula expresses the fact that the $\mathscr{D}^j_{m'm}(\alpha,\beta,\gamma)$ constitute a system of orthogonal functions of α, β, and γ. It can be shown that this system is complete.

8. *Spherical harmonics*

Consider the case where j is an integer ℓ. Then the group composition law for the matrices $\mathscr{D}^j$ is written as

$$\sum_m \mathscr{D}^j_{m'm}(R_1)\, \mathscr{D}^j_{mm''}(R_2) = \mathscr{D}^j_{m'm''}(R_1 R_2). \tag{7.23}$$

Putting $m'' = 0$ in this equation, one sees from eq. (7.11) that $\mathscr{D}^j_{m0}(R_2)$ depends only on α_2 and β_2, but not on γ_2. Let us define $\alpha_2 = \varphi$, $\beta_2 = \theta$. In view of eq. (7.22) one notices that on putting

$$Y_\ell^m(\theta,\varphi) = \left(\frac{2\ell+1}{4\pi}\right)^{1/2} \mathscr{D}^j_{m0}(\varphi,\theta,0) \tag{7.24}$$

one gets

$$\int Y_\ell^{m*}(\theta,\varphi)\, Y_\ell^m(\theta,\varphi) \sin\theta\, d\theta\, d\varphi = 1. \tag{7.25}$$

The $Y_\ell^m(\theta,\varphi)$ are called spherical harmonics. According to equation (7.23) they constitute a basis for the representation $\{\ell\}$, since

$$\sum_m \mathscr{D}^\ell_{m'm}(R)\, Y_\ell^m(\theta,\varphi) = Y_\ell^{m'}(R\theta, R\varphi).$$

They are related in particular to the Legendre polynomials:

$$\mathscr{D}^\ell_{00}(\theta) = P_\ell(\cos\theta). \tag{7.26}$$

The group multiplication rule for the matrices $\mathscr{D}^j$ yields the following important formula:

$$\frac{4\pi}{2\ell+1}\sum_m Y^*_{\ell m}(\theta,\varphi)\, Y_{\ell m}(\theta'\varphi') = P_\ell(\cos\theta\cos\theta' + \sin\theta\sin\theta'\cos(\varphi-\varphi')). \tag{7.27}$$

9. *Properties of unitary representations*

It is possible to prove the following theorem, which formulates the fact that knowledge of the representations $\{j\}$ amounts to knowledge of all unitary representations.

Theorem

Every representation of the rotation group, finite or infinite, is a direct sum of irreducible representations. In other words, there always exists a basis in which the representation assumes block-diagonal form.

10. *Direct product of two representations*

As an example of the combination of two angular momenta, we can consider the eigenstates $|j_1m_1\rangle$ of a particle's orbital angular momentum and the eigenstates $|j_2m_2\rangle$ of its spin, and investigate the total angular momentum.

In the light of the above theorem, this is a meaningful problem. The vectors $|j_1m_1\rangle\,|j_2m_2\rangle$ constitute a unitary representation of the rotation group, characterised by

$$\mathscr{U}_R\,|j_1m_1\rangle\,|j_2m_2\rangle = \mathscr{D}^{j}_{m_1'm_1}(R)\,\mathscr{D}^{j}_{m_2'm_2}(R)\,|j_1m_1'\rangle\,|j_2m_2'\rangle. \tag{7.29}$$

It is important to note the order of the indices, which corresponds to the definition (7.9).

This representation is reducible. Hence there exists a basis $|jm\rangle$ such that

$$|j_1m_1\rangle\,|j_2m_2\rangle = \sum_{j,m}\langle jm\,|\,j_1m_1j_2m_2\rangle\,|jm\rangle. \tag{7.30}$$

It is easy to show that the only value of m which can occur in (7.30) is $m = m_1 + m_2$. Indeed, if equation (7.29) is restricted to rotations about the z axis, one obtains

$$\begin{aligned}\mathscr{U}_R\,|j_1m_1\rangle\,|j_2m_2\rangle &= e^{im_1\varphi}e^{im_2\varphi}\,|j_1m_1\rangle\,|j_2m_2\rangle\\ &= e^{i(m_1+m_2)\varphi}\,|j_1m_1\rangle\,|j_2m_2\rangle\end{aligned} \tag{7.31}$$

whence, because of

$$\mathscr{U}_R = e^{iJ_3\varphi} \tag{7.32}$$

there follows

$$m = m_1 + m_2\,. \tag{7.33}$$

It turns out, and this is peculiar to the rotation group, that the basis in which $\mathscr{U}_R$ is block diagonal is uniquely determined by the value of j and m. In other words, a given value of j appears only once when combining j_1 and j_2. More precisely, one can show that j assumes the values

$$|j_1 - j_2|,\ |j_1 - j_2| + 1, \ldots, j_1 + j_2 \tag{7.34}$$

each only once. The scalar products $\langle jm \mid j_1m_1j_2m_2\rangle$ are numbers which are determined uniquely if the phases of the vectors are chosen once and for all. They are called *Clebsch–Gordan coefficients*.

11. *Properties of the Clebsch–Gordan coefficients*

One has the orthogonality condition

$$\sum_{j,m} \langle j_1j_2m_1m_2 \mid jm\rangle\langle jm \mid j_1j_2m_1'm_2'\rangle = \delta_{m_1m_1'} \cdot \delta_{m_2m_2'} \tag{7.35}$$

which is simply the self evident relation

$$\langle j_1m_1 \mid \langle j_2m_2 \mid I \mid j_1m_1'\rangle \mid j_2m_2'\rangle = \delta_{m_1m_1'} \cdot \delta_{m_2m_2'}$$

and where one has exploited the fact that the system $|jm\rangle$ is orthonormal and complete in order to replace the unit operator I by

$$\sum_{j,m} \mid jm\rangle\langle jm \mid .$$

Occasionally we shall substitute for $\langle jm \mid j_1j_2m_1m_2\rangle$ the notation

$$C(j_1\,j_2m_1m_2\,;jm)$$

or

$$C\begin{pmatrix} j_1j_2j \\ m_1m_2m \end{pmatrix}.$$

12. *Further properties of rotation matrices*

The self-evident relation

$$\langle j_1m_1' \mid \langle j_2m_2' \mid \mathscr{U}_R \mid j_1m_1\rangle \mid j_2m_2\rangle$$
$$= \sum_{jm} \langle j_1\,j_2m_1'm_2' \mid jm'\rangle\langle jm' \mid \mathscr{U}_R \mid jm\rangle\langle jm \mid j_1\,j_2m_1m_2\rangle$$

can be written as

$$\mathscr{D}^{j_1}_{m_1'm_1}(R)\,\mathscr{D}^{j_2}_{m_2'm_2}(R) = \sum_{jm} C(j_1\,j_2m_1'm_2';jm')\,\mathscr{D}^{j}_{m'm}(R)\,C(j_1\,j_2m_1m_2\,;jm). \tag{7.36}$$

13. *Numerical values of Clebsch–Gordan coefficients*

We give the values of the Clebsch–Gordan coefficients for the cases where j_2 is $\frac{1}{2}$ or 1. Then one has

$C(j_1\,\frac{1}{2}m_1m_2\,;jm)$ (7.37)

$j =$	$m_2 = 1/2$	$-1/2$
$j_1 + j_2$	$\sqrt{\dfrac{j_1 + m + 1/2}{2j_1 + 1}}$	$\sqrt{\dfrac{j_1 - m + 1/2}{2j_1 + 1}}$
$j_1 - 1/2$	$-\sqrt{\dfrac{j_1 - m + 1/2}{2j_1 + 1}}$	$\sqrt{\dfrac{j_1 + m + 1/2}{2j_1 + 1}}$

$C(j_1\,1 m_1 m_2\,; jm)$ (7.38)

j \ m_2	1	0	−1
$j+1$	$\sqrt{\frac{(j_1+m)(j_1+m+1)}{(2j_1+1)(2j_1+2)}}$	$\sqrt{\frac{(j_1-m+1)(j_1+m+1)}{(2j_1+1)(j_1+1)}}$	$\sqrt{\frac{(j_1-m)(j_1-m+1)}{(2j_1+1)(2j_1+2)}}$
j_1	$\sqrt{\frac{(j_1+m)(j_1-m+1)}{2j_1(j_1+1)}}$	$\sqrt{\frac{m}{j_1(j_1+1)}}$	$\sqrt{\frac{(j_1-m)(j_1+m+1)}{2j_1(j_1+1)}}$
$j-1$	$\sqrt{\frac{(j_1-m)(j_1-m+1)}{2j_1(2j_1+1)}}$	$\sqrt{\frac{(j_1-m)(j+m)}{j_1(2j+1)}}$	$\sqrt{\frac{(j_1+m+1)(j_1+m)}{2j(2j_1+1)}}$

14. *Tensor operators*

A tensor operator is a set of operators $T_M{}^J$ transforming under rotations like a vector of the irreducible representation $\{J\}$. In other words, if one applies a rotation R to the co-ordinate axes, then the states transform according to

$$|\alpha'\rangle = \mathscr{U}_R\,|\alpha\rangle$$

so that the matrix elements of any arbitrary operator A become

$$\langle\beta'\,|\,A\,|\,\alpha'\rangle = \langle\beta\,|\,\mathscr{U}_R{}^+A\mathscr{U}_R\,|\,\alpha\rangle = \langle\beta\,|\,A'\,|\,\alpha\rangle$$

where

$$A' = \mathscr{U}_R{}^+A\mathscr{U}_R\,. \tag{7.39}$$

For a tensor operator, one has by definition

$$\mathscr{U}_R{}^+T_M{}^J\mathscr{U}_R^+ = \sum_{M'}\mathscr{D}^J_{MM'}(R)\,T^J_{M'}\,. \tag{7.40}$$

Examples. The momentum is a tensor operator of rank 1, with

$$P^1_0 = P_z \qquad P^1_1 = P_x + iP_y \qquad P^1_{-1} = P_x - iP_y \tag{7.41}$$

The electric and magnetic dipole moments of an atom transform like vectors under rotations: they are tensor operators of rank 1. The quadrupole moment is a tensor operator of rank 2. The charge is a tensor operator of rank 0.

Such operators obey a fundamental theorem:

15. *Wigner–Eckart theorem*

Matrix element of a tensor operator of the type

$$\langle j_1m'\beta\,|\,T_M{}^J\,|\,j_2m\alpha\rangle$$

(where α and β are quantum numbers other than angular momentum) can be written in the form

$$C(j_1m'JM; j_2m)\langle\beta j_1 \| T^J \| j_2\alpha\rangle \tag{7.42}$$

where the reduced matrix elements $\langle\beta j_1 \| T^J \| j_2\alpha\rangle$ do not depend on m_1, m_2, and M.

This theorem has two important applications:

1. For given j_1, j_2, α, and β, it reduces the calculation of matrix elements of a tensor operator to the calculation of a single matrix element, which then determines the reduced matrix element.
2. It leads to selection rules: the Clebsch–Gordan coefficient occurring in equation (7.42) vanishes when j_1, j_2, and J fail to satisfy the triangle inequalities.

16. *Miscellaneous points*

We list some useful formulae in no particular order.

(a) *Spherical harmonics of ranks* 0, 1, *and* 2

$$Y_0^0(\theta, \varphi) = \frac{1}{\sqrt{4\pi}} \quad Y_1^0(\theta, \varphi) = \sqrt{\frac{3}{4\pi}}\cos\theta \quad Y_1^{\pm 1}(\theta, \varphi) = \mp\sqrt{\frac{3}{4\pi}}\sin\theta e^{\pm i\varphi}$$

$$Y_2^0(\theta, \varphi) = \sqrt{\frac{5}{16\pi}}(3\cos^2\theta - 1) \quad Y_2^{\pm}(\theta, \varphi) = \mp\sqrt{\frac{15}{8\pi}}\cos\theta\sin\theta\, e^{\pm i\varphi} \tag{7.43}$$

$$Y_2^{\pm 2}(\theta, \varphi) = \sqrt{\frac{15}{32\pi}}\sin^2\theta\, e^{\pm 2i\varphi}$$

(b) *Properties of the Pauli matrices*

$$\sigma_1\sigma_2 = i\sigma_3 \qquad \sigma_2\sigma_3 = i\sigma_1 \qquad \sigma_3\sigma_1 = i\sigma_2 \qquad \sigma_1^2 = \sigma_2^2 = \sigma_3^2 = 1 \tag{7.44}$$

$$(\boldsymbol{\sigma}\cdot\mathbf{a})(\boldsymbol{\sigma}\cdot\mathbf{b}) = \mathbf{a}\cdot\mathbf{b} + i(\boldsymbol{\sigma}, \mathbf{a}, \mathbf{b}) \tag{7.45}$$

$$(\boldsymbol{\sigma}\cdot\mathbf{a})(\boldsymbol{\sigma}\wedge\mathbf{b}) = \mathbf{a}\wedge\mathbf{b} - i\mathbf{a}(\boldsymbol{\sigma}\cdot\mathbf{b}) + i\boldsymbol{\sigma}(\mathbf{a}\cdot\mathbf{b}) \tag{7.46}$$

$$(\boldsymbol{\sigma}\wedge\mathbf{b})(\boldsymbol{\sigma}\cdot\mathbf{a}) = \mathbf{a}\wedge\mathbf{b} + i\mathbf{a}(\boldsymbol{\sigma}\cdot\mathbf{b}) - i\boldsymbol{\sigma}(\mathbf{a}\cdot\mathbf{b}) \tag{7.47}$$

$$\begin{aligned}(\boldsymbol{\sigma}\cdot\mathbf{a})(\boldsymbol{\sigma}\wedge\mathbf{b})(\boldsymbol{\sigma}\cdot\mathbf{c}) = {} & (\mathbf{a}\wedge\mathbf{b})(\boldsymbol{\sigma}\cdot\mathbf{c}) + i\mathbf{a}(\mathbf{b}\cdot\mathbf{c}) - \mathbf{a}(\boldsymbol{\sigma}, \mathbf{b}, \mathbf{c}) \\ & - i\mathbf{c}(\mathbf{a}\cdot\mathbf{b}) - (\boldsymbol{\sigma}\wedge\mathbf{c})(\mathbf{a}\cdot\mathbf{b})\end{aligned} \tag{7.48}$$

$$\begin{aligned}(\boldsymbol{\sigma}\cdot\mathbf{a})(\boldsymbol{\sigma}\cdot\mathbf{b})(\boldsymbol{\sigma}\cdot\mathbf{c}) = {} & (\boldsymbol{\sigma}\cdot\mathbf{a})(\mathbf{b}\cdot\mathbf{c}) \\ & - (\boldsymbol{\sigma}\cdot\mathbf{b})(\mathbf{a}\cdot\mathbf{c}) + (\boldsymbol{\sigma}\cdot\mathbf{c})(\mathbf{a}\cdot\mathbf{b}) + i(\mathbf{a}, \mathbf{b}, \mathbf{c})\end{aligned} \tag{7.49}$$

8. The concept of isotopic spin

Let us return to the problem of multiplets, and see what properties we can expect of the invariance group of the strong interactions:

1. The strong interactions are invariant under a continuous group G. The operations in this group transform a state of a particle into another state with the same strangeness and the same baryon number, but in general with a different charge.
2. The states corresponding to a multiplet of particles form a set of basis vectors for a representation of G.
3. G has unitary representations with the dimensionalities 1, 2, and 3 corresponding respectively to the Λ^0, the nucleon, and the pion.
4. There is only particle for a given representation and a given charge state (e.g. there is only one π^+).

Elie Cartan has classified all continuous groups having finite unitary representations. Only one of them satisfies condition 4. This condition implies that in the Lie algebra one cannot find two independent commuting operators. (The group is then said to be of rank 1.) If two such operators did exist, one would be the charge and the other an operator A diagonalisable simultaneously with the charge. Apart from very special cases that could arise by accident, a multiplet would then contain several particles with the same charge but belonging to different eigenvalues of A.

The only group satisfying condition 4 is $SU(2)$, or R_3, which amounts to the same thing in practice. This group also satisfies condition 3 since it does indeed have representations of dimensionalities 1, 2, and 3. It is remarkable that this is actually the only group satisfying 3.

In this particular case the generators of R_3 are denoted by I_1, I_2, and I_3. By analogy with angular momentum, I_1, I_2, and I_3 are called the *isotopic spin* operators and in this context R_3 is called the isotopic spin group or *charge independence* (or charge invariance). Each representation is specified by an eigenvalue $\mathbf{I}^2 = I_1{}^2 + I_2{}^2 + I_3{}^2 = I(I+1)$, and has the dimensionality $(2I + 1)$.

Isotopic spin is a misnomer, because the different members of a multiplet are not isotopes, having in fact different charges. On the contrary, they have the same mass, and form isobars. The more correct appellation of isobaric spin has been proposed, but is having difficulty in supplanting customary usage.*

Classification of particles

We note the isospin properties of various hadrons:

1. There are three π's, whence $I = 1$. By convention we shall attribute

* Translator's note: the expressions i-spin and isospin are frequently used in English.

increasing values of I_3 to particles with increasing charge, which gives

$$I_3 \mid \pi^+\rangle = \mid \pi^+\rangle \qquad I_3 \mid \pi^0\rangle = 0 \qquad I_3 \mid \pi^-\rangle = - \mid \pi^-\rangle \tag{8.1}$$

2. K. There are two K's, whence $I = \frac{1}{2}$

$$I_3 \mid K^+\rangle = 1/2 \mid K^+\rangle \qquad I_3 \mid K^0\rangle = -1/2 \mid K^0\rangle \tag{8.2}$$

3. $\bar{K}$.

$$I_3 \mid \bar{K}^0\rangle = 1/2 \mid \bar{K}^0\rangle \qquad I_3 \; K^-\rangle = -1/2 \mid K^-\rangle \tag{8.3}$$

4. *Nucleons:* $I = \frac{1}{2}$

$$I_3 \mid p\rangle = 1/2 \mid p\rangle \qquad I_3 \mid n\rangle = -1/2 \mid n\rangle \tag{8.4}$$

5. $\Lambda . I = 0$

$$I_3 \mid \Lambda\rangle = 0 \tag{8.5}$$

6. $\Sigma . I = 1$

$$I_3 \mid \Sigma^+\rangle = \mid \Sigma^+\rangle \qquad I_3 \mid \Sigma^0\rangle = 0 \qquad I_3 \mid \Sigma^-\rangle = - \mid \Sigma^-\rangle \tag{8.6}$$

7. $\Xi . I = \frac{1}{2}$

$$I_3 \mid \Xi^+\rangle = 1/2 \mid \Xi^0\rangle \qquad I_3 \mid \Xi^-\rangle = -1/2 \mid \Xi^-\rangle \tag{8.7}$$

8. $\bar{N} . I = \frac{1}{2}$

$$I_3 \mid \bar{n}\rangle = 1/2 \mid \bar{n}\rangle \qquad I_3 \mid \bar{p}\rangle = -1/2 \mid \bar{p}\rangle \tag{8.8}$$

The Gell-Mann–Nishijima formula

It is clear from the table above that the charge is connected with I_3, B, and S by

$$\boxed{Q = I_3 + \frac{B + S}{2}} \tag{8.9}$$

We have now arrived at a precise definition of isotopic spin, but do not as yet have any compelling reason to assume that this concept actually plays a role in nature. We shall need to investigate the consequences of assuming invariance under the isotopic spin group before we can tell whether this invariance is confirmed quantitatively by experiment.

9. Tests of isotopic spin conservation

1. *The πN system*

Consider a state consisting of a pion and a nucleon, for instance $|\pi^- p\rangle$. In order to study the consequences of isotopic spin conservation, we must first rewrite this state as a superposition of states with sharp isotopic spin I.

Since the π has isospin 1 and the nucleon has isospin $\frac{1}{2}$, the total isospin can be only $I = \frac{3}{2}$ or $\frac{1}{2}$. Hence one can write

$$\begin{aligned} | \pi^1 N^2 \rangle = C(1\tfrac{1}{2} I_3^1 I_3^2 \mid 3/2, I_3^1 + I_3^2) \mid 3/2, I_3^1 + I_3^2 \rangle \\ + C(1\tfrac{1}{2} I_3^1 I_3^2 \mid 1/2, I_3^1 + I_3^2) \mid \tfrac{1}{2}, I_3^1 + I_3^2 \rangle \end{aligned} \tag{9.1}$$

where the C's are Clebsch–Gordan coefficients. By the formula (7.37) for these coefficients one has

$$\begin{aligned} | \pi^+ p \rangle &= | 3/2\ 3/2 \rangle \\ | \pi^0 p \rangle &= \sqrt{\tfrac{1}{3}} \, | 3/2\ 1/2 \rangle - \sqrt{\tfrac{2}{3}} \, | 1/2\ 1/2 \rangle \\ | \pi^+ n \rangle &= \sqrt{\tfrac{2}{3}} \, | 3/2\ 1/2 \rangle + \sqrt{\tfrac{1}{3}} \, | 1/2\ 1/2 \rangle \\ | \pi^0 n \rangle &= \sqrt{\tfrac{1}{3}} \, | 3/2 - 1/2 \rangle + \sqrt{\tfrac{2}{3}} \, | 1/2 - 1/2 \rangle \\ | \pi^- p \rangle &= \sqrt{\tfrac{2}{3}} \, | 3/2 - 1/2 \rangle - \sqrt{\tfrac{1}{3}} \, | 1/2 - 1/2 \rangle \\ | \pi^- n \rangle &= | 3/2 - 3/2 \rangle . \end{aligned} \tag{9.2}$$

To say that isotopic spin is conserved means that the transition amplitude connects only states with the same isospin, and that it is independent of the charge. Let these amplitudes be $A^{(1/2)}$ and $A^{(3/2)}$.

$$\begin{aligned} A^{(3/2)} &= \langle 3/2 I_3 \mid A \mid 3/2 I_3 \rangle \\ A^{(1/2)} &= \langle 1/2 I_3 \mid A \mid 1/2 I_3 \rangle . \end{aligned} \tag{9.3}$$

By (9.2), one then has the following expressions for the probability amplitudes for the actually observable reactions:

$$\begin{aligned} A(\pi^+ p \to \pi^+ p) &= A^{3/2} \\ A(\pi^- p \to \pi^- p) &= 2/3 A^{3/2} + 1/3 A^{1/2} \\ A(\pi^- p \to \pi^0 n) &= \sqrt{\tfrac{2}{3}}\, A^{3/2} - \sqrt{\tfrac{2}{3}}\, A^{1/2} \end{aligned} \tag{9.4}$$

Comments

1. These last formulae are simply expressions of the Wigner–Eckart theorem, once one assumes that A is a tensor operator of rank 0, i.e. an invariant.

2. As we see, the hypothesis that isotopic spin is conserved implies, that instead of the 6 amplitudes which are possible a priori, only two are in fact linearly independent. This can and in practice often does provide a test of isospin conservation through a detailed analysis of sets of several experiments.

2. *Triangle inequalities*

Since cross-sections are proportional to the squares of amplitudes, one has the following relations between quantities of a given type (total or differential cross-sections):

$$\begin{aligned} \sigma_+ &= \sigma(\pi^+p \to \pi^+p) = |A^{3/2}|^2 \\ \sigma_0 &\equiv \sigma(\pi^-p \to \pi^0n) = \tfrac{2}{9}|A^{3/2} - A^{1/2}|^2 \\ \sigma_- &\equiv \sigma(\pi^-p \to \pi^-p) = |\tfrac{2}{3}A^{3/2} + \tfrac{1}{3}A^{1/2}|^2. \end{aligned} \tag{9.5}$$

$A^{1/2}$ and $A^{3/2}$ are complex numbers, and it is easy to check that

$$\tfrac{2}{3}A^{3/2} + \tfrac{1}{3}A^{1/2} = -\frac{1}{\sqrt{2}}\frac{\sqrt{2}}{3}(A^{3/2} - A^{1/2}) + A^{3/2}$$

If the complex amplitudes in this equation are interpreted geometrically as vectors in the Argand diagram, one sees that $\sqrt{\sigma_+}$, $\sqrt{\sigma_-}$, and $\sqrt{\sigma_0/2}$ form the three sides of a triangle. We shall comment later on the very remarkable experimental fact that at a pion energy of 280 MeV, this triangle is extremely elongated, and that for the total cross-sections one actually has (Figure 3.9.1):

$$\sigma_+ : \sigma_0 : \sigma_- \simeq 1 : \tfrac{2}{9} : \tfrac{4}{9} \tag{9.6}$$

which one interprets as due to a great difference in magnitude between the two amplitudes: $A^{3/2} \gg A^{1/2}$. Remarkably, at this energy, the condition (9.6) is satisfied by the differential cross-sections as well, which is strong

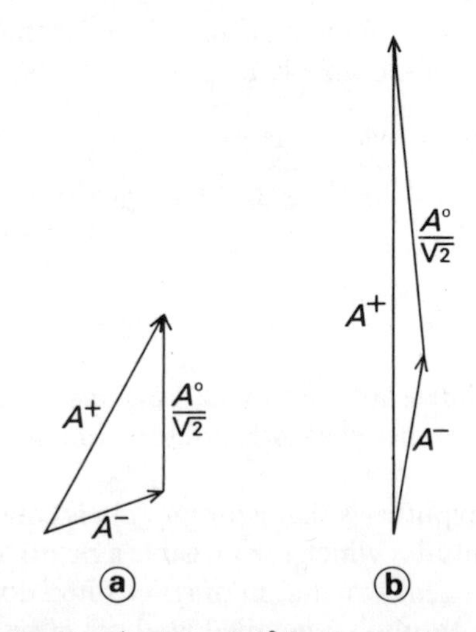

FIG. 3.9.1. The amplitudes A^+, A^-, and A^0 for the reactions $\pi^+p \to \pi^+p$, $\pi^-p \to \pi^-p$, and $\pi^-p \to \pi^0n$, (a) at some arbitrary energy; (b) at the resonance.

confirmation that isotopic spin is conserved. There are many similar verifications.

3. *The two-nucleon system*

The isospin of each nucleon being $\frac{1}{2}$, the total isospin of a system of two nucleons is 0 or 1. By (7.37) one has

$$\begin{aligned}
|pp\rangle &= |11\rangle \\
|pn\rangle &= \frac{1}{\sqrt{2}}(|1\ 0\rangle + |0\ 1\rangle) \\
|np\rangle &= \frac{1}{\sqrt{2}}(|1\ 0\rangle - |0\ 1\rangle) \\
|nn\rangle &= |1\ -1\rangle
\end{aligned} \tag{9.7}$$

Conversely, one has

$$\begin{aligned}
|1\ 1\rangle &= |pp\rangle \\
|1\ 0\rangle &= \frac{1}{\sqrt{2}}(|pn\rangle + |np\rangle) \\
|0\ 0\rangle &= \frac{1}{\sqrt{2}}(|pn\rangle - |np\rangle) \\
|1\ -1\rangle &= |nn\rangle.
\end{aligned} \tag{9.8}$$

We note that the state $|1\ 0\rangle$ is symmetric in p and n while the state $|0\ 0\rangle$ is antisymmetric.

4. *Isotopic spin and nuclear forces*

Consider the scattering of two protons in an S state. According to (9.7) they are in a state of isospin 1. Nuclear forces must depend only on I but not on I_3 if the assumption of isospin conservation is correct. It follows that the pp scattering amplitude must be equal to that of two neutrons, and also to that of a proton-neutron system in a state of i-spin 1 (i.e. symmetric in p and n). This is well satisfied by experiment. To see just what is meant by saying that a system is symmetric in proton and neutron, we must investigate the connection between symmetry and isotopic spin.

5. *Symmetry and isotopic spin*

According to the Pauli principle, the state vector of a two proton system is necessarily antisymmetric in space and spin. The isospin state $|1\ 0\rangle$ of two nucleons is obtainable by acting on the two proton states with the step-down operator $I_- = I_-^1 + I_-^2$, where the superfixes 1 and 2 refer to the two nucleons. This operator acts only on the charge indices of the nucleons and does not affect the dependence of the wavefunction on the space and spin variables. Hence the space and spin wavefunction in $|1\ 0\rangle$ is antisymmetric in these variables jointly, just as it was for two protons.

By (9.8) the wavefunction $|0\ 0\rangle$ is orthogonal to $|1\ 0\rangle$, and is therefore symmetric in the space and spin variables.

Hence the wavefunction has the following properties:

$I = 1$	the wavefunction is: symmetric in the charge indices	antisymmetric in space and spin variables
$I = 0$	antisymmetric in the charge indices	symmetric in space and spin variables

This result seems to suggest the following theorem:

Theorem

The charge indices can be placed on the same footing as the space and spin variables; when isotopic spin is conserved, one can say that the wavefunction of Fermi-Dirac particles must be antisymmetric in the space-, spin-, *and* isospin variables. (Symmetry instead of antisymmetry is demanded for Bose–Einstein particles.)

Though it is easy to verify this theorem in particular cases, we shall not give a proof applicable to an arbitrary number of particles with arbitrary isotopic spin.

Note that for two nucleons in an S state and in the isotopic state $|1\ 0\rangle$, the wavefunction must be antisymmetric in space and spin. Since it is symmetric in space (S-wave), it must be antisymmetric in spin, and the spin is 0 (singlet state). Therefore isotopic spin conservation demands that the nuclear forces be the same in the S states of *pp* and *nn*, and in the singlet S state of *pn*.

6. *Charge symmetry*

Charge symmetry is a rotation through π about the 2 axis in isospace ($e^{i\pi I_2}$); in general it changes the charge, because it inverts the direction of the axis along which I_3 is measured.

In particular,

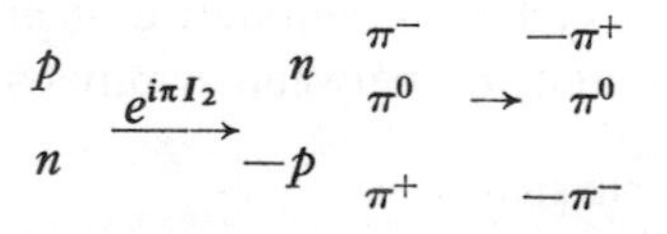

It has occasionally been suggested that invariance under this operation might apply to a higher degree of accuracy than does the conservation of isotopic spin; such suggestions having arisen mainly due to the historical order in which experimental information became available. It is difficult to find reasons for believing them.

The following is one reaction amongst several testing invariance under charge symmetry. The deuteron is a triplet state, $J = 1$, of proton and neutron (a mixture of S and D states). Therefore its isotopic wavefunction is antisymmetric, and it has isospin 0, comformably with the fact that there exists no other bound state of two nucleons. Hence the two reactions

$$\pi^+ + d \to p + p \tag{9.9}$$

$$\pi^- + d \to n + n \tag{9.10}$$

take place in a state of sharp i-spin, $I = 1$. Therefore their cross-sections must be equal, and this is well checked experimentally.

We shall actually consider this experiment as confirmation of isospin conservation, even though (9.10) and (9.9) are connected by charge symmetry. (Another example occurs in problem 10.)

7. *Violation of isotopic spin*

It is the electromagnetic interactions that are thought to be responsible for the splitting of multiplets, and in particular for the mass differences between neutral and charged pions, between neutron and proton, etc. In order to confirm this in detail one must use quantum electrodynamics, and most of the methods employed run into convergence problems.

Note finally that isotopic spin is not conserved by the weak interactions, as is sufficiently shown by the decay $K \to 2\pi$. We can therefore formulate the

Law: Isotopic spin is conserved by the strong interactions. It is not conserved by the electromagnetic nor by the weak interactions.

Problems

1. By counting parameters, show that $SU(2)$ is of order 3 and $SL(2C)$ of order 6.
2. Show that the Lie algebra of $SL(2C)$ consists of the 2×2 matrices with zero trace.
3. Prove the commutation relations (6.12). Use

$$[g_1, g_2] = \lim_{n\to\infty} n[ABA^{-1}B^{-1})^{1/n} - 1]$$

$$[G_1, G_2] = \lim_{n\to\infty} n\{\mathcal{U}_A \mathcal{U}_B \mathcal{U}_{A^{-1}} \mathcal{U}_{B^{-1}} - I\}$$

   where $[G_1, G_2]$ is the operator representing the element $[g_1, g_2]$ of the Lie algebra, and apply the formula (6.11).
4. Show that the Lie algebra of $SU(3)$ contains two independent commuting operators; ($SU(3)$ is said to be of rank 2).

5. What is the rank of $SL(2C)$?
6. Verify the theorem of paragraph (3.5) for the case of two pions.
7. Verify the same theorem for three π mesons. (A good knowledge of the permutation group of three objects seems useful in solving this problem.)
8. Show that on combining two spins $\frac{1}{2}$, the state with $J = 0$ is antisymmetric and the state with $J = 1$ is symmetric.
9. Consider a system of three pions, with isotopic spin 0, angular momentum 1, and negative parity. Given that pions have spin 0 and negative parity, show that each pair of pions is in a state with angular momentum 1.
10. Show that charge symmetry forbids the reaction

$$d + d \rightarrow He + \pi^0.$$

CHAPTER 4

RELATIVISTIC INVARIANCE

In this chapter we study the biggest *kinematic* group under which interactions are invariant, namely the Poincaré group, generated by Lorentz transformations and by translations in space and time. We show how the operations in this group act on single-particle states, and how it follows that as regards relativistic kinematics the only characteristics of a particle are its mass and its spin; and we shall discover how to treat these characteristics covariantly.

1. General orientation

(a) *Mass*

The mass of a particle is well defined by the conventions we have adopted. In other words, if we determine its energy E and momentum $\mathbf{p}$, then the quantity $m^2 = E^2 - \mathbf{p}^2$ is an invariant characteristic of the particle. This is possible only if particles move according to the laws of special relativity; that they do so to a very good approximation is confirmed, as we have seen already, by the very fact that accelerators work.

(b) *Spin*

Is the mass by itself enough to specify fully the physical state of a particle? Clearly not, because we know that electrons and protons at rest have a degree of freedom due to spin. The existence of electron spin is absolutely essential for our understanding of atomic spectra, of chemical reactions, of the structure of molecules, and of many other observable effects. It is evidenced directly in the Stern–Gerlach experiment and in electron magnetic resonance. The most direct evidence for the spin of the proton, and of atomic nuclei, arises similarly from experiments on nuclear magnetic resonance, and we will sketch it rapidly.

For instance, to confirm the proton spin, one puts hydrogen into a magnetic field $\mathbf{B}$. Then a proton with magnetic moment $\boldsymbol{\mu}$ has an energy $-\boldsymbol{\mu} \cdot \mathbf{B}$. If the proton has an intrinsic angular momentum, i.e. a spin $\mathbf{s}$, then in view of the fact that both $\mathbf{s}$ and $\boldsymbol{\mu}$ are axial vectors (pseudovectors), it is

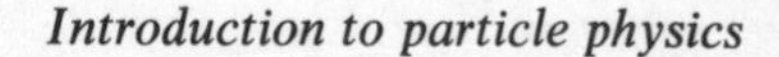

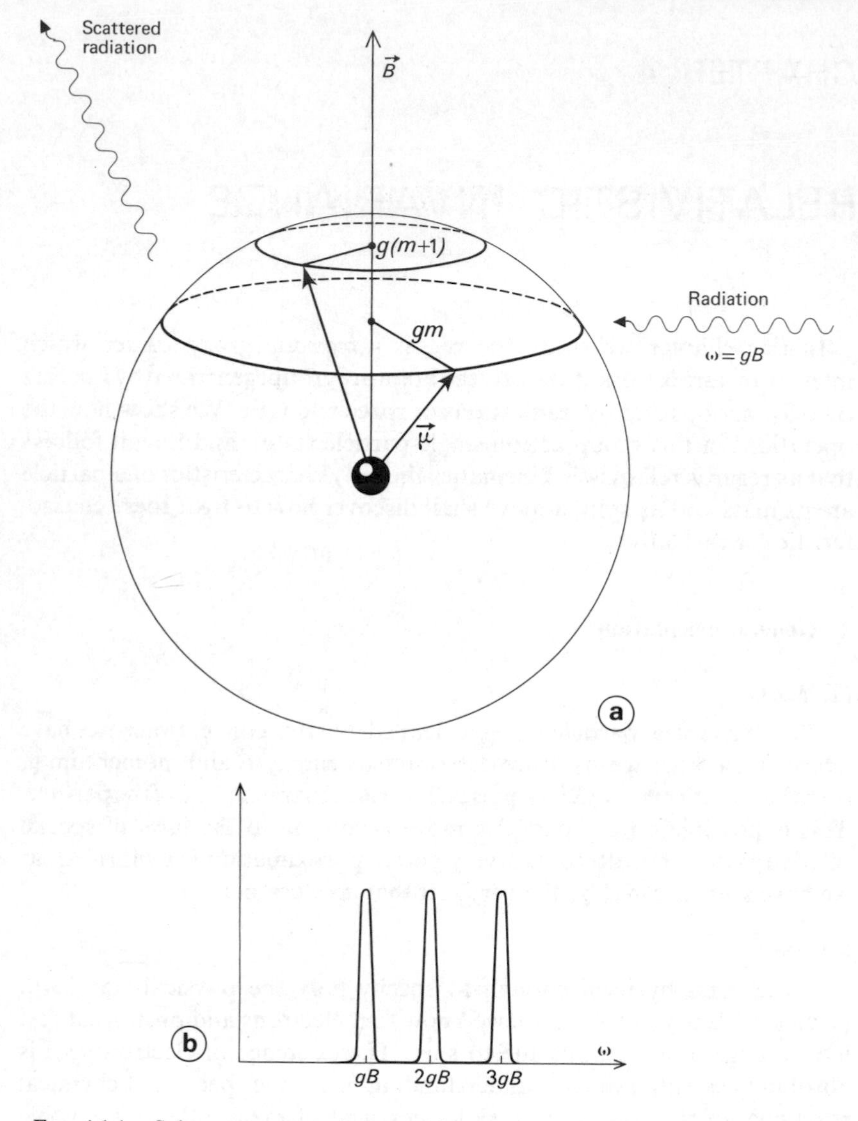

FIG. 4.1.1. Spin measurement by magnetic resonance. (a) Sketch; (b) Spectrum of scattered or observed frequencies in the case $s = 1$.

natural to assume that the magnetic moment is proportional to **s**, so that one has $\boldsymbol{\mu} = g\mathbf{s} \cdot g$ is called the gyromagnetic ratio (g-value). A proof of the collinearity of $\boldsymbol{\mu}$ and **s** is set as problem 1. If the quantisation axis for spin is chosen parallel to **B**, the energy of a proton in the spin state $s_3 = m$ is seen to be

$$-gm\mathbf{B}.$$

Hence a proton in a magnetic field has $(2s + 1)$ energy levels (Figure 4.1.1). An electromagnetic wave incident on the system will induce transitions between these levels, due to resonant absorption, if its frequency ω is equal to the energy difference between two levels. (Since we take $\hbar = 1$, we are free to identify frequency and energy.) Such values of ω are

$$\omega = gB, 2gB, \ldots 2sgB$$

and their number determines s. For protons one observes only a single resonance frequency, whence $s = \frac{1}{2}$.

(c) *General survey*

Some general questions arise naturally when considering the states of a particle:

1. Does a particle have kinematic properties other than its mass and spin?
2. We know how to describe the spin of a particle at rest or moving slowly. How are these ideas modified when the particle is relativistic?
3. How does one use the language of quantum mechanics to describe the states of a relativistic spinning particle?
4. Is there a way of describing particles mathematically from which mass and spin follow deductively? This would give us a better understanding of these concepts, and of their interrelation.

It is clear that the notion of mass is intimately connected with special relativity and therefore with Lorentz transformations, while the notion of spin derives from the rotation group. In order to synthetise these two concepts we must therefore study the connections which can interrelate rotations and Lorentz transformations within the framework of group theory.

(d) *Formulation of the problem*

We shall look for a group G of operations which embraces Lorentz transformations and rotations, and allows a definition of mass. Afterwards we shall try to answer the questions listed above by applying the general results of the last chapter concerning the action of a group on physical states.

2. The symmetry group

(a) *The principle of relativistic invariance*

We assume that accelerators and detectors can be described relative to any arbitrary inertial frame, and that the quantities characterising a given experiment, (for example, the measured values of a particle momentum), when referred to two frames S and S', are related by the Lorentz transformation which takes S into S'. This is a cumbrous but exact formulation of relativistic invariance. It implies a passive interpretation of Lorentz transformations, i.e. it postulates invariance under a change in description, not under a change in the experimental setup. As we have seen, these transformations will be represented by unitary or antiunitary operations acting on the state vectors of the system, and in particular, on the states of a single particle.

(b) *The homogeneous Lorentz group*

The transformations from one inertial frame S to another S' are linear, and therefore of the form

$$x'_\mu = \Lambda_{\mu\nu} x_\nu \tag{2.1}$$

They conserve the quantity

$$x^2 = g_{\mu\nu} x_\mu x_\nu \tag{2.2}$$

where the elements of the metric are defined by

$$\begin{aligned} g_{00} &= 1 \qquad g_{11} = g_{22} = g_{33} = -1 \\ g_{0i} &= g_{ij} = 0 \qquad \text{for} \quad i \neq j \end{aligned} \tag{2.3}$$

Hence we define

$$x^2 = x_0{}^2 - \mathbf{x}^2.$$

(Greek indices μ, ν, ··· will always run over 0, 1, 2, 3; latin indices i, j, ··· will run over 1, 2, 3.)

Example 1. For a Lorentz transformation with velocity v in the direction 1, the matrix Λ has the form

$$\Lambda_1(v) = \begin{bmatrix} \gamma & 0 & 0 & \gamma v \\ 0 & 1 & 0 & 0 \\ 0 & 0 & 1 & 0 \\ \gamma v & 0 & 0 & \gamma \end{bmatrix} \tag{2.4}$$

with $\gamma = (1 - v^2)^{-1/2}$.

Example 2. For a rotation through an angle θ about the 3 axis, the matrix Λ has the form

$$\Lambda_3(\theta) = \begin{bmatrix} \cos\theta & -\sin\theta & 0 & 0 \\ \sin\theta & \cos\theta & 0 & 0 \\ 0 & 0 & 1 & 0 \\ 0 & 0 & 0 & 1 \end{bmatrix} \tag{2.5}$$

The operations thus introduced clearly form a group: the group of all

4×4 matrices conserving x^2. The unit element of the group is the identity transformation $x_\mu' = x_\mu$. This group is called the *Lorentz group*; when one needs to distinguish it from certain other groups to be introduced later, it is called the *homogeneous Lorentz group*. We see that the rotation group R_3 is one of its subgroups.

The matrices Λ can be specified algebraically if one introduces the matrix for the metric

$$g = (g_{\mu\nu}) = \begin{bmatrix} -1 & 0 & 0 & 0 \\ 0 & -1 & 0 & 0 \\ 0 & 0 & -1 & 0 \\ 0 & 0 & 0 & +1 \end{bmatrix}. \tag{2.6}$$

Then the condition (2.2) can be written as

$$g_{\mu\nu}\Lambda_{\mu\kappa}\Lambda_{\nu\rho}\, x_\kappa x_\rho = g_{\kappa\rho}x_\kappa x_\rho$$

whence

$$g_{\mu\nu}\Lambda_{\mu\kappa}\Lambda_{\nu\rho} = g_{\kappa\rho}$$

or, in matrix form

$$\Lambda^T g \Lambda = g \tag{2.7}$$

where Λ^T is the transpose of the matrix Λ.

From equation (2.7) one can find the number of group parameters. Λ has 16 real matrix elements; g is a symmetric matrix, so that (2.7) equates two symmetric matrices and consequently imposes 10 conditions; hence in toto Λ depends on 6 real parameters. Therefore the Lorentz group is of order 6.

(c) *Infinitesimal generators of the Lorentz group*

We shall look for six independent elements of the Lie algebra of the Lorentz group, and shall adopt them as infinitesimal generators. To this end we consider the six matrices $\Lambda_1(v)$, $\Lambda_2(v)$, $\Lambda_3(v)$, $\Lambda_1(\theta)$, $\Lambda_2(\theta)$, $\Lambda_3(\theta)$ given by (2.4), (2.5), and the analogous matrices representing, respectively, pure Lorentz transformations along the three coordinate axes, and rotations about these axes. We introduce the corresponding elements of the Lie algebra

$$iJ_i = \lim_{n\to\infty} n[\Lambda_i^{1/n}(\theta) - I] \tag{2.8}$$

$$iK_i = \lim_{n\to\infty} n[\Lambda_i^{1/n}(v) - I]. \tag{2.9}$$

To begin with we calculate J_i. A rotation through an angle α, repeated n times, is simply a rotation through an angle $n\alpha$, whence one has

$$iJ_i = \lim_{n\to\infty} n[\Lambda_i(\theta/n) - I]$$

Therefore, using the explicit form (2.5) of the matrices $\Lambda_i(\theta)$,

$$J_3 = \begin{bmatrix} 0 & i & 0 & 0 \\ -i & 0 & 0 & 0 \\ 0 & 0 & 0 & 0 \\ 0 & 0 & 0 & 0 \end{bmatrix} \qquad J_2 = \begin{bmatrix} 0 & 0 & -i & 0 \\ 0 & 0 & 0 & 0 \\ i & 0 & 0 & 0 \\ 0 & 0 & 0 & 0 \end{bmatrix} \tag{2.10}$$

$$J_1 = \begin{bmatrix} 0 & 0 & 0 & 0 \\ 0 & 0 & i & 0 \\ 0 & -i & 0 & 0 \\ 0 & 0 & 0 & 0 \end{bmatrix}.$$

To calculate K_i, we need consider only small values of v; in that case one has $\Lambda_i^{1/n}(v) \approx \Lambda_i(v/n)$, because the changes in reference frame are now essentially non-relativistic and the velocities simply add. According to (2.4) it follows that

$$K_1 = \begin{bmatrix} 0 & 0 & 0 & -i \\ 0 & 0 & 0 & 0 \\ 0 & 0 & 0 & 0 \\ -i & 0 & 0 & 0 \end{bmatrix} \qquad K_2 = \begin{bmatrix} 0 & 0 & 0 & 0 \\ 0 & 0 & 0 & -i \\ 0 & 0 & 0 & 0 \\ 0 & -i & 0 & 0 \end{bmatrix} \tag{2.11}$$

$$K_3 = \begin{bmatrix} 0 & 0 & 0 & 0 \\ 0 & 0 & 0 & 0 \\ 0 & 0 & 0 & -i \\ 0 & 0 & -i & 0 \end{bmatrix}$$

Note that in (3.1) and (3.2) we have introduced a factor i to ensure that J_i and K_i are represented by Hermitean matrices (c.f. section 6 of chapter 3).

(d) *Commutation rules*

An explicit calculation establishes the following commutation relations:

$$[J_1, J_2] = iJ_3 \tag{2.12}$$

$$[J_1, K_1] = 0 \tag{2.13}$$

$$[J_1, K_2] = iK_3 \tag{2.14}$$

$$[K_1, K_2] = -iJ_3 . \tag{2.15}$$

The other commutation relations follow from these by a cyclic permutation of indices.

Equation (2.12) shows that the commutation relation for J_1, J_2, and J_3 are those of the infinitesimal generators of the rotation group. Indeed, we actually constructed them as generators of this group.

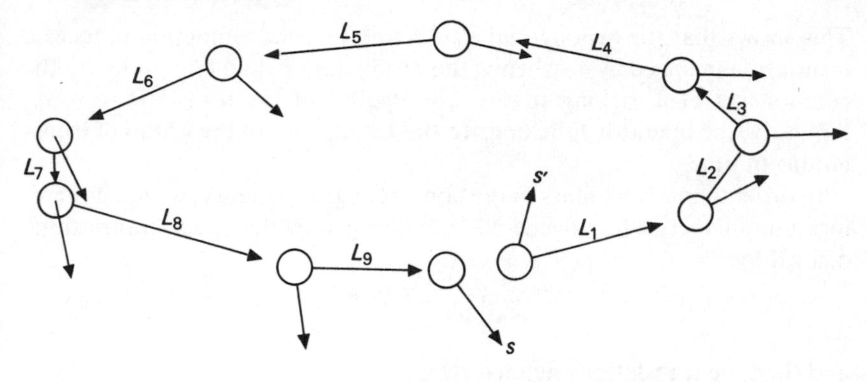

FIG 4.2.1. The Thomas precession: a sequence of Lorentz transformations, which bring the velocity of an object back to its initial value, but result in a rotation of the object itself ($\mathbf{S} \neq \mathbf{S}'$).

Equation (2.13) shows that a Lorentz transformation is unchanged by a rotation about the direction of the velocity involved, which is clear from the physics.

Equation (2.14) shows that K_1, K_2, and K_3 transform under rotations like the components of a tensor operator of rank 1, i.e. like a vector. They will often be written as $\mathbf{K}$.

Equation (2.15) shows that two pure Lorentz transformations, along the axes 1 and 2 respectively, generate a rotation on carrying out the operation $\Lambda_1\Lambda_2\Lambda_1^{-1}\Lambda_2^{-1}$. This result underlies the phenomenon known as the "Thomas precession", wherein the resultant of a sequence of pure Lorentz transformations applied to a spinning particle can leave its momentum unchanged while rotating its spin (Figure 4.2.1).

(e) *The inhomogeneous Lorentz group*

The fact that the rotation group is a subgroup of the Lorentz group will certainly enable us to describe spin. By contrast, mass can enter only if one takes account of energy and momentum, which are not generators of the Lorentz group.

Nevertheless, energy and momentum are known to have a simple interpretation in group theory. To see it, recall that in Schroedinger theory they are associated with the differential operators

$$\hat{E} = i\frac{\partial}{\partial t} \qquad \hat{\mathbf{p}} = -i\nabla \tag{2.16}$$

whence the Taylor expansion for a function of $f(x)$ gives

$$\begin{aligned} e^{i\mathbf{a}\cdot\hat{\mathbf{p}}} f(\mathbf{x}) = e^{\mathbf{a}\cdot\nabla} f(\mathbf{x}) &= f(\mathbf{x}) + \mathbf{a}\cdot\nabla f(\mathbf{x}) + \frac{1}{2!}\, a_i a_j \partial_i \partial_j f(\mathbf{x}) + \cdots \\ &= f(\mathbf{x} + \mathbf{a}). \end{aligned} \tag{2.17}$$

This shows that the exponential $e^{i\mathbf{a}\cdot\hat{\mathbf{p}}}$ acting on a wavefunction induces a translation in space by $\mathbf{a}$, whence the product $\mathbf{a}\cdot\hat{\mathbf{p}}$, and consequently the components of $\hat{\mathbf{p}}$, belong to the Lie algebra of the translation group. Similarly, the operator $\hat{E}$ belongs to the Lie algebra of the group of translations in time.

In order to interpret mass in the context of group theory, we must therefore adjoin to the homogeneous Lorentz group the space translations defined by

$$T_{\mathbf{a}}\begin{cases}\mathbf{x}' = \mathbf{x}+\mathbf{a}\\ t' = t\end{cases} \tag{2.18}$$

and the time translations defined by

$$T_{\tau}\begin{cases}\mathbf{x}' = \mathbf{x}\\ t' = t+\tau\end{cases} \tag{2.19}$$

All these transformations together constitute the inhomogeneous Lorentz group defined by

$$x_{\mu}' = \Lambda_{\mu\nu}x_{\nu} + a_{\mu}\,. \tag{2.20}$$

Denote by $\{a, \Lambda\}$ the transformation given by equation (2.20), where a is a four-vector, and Λ a matrix of the homogeneous Lorentz group. Thus, the identity element of the inhomogeneous Lorentz group is $\{0, I\}$. The group multiplication rule for the product of two elements $\{a_1, \Lambda_1\}$ and $\{a_2, \Lambda_2\}$ is given explicitly by the transformations

$$x_{\mu}' = \Lambda^{1}_{\mu\nu}x_{\nu} + a_{\mu}{}^{1}$$

$$x_{\mu}'' = \Lambda^{2}_{\mu\nu}x_{\nu}' + a_{\mu}{}^{2}$$

Hence

$$x_{\mu}'' = \Lambda^{2}_{\mu\rho}\Lambda^{1}_{\rho\nu}x_{\nu} + \Lambda^{2}_{\mu\nu}a_{\nu}{}^{1} + a_{\mu}{}^{2}$$

determines the multiplication rule:

$$\{a_2\,, \Lambda_2\}\cdot\{a_1\,, \Lambda_1\} = \{\Lambda_2 a_1 + a_2; \Lambda_2\Lambda_1\}. \tag{2.21}$$

The inhomogeneous Lorentz group can be represented as a matrix group by introducing a fifth coordinate s and a vector y_{μ} such that x_{μ} can be represented homogeneously by

$$x_{\mu} = y_{\mu}/s \tag{2.22}$$

while (2.20) can be written as a matrix transformation in a homogeneous 5 dimensional space with coordinates (y_1, y_2, y_3, y_4, s):

$$\begin{cases}y_{\mu}' = \Lambda_{\mu\nu}y_{\nu} + a_{\mu}s\\ s' = s.\end{cases} \tag{2.23}$$

(f) *Infinitesimal translations*

In the homogeneous space just defined, $T_{\mathbf{a}}$ is represented by the matrix

$$T_{\mathbf{a}} = \begin{vmatrix} 1 & 0 & 0 & 0 & a_1 \\ 0 & 1 & 0 & 0 & a_2 \\ 0 & 0 & 1 & 0 & a_3 \\ 0 & 0 & 0 & 1 & 0 \\ 0 & 0 & 0 & 0 & 1 \end{vmatrix}$$

and evidently we have $(T_{\mathbf{a}})^{1/n} = T_{\mathbf{a}/n}$. Hence the element of the Lie algebra corresponding to $T_{\mathbf{a}}$ is given by

$$\alpha_{\mathbf{a}} = \lim_{n\to\infty} n[T_{\mathbf{a}/n} - I] = \begin{vmatrix} 0 & 0 & 0 & 0 & a_1 \\ 0 & 0 & 0 & 0 & a_2 \\ 0 & 0 & 0 & 0 & a_3 \\ 0 & 0 & 0 & 0 & 0 \\ 0 & 0 & 0 & 0 & 0 \end{vmatrix} \tag{2.24}$$

One can introduce three infinitesimal generators P_1, P_2, and P_3 such that $\alpha_{\mathbf{a}}$ is written as

$$\alpha_{\mathbf{a}} = ia_1P_1 + ia_2P_2 + ia_3P_3 \tag{2.25}$$

where

$$iP_1 = \begin{vmatrix} 0 & 0 & 0 & 0 & 1 \\ 0 & 0 & 0 & 0 & 0 \\ 0 & 0 & 0 & 0 & 0 \\ 0 & 0 & 0 & 0 & 0 \\ 0 & 0 & 0 & 0 & 0 \end{vmatrix} \qquad iP_2 = \begin{vmatrix} 0 & 0 & 0 & 0 & 0 \\ 0 & 0 & 0 & 0 & 1 \\ 0 & 0 & 0 & 0 & 0 \\ 0 & 0 & 0 & 0 & 0 \\ 0 & 0 & 0 & 0 & 0 \end{vmatrix}$$

$$iP_3 = \begin{vmatrix} 0 & 0 & 0 & 0 & 0 \\ 0 & 0 & 0 & 0 & 0 \\ 0 & 0 & 0 & 0 & 1 \\ 0 & 0 & 0 & 0 & 0 \\ 0 & 0 & 0 & 0 & 0 \end{vmatrix} \tag{2.26}$$

It is easy to show in the same way that a time translation T_τ corresponds to the element $i\tau P_0$ of the Lie algebra, where

$$iP_0 = \begin{vmatrix} 0 & 0 & 0 & 0 & 0 \\ 0 & 0 & 0 & 0 & 0 \\ 0 & 0 & 0 & 0 & 0 \\ 0 & 0 & 0 & 0 & 1 \\ 0 & 0 & 0 & 0 & 0 \end{vmatrix} \tag{2.27}$$

(g) *Commutation relations of the inhomogeneous Lorentz group*

It is easy to write down the expressions for **J** and **K** in the homogeneous 5-dimensional space; indeed all that is necessary is to add a fifth row and a fifth column of zeros to the matrices (2.10) and (2.11). In these circumstances it is easy to find by direct calculation the commutation relations of the infinitesimal generators of the Lorentz group. Using (2.10), (2.11), (2.26), and (2.27), one finds

$$[P_i, P_0] = 0 \qquad [P_i, P_j] = 0 \tag{2.28}$$

$$[J_1, P_2] = iP_3 \tag{2.29}$$

$$[J_1, P_1] = 0 \tag{2.30}$$

$$[J_1, P_0] = 0 \tag{2.31}$$

$$[K_1, P_0] = iP_1 \tag{2.32}$$

$$[K_1, P_1] = iP_0 \tag{2.33}$$

$$[K_1, P_2] = 0 \tag{2.34}$$

$$[J_1, J_2] = iJ_3 \quad [J_1, K_2] = iK_3 \quad [J_1, K_1] = 0 \quad [K_1, K_2] = -iJ_3 \tag{2.35}$$

(2.28) shows that translations commute with each other; in other words they form an Abelian subgroup of the Lorentz group.

(2.29) shows that (P_1, P_2, P_3) are components of a tensor operator of rank 1, whence they transform under rotations like the three components of a vector, which is evident geometrically.

(2.30) shows that a rotation is unaffected by a translation along its axis.

(2.31) shows that a rotation is unaffected by a translation of the origin in time.

(2.32) shows how the energy changes under a Lorentz transformation.

(2.33) and (2.34) show how the momentum changes under a Lorentz transformation: only the momentum component along the relative velocity of the two frames is changed, as is seen from (2.34).

(2.35) has been considered already.

(h) *The pieces of the inhomogeneous Lorentz group*

As we saw already in the last chapter, there is a great difference between the treatment of the general case, and of the special case of groups all of whose elements can be generated by continuous variation starting from the identity element.

It is clear that the operations of space and time reflection:

$$x_\mu' = \Pi_{\mu\nu} x_\nu \qquad x_\mu' = \Theta_{\mu\nu} x_\nu \tag{2.36}$$

where

$$\Pi = \begin{vmatrix} -1 & 0 & 0 & 0 \\ 0 & -1 & 0 & 0 \\ 0 & 0 & -1 & 0 \\ 0 & 0 & 0 & 1 \end{vmatrix} \qquad \Theta = \begin{vmatrix} 1 & 0 & 0 & 0 \\ 0 & 1 & 0 & 0 \\ 0 & 0 & 1 & 0 \\ 0 & 0 & 0 & -1 \end{vmatrix} \tag{2.37}$$

conserve x_μ^2 and therefore belong to the Lorentz group. But they cannot be generated continuously from the identity.

Indeed, by taking the determinant of equation (2.7), one obtains

$$(\det \Lambda^T)(\det g)(\det \Lambda) = \det g$$

whence, the determinant of Λ^T being equal to that of Λ,

$$(\det \Lambda)^2 = 1$$

admitting the possibilities

$$\det \Lambda = +1 \tag{2.38a}$$

$$\det \Lambda = -1 \tag{2.38b}$$

Lorentz transformations can therefore be classified into two categories, according to whether their determinants are $+1$ or -1. (Figure 4.2.2.) These two categories are evidently disjoint. The matrices which can be generated continuously from the identity clearly have determinant $+1$. The operations Π and Θ have determinants -1. Hence the operation $\Pi\Theta$ has the determinant 1.

Next, we show that $\Pi\Theta$ cannot be joined continuously to the identity.

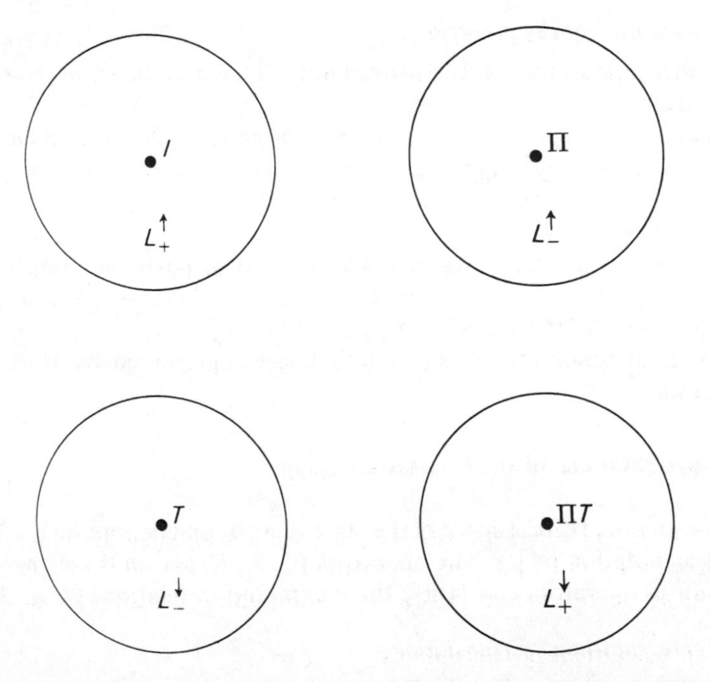

FIG. 4.2.2. The pieces of the Lorentz group. The sign $\pm$ represents the sign of the determinant of Λ. The arrow ↑↓ shows whether or not the transformations are orthochronous.

To this end, consider the component $\mu = 0 = \nu$ of equation (2.7), which gives

$$g_{00} = 1 = (\Lambda_{00})^2 - \sum_{j=1}^{3} (\Lambda_{0j})^2$$

whence

$$(\Lambda_{00})^2 \geqslant 1$$

This leads to a new classification of Lorentz transformations into two categories:

$$\Lambda_{00} \geqslant 1 \tag{2.39a}$$

$$\Lambda_{00} \leqslant -1. \tag{2.39b}$$

Transformations which can be joined continuously to the identity obey (2.39a), while $\Pi\Theta$ obeys (2.39b).

Lorentz transformations obeying (2.38a) are said to be *proper*, and those obeying (2.39a) are *orthochronous* (equation (2.39a) entails that t' and t have the same sign). Thus, the operations generated continuously from the identity constitute the proper orthochronous inhomogeneous Lorentz group.

In the remainder of the chapter we shall be interested only in this group, and shall call it the Poincaré group for brevity.

(i) *Formulation of the problem*

We give now a more precise formulation of the hypothesis of relativistic invariance.

Fundamental hypothesis: the properties of particles are describable similarly before and after applying to the reference frame a transformation belonging to the proper orthochronous Lorentz group, and a translation in space or time. In other words, they are invariant under the Poincaré group.

It follows that the states of a system, and in particular those of an isolated particle, are described by the vectors of a Hilbert space on which the operations of the Poincaré group act as unitary operators. Then the infinitesimal generators $\mathbf{P}$, P_0, $\mathbf{J}$ and $\mathbf{K}$ are represented by Hermitean operators.

3. Representations of the translation group

Consider the Hilbert space of the state vectors, and denote such a vector in Dirac notation by $| \rangle$. The operators P_μ, J_i, K_i act on these vectors as Hermitean operators and satisfy the commutation relations (2.28–35).

(a) *Representations of translations*

Since the operators P_0, P_1, P_2, P_3 commute, they can be diagonalised simultaneously. Therefore one can define vectors $|p\rangle$ corresponding to such states i.e. eigenvectors of the components of P^μ:

$$P_\mu \mid p\rangle = p_\mu \mid p\rangle. \tag{3.1}$$

By the definition of the infinitesimal generators, every element of the Lie algebra can be written in the form $ia_\mu P_\mu$, and every translation in the form $e^{ia_\mu P_\mu}$; if we denote by $U\{a, \Lambda\}$ the unitary operation representing a general Poincaré transformation, then we have

$$U\{a, I\} \mid p\rangle = e^{ia_\mu P_\mu} \mid p\rangle = e^{ia_\mu p_\mu} \mid p\rangle = e^{ip\cdot a} \mid p\rangle. \tag{3.2}$$

Comparison of this expression with (2.17), which applies in Schroedinger theory, suggests that p is the four-momentum of the physical system whose state vector is $|p\rangle$.

(b) *Effects of rotations and Lorentz transformations*

How does a rotation act on $|p\rangle$? To discover this, we begin by asking what values of the quantum number p are connected by this operation. In other words, if $U\{0, R\}$ represents a rotation R, we investigate whether $U\{0, R\} \, |p\rangle$ is an eigenvector of the P_μ, and if so, to what eigenvalues it belongs. We must therefore consider the vector

$$P_\mu U\{0, R\} \mid p\rangle. \tag{3.3}$$

We have noted already the commutation relations between P_μ and $\mathbf{J}$, which express the fact that P_0 is invariant under rotations while $\mathbf{P}$ transforms like a vector; in other words

$$U^{-1}\{0, R\} \, P_0 U\{0, R\} = P_0 \tag{3.4}$$

$$U^{-1}\{0, R\} \, P_i U\{0, R\} = R_{ij} P_j \tag{3.5}$$

where R_{ij} is the 3×3 matrix of the rotation R. So far, equation (3.5) is based on the Wigner–Eckart theorem; it is of interest to establish it more directly. To this end we introduce a representation of R as a function on the Lie algebra

$$U\{0, R\} = e^{i(\theta_1 J_1 + \theta_2 J_2 + \theta_3 J_3)} \tag{3.6}$$

and exploit the fact that

$$U^{-1} P_i U = [U^{-1/n}]^n P_i [U^{1/n}]^n \tag{3.7}$$

where n is an integer which we take to be very large. Writing $U = U\{0, R\}$, it then follows from (3.6) that

$$\begin{aligned} U^{-1/n} P_i U^{1/n} &= \left(I - i\frac{\theta_1}{n} J_1 - i\frac{\theta_2}{n} J_2 - i\frac{\theta_3}{n} J_3\right) \\ &\quad \times P_i \left(I + i\frac{\theta_1}{n} J_1 + i\frac{\theta_2}{n} J_2 + i\frac{\theta_3}{n} J_3\right) \\ &= P_i + \frac{i}{n} [P_i \, , \theta_1 J_1 + \theta_2 J_2 + \theta_3 J_3] \end{aligned} \tag{3.8}$$

But the commutation relations of the operators **P** with **J** are identical to those of the operators **J** with each other; according to the group multiplication rule,

$$\frac{i}{n}[J_i, \theta_1 J_1 + \theta_2 J_2 + \theta_3 J_3] = (R_{ij}^{1/n} - I)\, J_j \tag{3.9}$$

whence

$$U^{-1/n} P_i U^{1/n} = R_{ij}^{1/n} P_j \tag{3.10}$$

which entails (3.5).

Therefore we have the eigenvalue relation

$$P_i U\{0, R\} \mid p\rangle = U\{0, R\}\, R_{ij} P_j \mid p\rangle = R_{ij} p_j U\{0, R\} \mid p\rangle \tag{3.11}$$

which shows that $U\{0, R\} \, |p\rangle$ is an eigenstate of **P** belonging to the eigenvalue $R\mathbf{p}$. Similarly, it is an eigenstate of P_0 belonging to the eigenvalue p_0.

(c) *Lorentz transformations*

Consider now how p changes under a Lorentz transformation along the 1 axis. It is simplest to consider an infinitesimal transformation

$$\Lambda = I + i\epsilon K_1$$

where K_1 is a 5×5 matrix in the homogeneous space.

Λ will be represented by

$$U\{0, \Lambda\} = I + i\epsilon K_1 \tag{3.12}$$

where K_1 is a Hermitean operator in Hilbert space. We need to study the vector

$$P_\mu U\{0, \Lambda\} \mid p\rangle \tag{3.13}$$

(α) Begin by considering the energy, i.e. the case $\mu = 0$. The commutation relation

$$[K_1, P_0] = iP_1 \tag{3.14}$$

entails

$$P_0 K_1 = K_1 P_0 + iP_1 \tag{3.15}$$

whence, to first order in ϵ, we have

$$P_0(I + i\epsilon K_1) = (I + i\epsilon K_1)\, P_0 + \epsilon P_1 = (I + i\epsilon K_1)(P_0 + \epsilon P_1) \tag{3.16}$$

This implies

$$P_0 U \mid p\rangle = U(P_0 + \epsilon P_1) \mid p\rangle = (p_0 + \epsilon p_1)\, U \mid p\rangle \tag{3.17}$$

Note (c.f. equation (2.4)) that $p_0 + \epsilon p_1$ is simply the 0 component of the vector p' arising from p under Λ:

$$p' = \Lambda p \tag{3.18}$$

(β) Consider the case $\mu = 1$. We have

$$[K_1, P_1] = iP_0 \tag{3.19}$$

whence

$$P_1(I + i\epsilon K_1) = (I + i\epsilon K_1)(P_1 + \epsilon P_0)$$

implying

$$P_1 U \mid p\rangle = (p_1 + \epsilon p_0)\, U \mid p\rangle \tag{3.20}$$

Note that $p_1 + \epsilon p_0 = p_1'$.

(γ) For $\mu = i = 2$ or 3, we have $[K_1, P_i] = 0$, whence

$$P_i U \mid p\rangle = p_i U \mid p\rangle \tag{3.21}$$

From (3.17), (3.20), and (3.21) we deduce

$$P_\mu U\{0, \Lambda\} \mid p\rangle = \Lambda_{\mu\nu} p_\nu U\{0, \Lambda\} \mid p\rangle. \tag{3.22}$$

This equation holds when Λ is an infinitesimal Lorentz transformation in the 1 direction. But by the group composition law it holds also for a finite Lorentz transformation in the 1 direction; further, we know already that (3.22) holds for a pure rotation. Given that every element of the homogeneous Lorentz group can be written as a product of rotations with a Lorentz transformation along the 1 axis, we see that (3.22) holds for every such element.

To sum up, equation (3.27) shows that $U\{0, \Lambda\} \mid p\rangle$ is an eigenvector of P_μ belonging to the eigenvalue Λp. Hence finally we obtain the important relation

$$U\{0, \Lambda\} \mid \alpha, p\rangle = \mid \alpha', \Lambda p\rangle \tag{3.23}$$

where the index α denotes the values of all quantum numbers other than p, for instance the spin. The reason why a value α' different from α appears on the right lies in our ignorance, at this time, of the action of the Poincaré group on these other quantum numbers.

(d) *Classification of four-vectors*

If there exists amongst the state vectors an eigenstate of P belonging to the eigenvalue p, then it follows from equation (3.23) that by acting on this state with the operators $U\{a, \Lambda\}$ we can generate eigenstates of P belonging to all eigenvalues of the form Λp.

The set of all such vectorial eigenvalues Λp is a 3 dimensional surface (Σ) in the 4 dimensional space of the vectors p. The geometric properties of this surface depend on the value of p^2 and possibly on the sign of p^0. Thus, we have the cases

(i) $p^2 > 0, p^0 > 0$. Σ is the upper branch of a two-sheeted hyperboloid: $p' = \Lambda p, p'^2 = p^2, p'^0 > 0$.

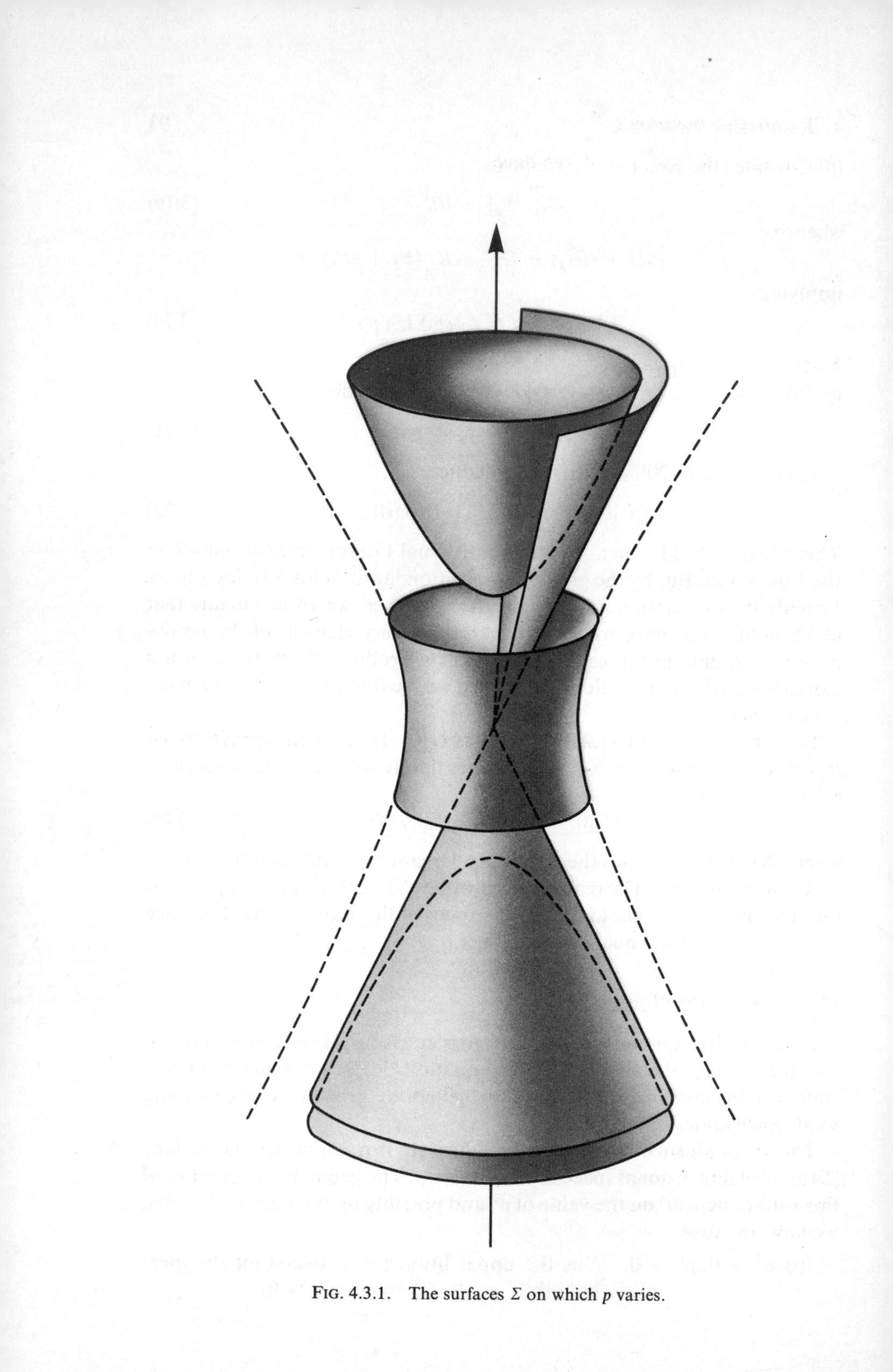

FIG. 4.3.1. The surfaces Σ on which p varies.

(ii) $p^2 > 0, p^0 < 0.$ Σ is the lower branch of the same two-sheeted hyperboloid.
(iii) $p^2 = 0, p^0 > 0.$ Σ is the upper half of the light cone $p'^2 = 0$.
(iv) $p^2 = 0, p^0 < 0.$ Σ is the lower half of the same cone.
(v) $p^2 < 0.$ Σ is a single-sheeted hyperboloid (Figure 4.3.1).

Our interpretation of p as a four-momentum shows that only cases (i) and (iii) are physically relevant, case (iii) being realised by a particle of zero mass.

(e) *Algebraic formulation of the conservation of mass*

Equation (3.23) shows that Lorentz transformations leave invariant the eigenvalue of the expression

$$P^2 = P_0^2 - P_1^2 - P_2^2 - P_3^2. \tag{3.24}$$

Algebraically speaking this depends on the fact that the operator P^2 commutes with all the generators of the Poincaré group. Indeed, from the commutation relations (2.28–35) one verifies readily that

$$[P^2, P_0] = [P^2, P_i] = [P^2, J_i] = [P^2, K_i] = 0. \tag{3.25}$$

Hence the eigenvalue of P^2 is a quantum number which can always be diagonalised, and which for obvious reasons is called the (squared) *mass* of the state

$$P^2 = M^2.$$

Note that in general a physical state can consist of several particles, so that the mass of the system is actually its total energy as measured in the centre of mass frame.

4. Unitary representations of the Poincaré group. Particles with nonzero mass

(a) *The problem*

At this point we confine ourselves to the case of a single particle whose mass is well defined, and assumed for the time being to be nonzero. Then p varies on one branch of the hyperboloid Σ_M. For any given value of the four-momentum p, there exists a Lorentz transformation with velocity parallel to $\mathbf{p}$ which transforms the four-vector $(\mathbf{0}, M) = p^{(0)}$ into p. We call this transformation $L(p)$ the boost of p, since it sends a vector at rest into the four-momentum in which we are interested. Its inverse $L^{-1}(p)$ reduces the particle to rest. Explicitly, if

$$p = [p, 0, 0, \sqrt{M^2 + p^2}],$$

then one has

$$L(p) = \begin{vmatrix} \gamma & 0 & 0 & \gamma v \\ 0 & 1 & 0 & 0 \\ 0 & 0 & 1 & 0 \\ \gamma v & 0 & 0 & \gamma \end{vmatrix} \tag{4.1}$$

where

$$\gamma = \frac{(M^2 + p^2)^{1/2}}{M} \qquad v = \frac{p}{(M^2 + p^2)^{1/2}} \qquad \gamma v = \frac{p}{M}.$$

Our problem now is to make explicit the role of the quantum numbers α, α' in the equation

$$U\{0, \Lambda\} \mid \alpha, p\rangle = \mid \alpha', \Lambda p\rangle, \tag{4.2}$$

or in other words to find the transformation rule for the spin states.

(b) *A special case*

Consider the special case where $p = p^{(0)}$ (we always write $p^{(0)} = (\mathbf{0}, M)$) and where Λ is a rotation R. Then equation (4.2) can be written as

$$U\{0, \Lambda\} \mid \alpha, p^{(0)}\rangle = \mid \alpha'; p^{(0)}\rangle \tag{4.3}$$

since $Rp^{(0)} = p^{(0)}$. Thus equation (4.3) expresses the fact that $U\{0, R\}$ is a unitary representation of the rotation group for which $p^{(0)}$ is a dummy index. But we know that every representation of the rotation group is a direct sum of irreducible representations, whose basis vectors are specified by the eigenvalues $j(j + 1)$ and m of $\mathbf{J}^2$ and J_3. Thus we can write

$$\mid \alpha; p^{(0)}\rangle = \mid \beta; jm; p^{(0)}\rangle \tag{4.4}$$

where β represents all quantum numbers needed to specify the state, other than j, m, and $p^{(0)}$. Then equation (4.3) becomes

$$U\{0, R\} \mid \beta; j, m; p^{(0)}\rangle = \mathscr{D}^j_{m'm}(R) \mid \beta; j, m'; p^{(0)}\rangle. \tag{4.5}$$

(c) *Definition of the state vectors*

Abandoning the vague notation $|\alpha p\rangle$ for the state vectors, we now adopt the definition

$$\mid \beta; j, m; p\rangle = U\{0, L(p)\} \mid \beta; j, m; p^{(0)}\rangle. \tag{4.6}$$

Our problem is to rewrite equation (4.2) explicitly in this notation, which shows more clearly how the states depend on their quantum numbers.

(d) *Action of an arbitrary Lorentz transformation*

We need to evaluate $U\{a, \Lambda\} \, |\beta; j, m; p\rangle$. Begin by putting $a = 0$; then from (4.6) one has

$$U\{0, \Lambda\} \mid \beta; j, m; p\rangle = U\{0, \Lambda\} \, U\{0, L(p)\} \mid \beta; j, m; p^{(0)}\rangle. \tag{4.7}$$

Putting $p' = \Lambda p$, one has from the group composition law

whence

$$U\{0, \Lambda\} \mid \beta; j, m; p\rangle = U\{0, L(p')\}\, U\{0, L^{-1}(p')\,\Lambda L(p)\} \mid \beta; j, m; p^{(0)}\rangle \quad (4.8)$$

Consider the transformation $L^{-1}(p')\Lambda L(p) = T$. The transformation $L(p)$ transforms $p^{(0)}$ into p; Λ transforms p into p'; and $L^{-1}(p')$ transforms p' into $p^{(0)}$ (Figure 4.4.1). Consequently T is a Lorentz transformation conserving $p^{(0)} = (\mathbf{0}, M)$. Hence it is a rotation, and equation (4.5) shows how it acts on $|\beta; j, m; p_0\rangle$, namely

$$U\{0, L^{-1}(p')\,\Lambda L(p)\} \mid \beta; j, m; p^{(0)}\rangle$$
$$= \sum_{m'} \mathscr{D}^{j}_{m'm}(L^{-1}(p')\,\Lambda L(p)) \mid \beta; j, m'; p^{(0)}\rangle \quad (4.9)$$

In view of the definition (4.6) one has

$$U\{0, L(p')\} \mid \beta; j, m'; p^{(0)}\rangle = \mid \beta; j, m'; \Lambda p\rangle \quad (4.10)$$

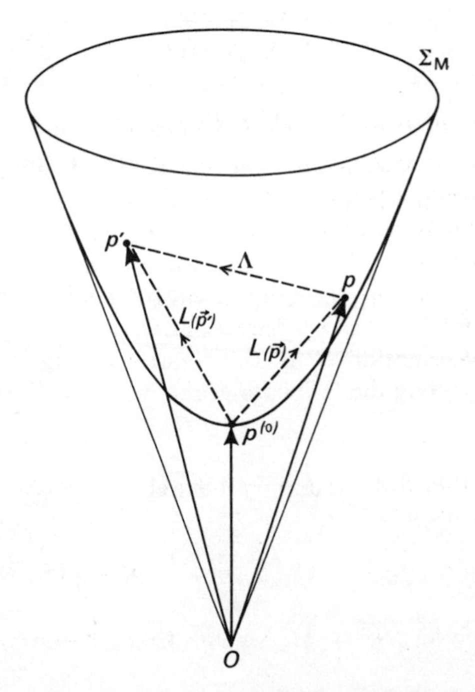

FIG. 4.4.1. The mass hyperboloid Σ_M of a particle. The effects of boosts $L(\mathbf{p})$, $L(\mathbf{p})$, and of a Lorentz transformation Λ. One sees that the transformation $L^{-1}(\mathbf{p})\Lambda L(\mathbf{p})$ leaves $p^{(0)}$ unchanged.

which together with (4.8) gives

$$U\{0, \Lambda\} \mid \beta; j, m; p\rangle = \sum_{m'} \mathscr{D}^j_{m'm}(L^{-1}(p')\,\Lambda L(p) \mid \beta; j, m'; \Lambda p\rangle. \tag{4.11}$$

It is now easy to determine the effects of $U\{a, \Lambda\}$. Indeed, by equation (2.21):

$$U\{a, \Lambda\} = U\{a, I\}\; U\{0, \Lambda\} \tag{4.12}$$

and by (4.2):

$$U\{a, I\} \mid \beta; j, m; p\rangle = e^{ip\cdot a} \mid \beta; j, m; p\rangle \tag{4.13}$$

whence

$$U\{a, \Lambda\} \mid \beta; j, m; p\rangle$$
$$= e^{ia\cdot p'} \sum_{m'} \mathscr{D}^j_{m'm}(L^{-1}(p')\Lambda L(p))\; |\beta, j, m'; p'\rangle, \tag{4.14}$$

where $p' = \Lambda p$.

Finally one notes that the quantum numbers β play no part in this relation; they can be suppressed if one is interested in a given irreducible representation. *More precisely, we shall assume that every particle is associated with an irreducible representation, and that consequently the quantum numbers j, m, and p suffice to specify the state of a particle of mass M*. Hence

$$\boxed{U\{a, \Lambda\} \mid jmp\rangle = e^{ia\cdot p'}\, \mathscr{D}^j_{m'm}(L^{-1}(p')\,\Lambda L(p)) \mid jmp'\rangle;\; p' = \Lambda p} \tag{4.15}$$

(e) *Properties of representations of the Poincaré group*

The formula (4.15) is seen to involve only the irreducible representations of the rotation group. The only quantum numbers to appear are the mass M and the spin quantum numbers j and m, which shows that a particle is fully specified by its mass, its *spin* j, and the orientation m of its spin. This identification of (j, m) with the spin is clear from subsection b, since (j, m) specify the rotational states of a particle at rest.

Since the representation $\mathscr{D}^j$ of the rotation group is irreducible, the representation (4.15) of the Poincaré group is also irreducible.

(f) *Unitarity*

The representation (4.15) is unitary if one choses an invariant normalisation for the state vectors:

$$\langle j'm'p' \mid jmp\rangle = \delta_{m'm}\, \delta_{jj'}\, \delta(\mathbf{p}' - \mathbf{p})\, \sqrt{\mathbf{p}^2 + M^2} \tag{4.16}$$

Note that $\delta(\mathbf{p}' - \mathbf{p})\sqrt{\mathbf{p}^2 + M^2}$ is an invariant, because $d^3p/\sqrt{\mathbf{p}^2 + M^2}$ is invariant and

$$\int \delta(\mathbf{p}' - \mathbf{p})\, \sqrt{\mathbf{p}^2 + M^2}\, \frac{d^3p'}{\sqrt{\mathbf{p}'^2 + M^2}} = 1.$$

The invariance of $d^3p/\sqrt{\mathbf{p}^2 + M^2}$ is evident from the fact that d^3p and $\sqrt{\mathbf{p}^2 + M^2}$ are both multiplied by γ'/γ under a Lorentz transformation which sends p into Λp (defining $\gamma = p^0/M, \gamma' = (\Lambda p)^0/M$).

Verification of unitarity: we must show that $U^+U = I$ in Hilbert space, i.e.

$$\langle p'j'm' \mid U^+\{a, \Lambda\}\, U\{a, \Lambda\} \mid pjm\rangle = \langle p'j'm' \mid pjm\rangle. \tag{4.17}$$

By (4.15) the left hand side equals

$$e^{-i\Lambda p'\cdot a}\, e^{i\Lambda p\cdot a}\, \mathscr{D}^{j'^*}_{m''m'}(R)\, \mathscr{D}^{j}_{m'''m}(R) \langle \Lambda p', j'm''' \mid \Lambda p, j, m''\rangle, \tag{4.18}$$

while one has from (16) that

$$\begin{aligned}\langle \Lambda p', j', m''' \mid \Lambda p, j, m''\rangle \\ &= \delta_{j'j}\, \delta_{m'''m''}\, \delta(\Lambda p' - \Lambda p) \times \sqrt{(\Lambda \mathbf{p})^2 + M^2} \\ &= \delta_{j'j}\, \delta_{m''m'''}\, \delta(p' - p) \times \sqrt{\mathbf{p}^2 + M^2}\end{aligned} \tag{4.19}$$

where the last step derives from the invariance of (4.16).

From the unitarity of the representation $\mathscr{D}^j$ of the rotation group, and from $m'' = m'''$, it follows that

$$\begin{aligned}\mathscr{D}^{j^*}_{m''m'}(R)\, \mathscr{D}^{j}_{m''m}(R) &= \mathscr{D}^{j+}_{m'm''}(R)\, \mathscr{D}^{j}_{m''m}(R) \\ &= \mathscr{D}^{j}_{m'm''}(R^{-1})\, \mathscr{D}^{j}_{m''m}(R) = \mathscr{D}^{j}_{m'm}(I) = \delta_{mm'}\,.\end{aligned} \tag{4.20}$$

But equations (4.19) and (4.20) lead to (4.17), which is what we had to show.

(g) *Scalar products*

Recall that by chosing a scalar product of the basis vectors as in (4.16), one fixes the coefficient A in the resolution of the identity

$$I = \sum_{m,j} \int \mid jmp\rangle\langle jmp \mid A d^3p. \tag{4.21}$$

Indeed, substituting (4.21) into (4.16), and setting $E = \sqrt{\mathbf{p}^2 + M^2}$, one obtains

$$\begin{aligned}\langle j'm'p' \mid jmp\rangle &= \langle j'm'p' \mid I \mid jmp\rangle \\ &= \int \langle j'm'p' \mid j_1m_1p_1\rangle\, A d^3p_1 \langle j_1m_1p_1 \mid jmp\rangle \\ &= \int \delta_{m'm_1}\, \delta_{mm_1}\, \delta_{j'j_1}\, \delta_{jj_1}\, \delta(\mathbf{p}' - \mathbf{p}_1)\, \delta(\mathbf{p} - \mathbf{p}_1)\, EE_1 A d^3p_1 \\ &= AEE'\, \delta(\mathbf{p}' - \mathbf{p})\, \delta_{mm'}\, \delta_{jj'}\end{aligned}$$

whence $A = E^{-1}$. Thus,

$$I = \sum_{m,j} \int \mid jmp\rangle \frac{d^3p}{E} \langle jmp \mid. \tag{4.22}$$

Let us consider more generally two state vectors $|\varphi\rangle$ and $|\psi\rangle$, (two wavepackets for instance), whose wavefunctions in momentum space are given by

$$\langle jm\mathbf{p} \mid \varphi\rangle = \varphi_m^j(\mathbf{p}) \qquad \text{et} \qquad \langle jm\mathbf{p} \mid \psi\rangle = \psi_m^j(\mathbf{p})$$

Then their scalar product is

$$\langle\varphi \mid \psi\rangle = \int \sum_{j,m} \langle\varphi \mid jm\mathbf{p}\rangle \frac{d^3\mathbf{p}}{E} \langle jm\mathbf{p} \mid \psi\rangle$$

$$= \sum_{j,m} \int \varphi_m^{*j}(\mathbf{p})\, \psi_m^j(\mathbf{p}) \frac{d^3p}{E}. \tag{4.23}$$

5. The theory of spin operators

We have seen that a state of nonzero mass is specified by the three quantum numbers M^2, j, and m. We know already that the squared mass M^2 is an eigenvalue of the operator

$$P^2 = P_0^2 - P_1^2 - P_2^2 - P_3^2. \tag{5.1}$$

Thus one is led to ask, what are the operators having j and m as their eigenvalues?

To see this, introduce the Kronecker tensors*

$\epsilon_{ijk} = +1$ if $i \neq j \neq k \neq i$ and if (ijk) is an even permutation of (1, 2, 3)
$\epsilon_{ijk} = -1$ if $i \neq j \neq k \neq i$ and if (ijk) is an odd permutation of (1, 2, 3)
$\epsilon_{ijk} = 0$ if two indices are equal
$\epsilon_{\mu\nu\rho\sigma} = \pm 1$ if no two indices are equal and if $(\mu\nu\rho\sigma)$ is an even (odd) permutation of (0, 1, 2, 3)
$\epsilon_{\mu\nu\rho\sigma} = 0$ in all other cases.

We can then define an antisymmetric tensor of rank two to replace the J_i and K_i by

$$M_{jk} = \epsilon_{ijk} J_i \qquad M_{0i} = -M_{i0} = K_i \qquad M_{00} = 0. \tag{5.2}$$

After a little algebra the commutation relations of the J_i and K_i can be expressed in the covariant form

$$[M_{\mu\nu}, M_{\kappa\rho}] = -i(g_{\mu\kappa}M_{\nu\rho} - g_{\mu\rho}M_{\nu\kappa} + g_{\nu\rho}M_{\mu\kappa} - g_{\nu\kappa}M_{\mu\rho}) \tag{5.3}$$

$$[M_{\mu\nu}, P_\rho] = -i(g_{\mu\rho}P_\nu - g_{\nu\rho}P_\mu). \tag{5.4}$$

We can then introduce a four dimensional vector operator W_μ defined by

$$W_\mu = \tfrac{1}{2}\epsilon_{\mu\nu\rho\sigma}P_\nu M_{\rho\sigma} \tag{5.5}$$

* Translator's note: in English these are more commonly known as the Levy–Civita symbols.

or, more explicitly

$$W^0 = -\mathbf{P} \cdot \mathbf{J} = -P_0\mathbf{J} + \mathbf{P} \times \mathbf{K}. \tag{5.6}$$

where the symbol $\times$ denotes the vector product. The antisymmetry of the tensor entails

$$W_\mu P_\mu = 0 \tag{5.7}$$

This shows that when W_μ acts on an eigenvector of P_μ, only its three components orthogonal to p are independent. Indeed, in the rest frame of the physical system (always assuming nonzero mass) one has

$$W^0 = 0 \qquad \mathbf{W} = -M\mathbf{J} \tag{5.8}$$

which shows that up to a factor $-M$ the independent components of W are just the spin $\mathbf{J}$ of the particle. This can be seen equally well in an invariant way by appeal to the commutation relations between W_μ and the generators of the Poincaré group

$$[W_\mu , P_\nu] = 0 \tag{5.9}$$

$$[W_\mu , M_{\nu\rho}] = i(g_{\mu\nu}W_\rho - g_{\mu\rho}W_\nu) \tag{5.10}$$

whence $W^2 = g_{\mu\nu}W_\mu W_\nu$ is an invariant. Indeed,

$$[W^2, P_\nu] = 0 \tag{5.11}$$

$$[W^2, M_{\mu\nu}] = 0. \tag{5.12}$$

From (5.8) we get

To conclude, we see that $j(j + 1)$ is the eigenvalue of W^2 up to a factor $-M^2$. It is also easy to see that Mm is the eigenvalue of the scalar product (W, n), where n is a spacelike unit vector orthogonal to p, obtainable from the unit vector along the z axis by the boost of p.

Spin of a particle with zero mass

When a particle has zero mass, there exists no reference frame in which it is at rest, and the whole theory of spin must be rewritten from the start. To begin with, the state of a particle with zero mass can be specified by its four-momentum

$$P_\mu \,|\, p\rangle = p_\mu \,|\, p\rangle \tag{5.14}$$

with

$$p^2 = 0. \tag{5.15}$$

Then from equation (7) one obtains

$$W_\mu p_\mu \,|\, p\rangle = 0 \tag{5.16}$$

and from (13)

$$W^2 \,|\, p\rangle = 0 \tag{5.17}$$

whence

$$\langle p \mid W_\mu \cdot W_\mu \mid p \rangle = 0 \tag{5.18}$$

Thus, $W_\mu \, |p\rangle$ is a vector of zero length (in 4 dimensional space) orthogonal to p. Since p itself is of zero length, these two vectors are necessarily collinear, and

$$W_\mu \mid p \rangle = \sigma p_\mu \mid p \rangle \tag{5.19}$$

where σ is an arbitrary real number. $|\sigma|$ is called the *spin* of the zero-mass particle. Actually a state has physical meaning only if σ is an integer or half integer. By writing (5.19) in the form

$$W_\mu \mid \sigma p \rangle = \sigma p_\mu \mid \sigma p \rangle . \tag{5.20}$$

one can easily see that, $\mathbf{p}$ being a polar and $\mathbf{J}$ an axial vector, $\mathbf{W}$ is an axial vector; it follows that on reflecting the space axes, and writing the parity operator as Π, one has

$$\Pi W_i \Pi^{-1} = W_i$$

$$\Pi P_i \Pi^{-1} = -P_i$$

whence

$$\Pi \mid \sigma, p \rangle = \mid -\sigma, p \rangle . \tag{5.22}$$

For a zero mass particle there are consequently two spin states, $|\sigma, p\rangle$ and $|-\sigma, p\rangle$, having definite spin and parity. Thus the photon has the two spin states $|1, p\rangle$ and $|-1, p\rangle$, corresponding respectively to photons with left and right circular polarisation.

Note that for a system of particles, W^2/M^2 represents the total angular momentum in the centre of mass frame.

6. Further comments

(a) *Nonrelativistic limit*

At low energies one always has $p^0 \approx M$, $\gamma \approx 1$, $v \ll 1$, whence by (4.1) the boost differs very little from the identity. A reaction where all the particles have low energies can be described in terms of rotations R alone; then the expression $L^{-1}(p')RL(p)$ appearing in equation (4.15) is essentially the same as R, so that

$$U\{a, R\} \mid \mathbf{p}, j, m \rangle = e^{i(p^0 a^0 - \mathbf{aRp})} \mathscr{D}^j_{m'm}(R) \mid R\mathbf{p}, j, m' \rangle . \tag{6.1}$$

In equation (6.1) we recognise the usual representation of the spin of nonrelativistic particles, i.e. the Pauli formalism in the case of spin $\frac{1}{2}$ particles.

(b) *Dirac equation*

There exist methods different from the above for treating relativistic particles of spin $\frac{1}{2}$ or 1 (the Dirac equation for spin $\frac{1}{2}$ particles, the polarisation vector for spin 1), to which we shall return later. We stress, however, that all particles can be treated by the formalism of this chapter; there is no necessity for the Dirac equation, nor is it always the most convenient to use.

(c) *Reducibility*

Further, one can prove that every representation of the Poincaré group in Hilbert space has a spectrum of eigenvalues of P^2, called the mass spectrum. When this spectrum includes only positive values, the representation is a direct sum of irreducible representations of the kind (4.15), given a suitable choice of basis vectors in Hilbert space.

(d) *Parity and time reversal*

Recall finally that there exist Lorentz transformations (Π, Θ) which do not belong to the proper orthochronous Lorentz group. It is not evident a priori whether they should be treated on the same footing as the ordinary Lorentz transformations. We shall return to this problem later.

Problems

1. A statistical system is known to be described by a density matrix ρ. In particular, for a system of particles with spin s, ρ is a $(2s + 1) \times (2s + 1)$ matrix. Denote by S_1, S_2, S_3 the spin operators satisfying $[S_1, S_2] = iS_3$, etc; then the *polarisation* of the system of particles is by definition equal to

$$\mathbf{P} = \text{Trace}\,(\rho \mathbf{S}) \qquad \text{with} \quad \text{Trace}\,\rho = 1.$$

Similarly, denoting by $\boldsymbol{\mu}$ the set of three operators for the components of the magnetic moment of the particle, we shall call

$$\mathbf{M} = \text{Trace}\,(\rho\boldsymbol{\mu})$$

the magnetisation of the system.
By using the Wigner–Eckart theorem, show that $\mathbf{M}$ and $\mathbf{S}$ are necessarily parallel. To this end it is useful to investigate the transformation law for the density matrix under rotations.

2. A particle is circulating uniformly in an accelerator. Its motion can be interpreted as a succession of Lorentz transformations

$$\Lambda(\mathbf{v}_1)\Lambda(\mathbf{v}_2 - \mathbf{v}_1)\Lambda(\mathbf{v}_3 - \mathbf{v}_2) \cdots \Lambda(\mathbf{v}_n - \mathbf{v}_{n-1}),$$

where $\mathbf{v}_n$ is the velocity of the particle at time t_n, the time interval $t_n - t_{n-1}$ being very small. Investigate the precession of the particle spin during one revolution.

3. By using equation (2.7), prove equations (2.38) and (2.39).

CHAPTER 5

THE DETERMINATION OF SPINS

We show how the spin of a particle can be determined experimentally.

1. Orientation

We have seen that under the assumption of relativistic invariance, every particle is specified kinematically by its mass and its spin. Therefore it is a central problem to determine experimentally the spins of the various particles.

To this end, we shall analyse some basic experiments with the help of two theoretical tools:

1. The Golden Rule of Fermi;
2. The selection rules for angular momentum.

One could object that the Golden Rule depends on first order perturbation theory, and that we do not know whether for elementary particles an interaction Hamiltonian can be defined at all, nor, a fortiori, whether it can be treated by first order perturbation theory. But actually we shall exploit only some formal properties of the Golden Rule, which will moreover be justified quite generally when we come to deal with the general theory of collisions. Our only reason for discussing particle spin before general collision theory lies in the need to re-establish contact with experiment as soon as possible after acquiring a new theoretical concept.

We shall take it as read that electrons, positrons, protons and neutrons have spin $\frac{1}{2}$ and that photons have spin 1. At the beginning of the last chapter we discussed the magnetic resonance experiments which allow one to determine the electron and nucleon spins. As regards the spins of positrons and photons, it is more natural to treat the problem in the framework of electrodynamics. We note, however, and shall demonstrate it later when studying parity and charge conjugation, that particle and antiparticle must have the same spin, which immediately fixes the positron spin as $\frac{1}{2}$.

Note also that it is easy to decide whether the spin of a particle is integer or half-integer. Thus,

(a) the decay $n \to p + e + \bar{\nu}_e$ of the neutron implies that the antineutrino $\bar{\nu}_e$ has half-integer spin. The same is therefore true for the neutrino ν_e.

(b) the π^0 has integer spin, in view of $\pi^0 \to 2\gamma$.

(c) the (rare) reaction $\pi^- \to e^- + \bar{\nu}_e$ shows that π^-, and hence π^+, have integer spin.

(d) μ and ν_μ always appear together (except when μ appears with $\bar{\mu}$). Therefore it is difficult to tell a priori whether their spin is integer or half-integer. Nevertheless there is a strong indication that electrons and muons have the same spin, in view of the extraordinary similarity between their electromagnetic properties, and in particular between their gyromagnetic ratios, which are measured independently and without knowledge of the spin; thus, the muon has spin $\frac{1}{2}$.

(e) The decay $K \to 2\pi$ implies that K has integer spin.

(f) The decay $\Lambda \to p + \pi$ shows that Λ has half-integer spin.

Recall finally that the concept of isotopic spin makes sense only if the different members of a given multiplet have identical spins (whence (π^0 spin) = (π^- spin) = (π^+ spin), (K^- spin) = (K^0 spin) = ($\bar{K}^0$ spin) = (K^+ spin), etc.).

2. Pion spin

(a) *General argument*

Observing the decays

$$\pi^+ \to \mu^+ + \nu_\mu, \qquad \pi^- \to \mu^- + \bar{\nu}_\mu, \qquad \pi^0 \to 2\gamma$$

in the rest frame of the pion, one is struck by the fact that the angular distributions of μ's and γ's are always isotropic, which means that there is no preferred direction in this frame. This can be the case only if the pion has spin 0, or if it has nonzero spin but is always produced with very little polarisation. It would be difficult to understand how the second alternative could be realised so systematically, and it seems likely that the pion spin is zero.

(b) *π^+ spin*

The π^+ spin can be determined exactly by measuring the cross-sections of the two mutually inverse reactions

$$\pi^+ + d \to p + p \tag{2.1}$$

$$p + p \to \pi^+ + d. \tag{2.2}$$

Denote by α, β, i, j the spin components along a fixed axis of the two protons, of the pion, and of the deuteron respectively; then at low energies, when all the particles are nonrelativistic, the probabilities per unit time for these reactions are

$$P_1 = |\langle p_\alpha p_\beta \mid H \mid \pi_i^+ d_j\rangle|^2 \rho_f(p, p, E) \tag{2.3}$$

$$P_2 = |\langle \pi_i^+ d_j \mid H \mid p_\alpha p_\beta\rangle|^2 \rho_f(\pi, d, E) \tag{2.4}$$

In the centre of mass frame, the densities of states are given by

$$\begin{aligned}\rho_f(p, p, E) &= \int \delta\left(E - 2\frac{\mathbf{p}^2}{2m}\right)\frac{d^3\mathbf{p}}{(2\pi)^6} = \int \delta\left(E - \frac{2\mathbf{p}^2}{2m}\right)\frac{p^2\, dp\, d\Omega_p}{(2\pi)^6} \\ &= \int \delta\left(E - \frac{2p^2}{2m}\right) d\left(\frac{2p^2}{2m}\right)\frac{mp}{2}\frac{d\Omega_p}{(2\pi)^6} = \frac{mp}{2}\frac{d\Omega_p}{(2\pi)^6}\end{aligned} \tag{2.5}$$

$$\begin{aligned}\rho_f(\pi, d, E) &= \int \delta(E - \sqrt{q^2 + \mu^2} - q^2/2M)\frac{d^3q}{(2\pi^6)} \\ &\simeq \int \delta(E - \omega)\frac{q^2\, dq\, d\Omega_\pi}{(2\pi)^6} \\ &= \int \delta(E - \omega)\, q\omega\, d\omega \frac{d\Omega_\pi}{(2\pi)^6} = q\omega\frac{d\Omega_\pi}{(2\pi)^6}\end{aligned} \tag{2.6}$$

where $\omega = \sqrt{q^2 + \mu^2}$, and m, μ, M are the masses of proton, pion, and deuteron.

Let the pion spin be s. We know from nuclear physics that the deuteron spin is 1. If the final state particle polarisations are not observed in the reactions (2.1) and (2.2), then the probabilities are

$$P_1 = \frac{1}{3(2s+1)}\sum_{ij\alpha\beta} |\langle p_\alpha p_\beta \mid H \mid \pi_i' d_j\rangle|^2 \frac{mp}{2}\frac{d\Omega_p}{(2\pi)^6} \tag{2.7}$$

$$P_2 = \frac{1}{4}\sum_{ij\alpha\beta} |\langle \pi_i d_j \mid H \mid p_\alpha p_\beta\rangle|^2 q\omega \frac{d\Omega_\pi}{(2\pi)^6}\,. \tag{2.8}$$

The cross-sections are given by

$$\sigma_1 = P_1/\Phi_1 \qquad \sigma_2 = P_2/\Phi_2 \tag{2.9}$$

where the fluxes Φ are, respectively,

$$\Phi_1 = q/\omega \qquad \Phi_2 = p/m. \tag{2.10}$$

From the Hermitecity of the Hamiltonian, one has

$$\langle p_\alpha p_\beta \mid H \mid \pi_i d_j\rangle = \langle \pi_i d_j \mid H \mid p_\alpha p_\beta\rangle^* \tag{2.11}$$

whence, by (2.7), (2.8), (2.9), and (2.10)

$$4\frac{d\sigma_1}{d\Omega_\pi}\frac{m}{qp\omega} = \frac{d\sigma_2}{d\Omega_p}\frac{3(2s+1)}{pm}\frac{\omega}{q}\,. \tag{2.12}$$

A measurement of the cross-sections determines s directly, and one finds that it is zero.

Note that the argument hinges on the equality (2.11). We shall see later that this is due to the invariance of the strong interactions under time reversal. In this context, the equality between the probability amplitudes for mutually inverse reactions is called 'detailed balance'.

(c) *Spins of π^+ and π^0*

The equality, to 1 part in 10^3, between the cross-sections of π^+ and π^- on carbon evidences the total similarity between these particles, which are moreover particle and antiparticle. Hence the π^- spin is also zero. In the same way, the well confirmed conservation of isotopic spin shows the π^0 spin to be zero.

3. Spins of strange particles

(a) Λ

We exploit the fact that in reactions like

$$K^- + p \to \Lambda + \pi^0$$

the Λ is polarised at certain energies.

Suppose that the Λ is slow; and suppose for simplicity that its spin ($s = \frac{1}{2}, \frac{3}{2}, \ldots ?$) is oriented completely along the z axis. Then the system resulting from the decay

$$\Lambda \to p + \pi^-$$

must be in a state of angular momentum $|j = s, m = s\rangle$. Denoting the proton spin by $\boldsymbol{\sigma}$ and the orbital angular momentum by $\mathbf{l}$, we have $\mathbf{j} = \mathbf{l} + \boldsymbol{\sigma}$, whence $l = j \pm \frac{1}{2}$. Therefore the matrix element $\langle p\pi^- | H | \Lambda\rangle$ has the form

$$\alpha Y^{s\pm 1/2}_{s+1/2}(\hat{\mathbf{p}}) + \beta Y^{s\pm 1/2}_{s-1/2}(\hat{\mathbf{p}})$$

where $\hat{\mathbf{p}}$ is the direction of emission of the proton, and the Y's are spherical harmonics. After integrating $|\langle p\pi^- | H | \Lambda\rangle|^2$ over the azimuthal angle, we see that the term of highest degree in $\cos\theta$ is $P_{2s+1}(\cos\theta)$. Experimentally, the probability as a function of θ is given by (Figure 5.3.1):

$$\frac{dP_{\Lambda\to p\pi}}{d\cos\theta} \simeq A + B\cos\theta + C\cos^2\theta, \quad \text{whence } s_\Lambda = \tfrac{1}{2} \qquad (3.1)$$

This method can be refined further.

(b) The same method yields $s_{\Sigma^+} = s_{\Sigma^-} = \frac{1}{2}$.

(c) The Σ^0 spin has not been measured directly. Isotopic spin requires $s_{\Sigma^0} = \frac{1}{2}$.

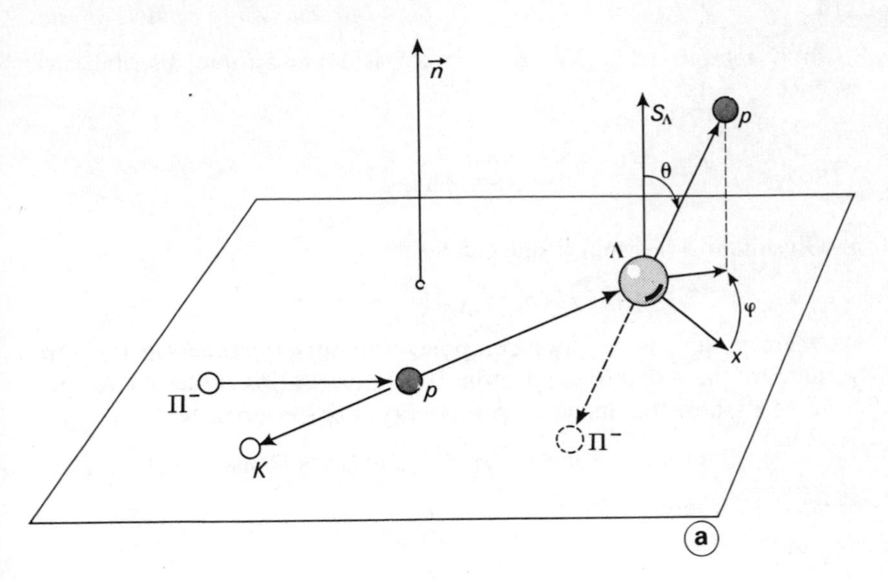

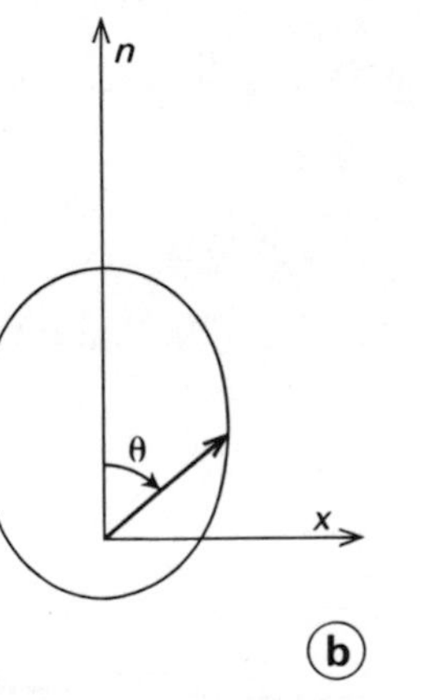

FIG. 5.3.1. Measurement of the Λ spin. (a) Production and decay kinematics of the Λ. (b) Decay spectrum of polarised Λ.

TABLE 1.

PARTICLE SPINS

Particle	γ	ν_e	ν_μ	μ	π	K	$\bar{K}$	p	n	$\bar{p}$	$\bar{n}$
Spin	1	1/2	1/2	1/2	0	0	0	1/2	1/2	1/2	1/2
Particle	Λ^0	Σ	Ξ	Ω							
Spin	1/2	1/2	1/2	3/2 ?							

(d) K mesons ($K^{\pm}$, K^0, $\overline{K}^0$) never yield anisotropic distributions, whence $s_K = 0$.

Problem

1. Given that Λ has spin $\frac{1}{2}$, one can write

$$\langle \Lambda\pi \mid H \mid p \rangle = \chi_\Lambda{}^*(A + B\boldsymbol{\sigma} \cdot \mathbf{p})\, \chi_p$$

where χ_Λ and χ_p are two-component spinors representing the spin states of the Λ and of the proton. If the polarisation of the Λ is known to be $\mathbf{P}$, show that its decay probability is proportional to

$$| A |^2 + | B |^2 p^2 \cos^2 \theta + 2\text{Re}(AB^*)\, pP \cos \theta$$

where θ is the angle between $\mathbf{p}$ and $\mathbf{P}$.

CHAPTER 6

SPACE REFLECTION, CHARGE CONJUGATION, TIME REVERSAL

We introduce certain discrete operations of fundamental importance: space reflection, which inverts the directions of the coordinate axes in space; charge conjugation, which changes particles into antiparticles; and time reversal. We show that the quantum numbers associated with these operations are not conserved by the weak interactions, though they are conserved by the strong and the electromagnetic interactions.

We indicate how the parity of a particle can be determined experimentally.

1. Space reflection and time reversal

When we were studying the action on a quantum state of the transformations of the Poincaré group, we did not concern ourselves with the transformations that cannot be generated continuously from the identity, nor in particular with space reflection Π and time reversal T:

$$\Pi: \qquad x'^0 = x^0 \qquad \mathbf{x}' = -\mathbf{x} \tag{1.1}$$

$$T: \qquad x'^0 = -x^0 \qquad \mathbf{x}' = \mathbf{x} \tag{1.2}$$

One can write down immediately the commutation relations between the matrices Λ_Π and Λ_T

$$\Lambda_\Pi = \begin{bmatrix} -1 & 0 & 0 & 0 \\ 0 & -1 & 0 & 0 \\ 0 & 0 & -1 & 0 \\ 0 & 0 & 0 & 1 \end{bmatrix} \qquad \Lambda_T = \begin{bmatrix} 1 & 0 & 0 & 0 \\ 0 & 1 & 0 & 0 \\ 0 & 0 & 1 & 0 \\ 0 & 0 & 0 & -1 \end{bmatrix} \tag{1.3}$$

and the infinitesimal generators of the Poincaré group:

$$[\Pi, \mathbf{J}] = 0 \qquad \{\Pi, \mathbf{K}\} = 0 \qquad [\Pi, P_0] = 0 \qquad \{\Pi, \mathbf{P}\} = 0 \tag{1.4}$$

$$[T, \mathbf{J}] = 0 \qquad \{T, \mathbf{K}\} = 0 \qquad \{T, P_0\} = 0 \qquad [T, \mathbf{P}] = 0 \tag{1.5}$$

In order to do this we need merely introduce a fifth homogeneous parameter s, invariant under Π and T, and calculate the matrix products explicitly. Actually (1.4) emerges immediately from the following argument:

Π changes the sign of the momentum, whence $\{\Pi, \mathbf{P}\} = 0$; it also changes the sign of the position vector $\mathbf{x}$; therefore it does not change $\mathbf{x} \times \mathbf{p}$, whence $[\Pi, \mathbf{J}] = 0$. It changes the sign of the relative velocity $\mathbf{v}$ in a Lorentz transformation, whence $\{\Pi, \mathbf{K}\} = 0$; but it does not change the energy, whence $[\Pi, P_0] = 0$.

Accordingly, for a single-particle state represented by $|\mathbf{p}, m\rangle$ one expects

$$\Pi \mid \mathbf{p}, m\rangle = \epsilon \mid -\mathbf{p}, m\rangle \tag{1.6}$$

It is easy to check that this transformation satisfies (1.4), once one has noticed that the relation $\{\Pi, \mathbf{K}\} = 0$ implies

$$\Pi L(\mathbf{p}) \Pi^{-1} = L(-\mathbf{p})$$

where $L(\mathbf{p})$ is the boost of $\mathbf{p}$. Given that $\Pi^2 = I$, one has $\epsilon^2 = 1$, whence $\epsilon = \pm 1$. (Indeed, at this point there would be some profit in reconsidering the representation of states up to a phase, but this is a delicate subject which we prefer to pass over.) The coefficient ϵ is called the intrinsic *parity* of the particle.

For time reversal one might be tempted to adopt the unitary transformation

$$T \mid \mathbf{p}m\rangle = \eta \mid \mathbf{p}m\rangle \tag{1.7}$$

which clearly satisfies the commutation relations $[T, \mathbf{J}] = 0$, $[T, \mathbf{P}] = 0$; but this is unsatisfactory for several reasons:

(a) (1.7) does not satisfy the relation $\{T, \mathbf{K}\} = 0$;

(b) according to (1.7) time reversal would simply be a non-transformation;

(c) (1.7) is contrary to common sense, which requires that momenta, and angular momenta $\mathbf{J} = \mathbf{x} \times \mathbf{p}$, change sign under time reversal. Furthermore, we expect the energy to remain unchanged.

On the other hand (1.7) is the only unitary transformation which has the correct commutation relations with $\mathbf{J}$ and $\mathbf{P}$. Therefore we must try an antiunitary transformation. We shall show that all the conditions are met by the transformation

$$T \mid \mathbf{p}m\rangle = \eta K \mid -\mathbf{p} - m\rangle \tag{1.8}$$

where K is the complex conjugation operator, having the property

$$Ki = -iK$$

To this end it sufficies to note that the operators of the Poincaré group are defined in terms of the infinitesimal generators by e^{iG}, whence K leads to a sign change of all generators:

$$KJK = -\mathbf{J} \qquad K\mathbf{K}K = -\mathbf{K} \qquad KP_0K = -P_0 \qquad K\mathbf{P}K = -\mathbf{P}$$

It is then immediately obvious that (1.8) satisfies (1.5). Further, it is satisfactory that according to (1.8) T changes the signs of $\mathbf{p}$ and m. Lastly, the relative phase of $|\mathbf{p}, m\rangle$ and $|-\mathbf{p}, -m\rangle$ is arbitrary and can always be chosen so that $\eta = 1$. (The same cannot be done for ϵ, because no arbitrary phase is involved in (1.6) when $\mathbf{p} = 0$.)

2. Parity and interactions

(a) *Strong and electromagnetic interactions*

The best proof that strong and electromagnetic interactions conserve parity is furnished by the selection rules for the emission of photons from nuclei, which we shall not discuss in detail.

(b) *Weak interactions*

It is a most remarkable fact that the weak interactions violate parity; it was established as follows. The nucleus of Cobalt 60 beta-decays according to

$$\mathrm{Co}^{60} \rightarrow \mathrm{Ni}^{60} + e + \bar{\nu} \tag{2.1}$$

The spin of Co^{60} has the large value 5. This makes it possible to polarise a sample of Cobalt by placing it into a magnetic field $\mathbf{B}$ at a low temperature. With such an arrangement, let us now ask what are the probabilities for emitting electrons into the two hemispheres separated by a plane Q normal to $\mathbf{B}$. Note in this connection that a reflection in the plane Q is the product of a rotation R through 180° about $\mathbf{B}$ and a space reflection Π. Since the rotation axis is parallel to the spin, R does not change the spin state of the nucleus, and since Π commutes with the spin, Q as characterised by its normal vector is invariant if Π is conserved. Hence, if the decay conserves parity, one must observe equal numbers of electrons emitted into the upper and the lower hemispheres (Figure 6.2.1).

But in the actual experiment one observes a great asymmetry, showing that parity is not conserved by the weak interactions.

Equivalently, one could note that this asymmetry implies the presence of a term proportional to the scalar product $\mathbf{B} \cdot \mathbf{p}$ in the decay probability. Since $\mathbf{B}$ is an axial vector, this term is a pseudoscalar and as such incompatible with the conservation of parity. Accordingly, this experiment establishes the following law:

Law: Electromagnetic and strong interactions conserve parity. The weak interactions do not conserve it.

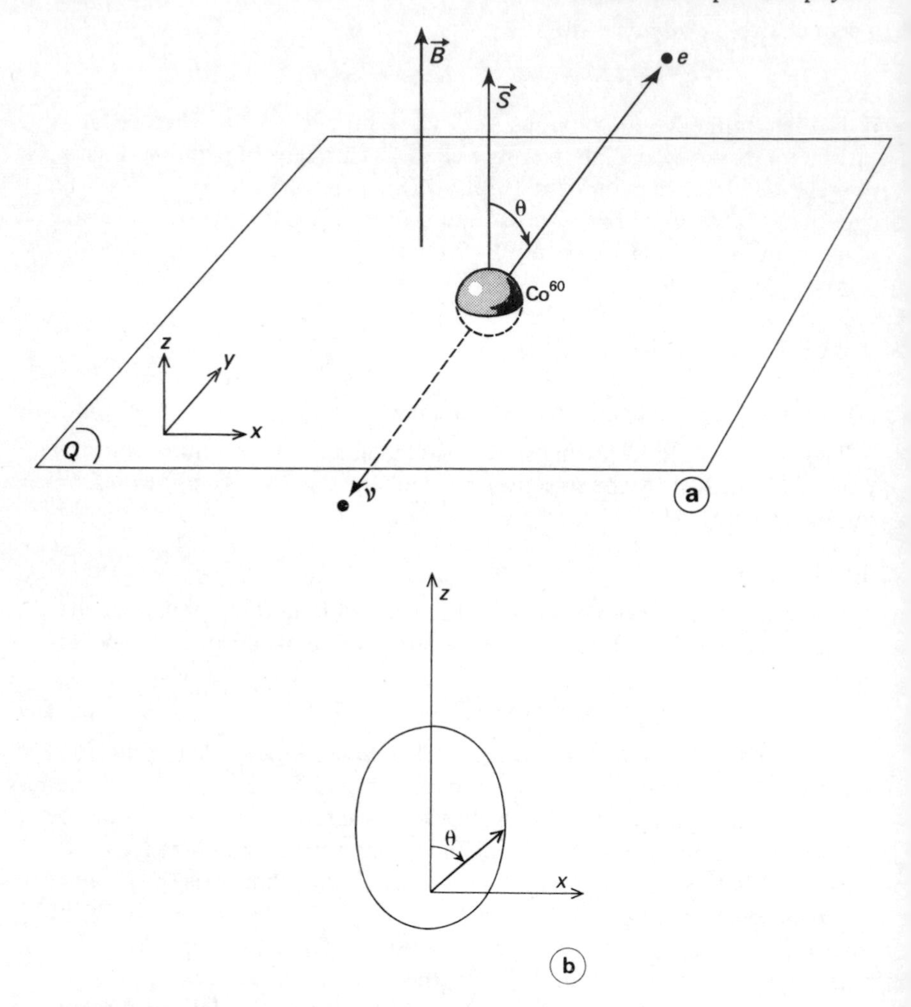

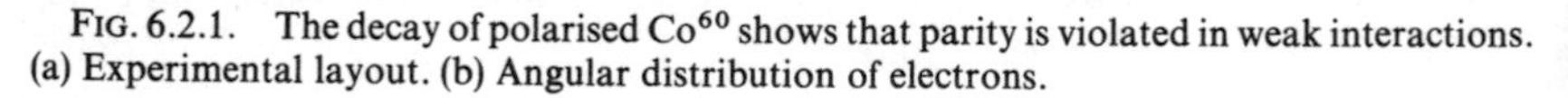

FIG. 6.2.1. The decay of polarised Co^{60} shows that parity is violated in weak interactions. (a) Experimental layout. (b) Angular distribution of electrons.

(c) *Parity and polarisation*

Parity conservation has interesting consequences for the polarisation of particles emerging from a collision, which we shall now examine.

Consider a collision between an unpolarised particle of momentum $\mathbf{p}$, and a particle at rest, also unpolarised. The collision results in two particles with momenta $\mathbf{q}$ and $\mathbf{q}'$. Let (Q) be the collision plane containing the vectors $\mathbf{p}$, $\mathbf{q}$, and $\mathbf{q}'$. Suppose that we observe the spin $\boldsymbol{\sigma}$ of particle $\mathbf{q}$ but not that of particle $\mathbf{q}'$. What does parity conservation tell us about $\boldsymbol{\sigma}$?

Denote by Σ the symmetry operation $R\Pi$, where R is a rotation through $180°$ about the normal $\mathbf{n}$ to (Q). Then we note that the initial state is unchanged by Σ; so are the vectors $\mathbf{q}$ and $\mathbf{q}'$. The spin $\boldsymbol{\sigma}$ is unchanged under Π. Under R, the component of $\boldsymbol{\sigma}$ normal to (Q) does not change, while the component parallel to (Q) changes sign. If parity is conserved, then the final state, like the initial state, must be unchanged by Σ. It follows that $\boldsymbol{\sigma}$ *must be normal to* (Q). Thus:

In a collision between two unpolarised particles leading to a two-body final state, the polarisation of an emergent particle is normal to the collision plane if parity is conserved in the reaction.

If the particle $\mathbf{q}$ is unstable, its polarisation can be measured by studying its decay. In effect, its polarisation singles out a preferred direction as regards the decay; and since the decay proceeds by the weak interactions which violate parity, the direction of $\boldsymbol{\sigma}$ can be ascertained through the presence in the decay probability of a term proportional to $\boldsymbol{\sigma} \cdot \mathbf{p}$. If the particle is stable, like a proton, then its polarisation can be measured through secondary collisions. Such measurements confirm the rule stated above, allowing one to check that parity is conserved by the strong and by the electromagnetic interactions. However, their accuracy is much lower than that of the electromagnetic selection rules. But the results are useful in practice since they allow one to determine uniquely the direction of polarisation.

3. Superselection rules

Before we consider how to determine the intrinsic parities ϵ of elementary particles, we must ask for which particles ϵ is arbitrary, and for which it can be determined by experiment.

Thus, the parity of the vacuum state is completely arbitrary, but by universal convention it is taken as $+1$. Then all particles whose internal quantum numbers are those of the vacuum have well determined parities; so for instance the photon and the π^0.

The parity of π^+ is arbitrary, because the charge prevents one from comparing it to that of the γ or of the π^0. There is no reaction like $\pi^+ \to \pi^0 \to \gamma + \gamma$ which might enable one to fix the parity of π^+ relative to known particles. In such cases one says that charge conservation leads to a superselection rule. This rule classifies states according to their charge, and no direct comparison is possible between states with different charges. Nevertheless, since the parity of π^0 is well determined, one choses the parity of π^+ to be the same as that of π^0, in order to simplify arguments based on isotopic spin. Similarly one choses the parity of π^- to be equal to that of π^+ and π^0.

Baryon number leads to another superselection rule, so that the parity of the proton is arbitrary. By convention it is chosen as $+1$. The parity of the neutron is then well determined, since the pion parity is known and since there exist reactions like

$$\pi^- + p \to \pi^0 + n + \pi^-$$

which allow one to compare proton and neutron directly. Actually, if isotopic spin is to make sense, it is necessary that the parity of the neutron should be $+1$ like that of the proton. We shall see later that this property is exploited in practice to determine the parity of the pion.

Strangeness is conserved by the strong and by the electromagnetic interactions, but not by the weak ones. But the weak interactions violate parity and cannot therefore be used in the determination of intrinsic parities; hence as regards parity the conservation of strangeness acts in practice like a superselection rule. Thus the parity of the Λ^0 is arbitrary, and by convention it is chosen to be $+1$. Thereafter the parity of the other particles is determined in principle through the reactions

$$\begin{aligned} K^+&: \quad \pi^0 + p \to \Lambda^0 + K^+ \\ K^0&: \quad \pi^- + p \to \Lambda^0 + K^0 \\ K^-&: \quad \pi^- + p \to p + K^0 + K^- \\ \bar{K}^0&: \quad \pi^+ + p \to p + K^+ + K^0 \\ \Sigma&: \quad \pi + p \to \Sigma + K \\ \Xi&: \quad \bar{K} + p \to \Xi + K \end{aligned}$$

In this way the parities of all hadrons are determined. Those of antiparticles can be determined through their annihilation; e.g. for the antiproton

$$\bar{p} + p \to \pi^+ + \pi^- + \pi^0.$$

It turns out that the proton and antiproton must have opposite parities if they can be described by a local field $\Psi(x)$ obeying the Dirac equation. Note that this condition is rather restrictive, and that it is useful to have an independent determination of the antiproton parity. Then the parities of other antiparticles can be determined through exchange reactions:

$$\bar{p} + p \to \bar{\Sigma} + \Sigma \cdots$$

The parities of the electron and of the μ^- are arbitrary in view of the superselection rules due to the conservation of electron number and of muon number. But the parity of e^+ is determined through the annihilation process

$$e^+ + e^- \to \gamma + \gamma$$

Since neutrinos take part only in weak interactions, there is no sense in which one can even think of their having a parity.

4. Parity of the pion

As one example of a parity determination, we discuss how one determines that of the π^+. One adopts the convention that the proton and neutron parities are the same, thus disposing of the two superselection rules due to the conservation of baryon number and of electric charge. Then the parity of the deuteron is well determined and that of the π^+ is found by studying the reaction

$$\pi^+ + d \rightarrow p + p \tag{4.1}$$

The deuteron parity is known from nuclear physics, which tells us that it is a superposition of S and D states containing a proton and a neutron, in a triplet spin state. Its total spin is 1 and its parity +.

At low energies the reaction (4.1) proceeds from an S state of the pion relative to the deuteron. This can be seen from the following heuristic argument (Figure 6.4.1). The deuteron has a spatial extension of R_0. When the pion impact parameter is b and its momentum $\mathbf{p}$, its angular momentum is $\ell = pb$. In practice the reaction will clearly not occur if $b \gg R_0$, i.e. if $\ell \gg pR_0$. If p is sufficiently small we see that the reaction can proceed only in the S state $\ell = 0$.

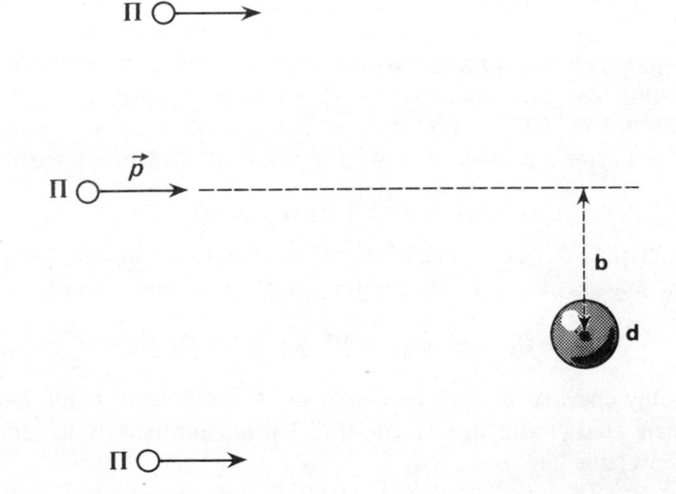

FIG. 6.4.1. The impact parameter b in a π-deuteron collision.

By vector addition, the initial state of (4.1) accordingly has total angular momentum $J = 1$. Evidently the same applies to the two protons in the final state. Hence our problem reduces to discovering what is the parity of a two proton state with total angular momentum 1.

Consider first the case when the two protons are in a triplet spin state, $\sigma = 1$. Then their spin wavefunction is symmetric. By the Pauli principle their orbital wavefunction must be antisymmetric. Hence their orbital angular momentum is $\ell = 1, 3, \ldots$. But only the state with $\ell = 1$ can be combined with a total spin of 1 to give the resultant $J = 1$. Therefore the parity is $(-1)^{\ell} = -1$.

If the two protons are in the singlet spin state, $\sigma = 0$, their spin wavefunction is antisymmetric, and their orbital wavefunction therefore symmetric: $\ell = 0, 2, \ldots$. Under these conditions it is impossible to have a resultant $J = 1$, and the reaction (4.1) cannot then proceed at low energies.

The fact that reaction (4.1) is observed implies that the parity of the π is $-$: the pion is pseudoscalar. One can verify experimentally that the reaction does indeed proceed from the S state.

It has been shown by similar methods that the K is pseudoscalar and that Σ has the same parity as Λ, i.e. $+$ by convention. The parity of the Ξ is still conjectural.

5. Charge conjugation

It is found experimentally that every particle appears to have an antiparticle, of equal mass, but with opposite signs of the electric charge, of baryon or lepton number, of strangeness, and of the third component I_3 of isotopic spin. Hence it seems natural to try to express this result formally, and to ascribe the equality in mass between particle and antiparticle to a conservation law; (in some sense they constitute a multiplet).

To this end we define a unitary operator C, called *charge conjugation*, which transforms a particle A into its antiparticle $\bar{A}$. More precisely,

$$C \mid A, \mathbf{p}, m\rangle = \eta_A \mid \bar{A}, \mathbf{p}, m\rangle. \tag{5.1}$$

The fact that C has no effect on the momentum, nor on the energy, (since the masses of A and $\bar{A}$ are equal), nor on the spin, entails

$$[C, \mathbf{P}] = [C, \mathbf{P}_0] = [C, \mathbf{K}] = [C, \mathbf{J}] = 0 \tag{5.2}$$

as is readily checked by direct calculation. On the other hand, denoting the electric charge and the baryon and lepton numbers by Q, B, and ℓ respectively, one has

$$\{C, Q\} = \{C, B\} = \{C, \ell\} = \{C, S\} = \{C, I_3\} = 0. \tag{5.3}$$

Our choice of C as unitary rather than antiunitary is dictated by quantum electrodynamics.

Note that evidently $C^2 = I$, whence $\eta_A = \pm 1$.

Next, we shall establish the following crucial property:*

Law: charge conjugation is conserved by the electromagnetic and strong interactions. It is not conserved by the weak interactions.

In the first instance this law stems from its necessity in quantum electrodynamics. It is evident that charge conjugation cannot be conserved by the electromagnetic interactions if it is not conserved by the strong ones. Further, in field theory one can prove that C conservation implies the exact equality of particle and antiparticle masses. Unfortunately, other and more general conservation laws, like that for the product ΠCT, can lead to the same result. Hence the strongest support for C conservation derives from the success of quantum electrodynamics applied to electrons. For hadrons, the exact limits of validity of C conservation are still uncertain. It has however been confirmed quite closely by all experiments that have tested it so far.

C violation in weak interactions is evidenced by pion decay:

$$\pi^+ \to \mu^+ + \nu_\mu \tag{5.4}$$

$$\pi^- \to \mu^- + \bar{\nu}_\mu \,. \tag{5.5}$$

If we were to assume that the weak interaction Hamiltonian H conserves C, i.e. that $[\mathrm{H}, C] = 0$, then we would have

$$\begin{aligned} &\langle \mu^+ \mathbf{p} m; \nu, -\mathbf{p}, -m \mid \mathrm{H} \mid \pi^+ \mathbf{0}, 0 \rangle \\ &= \langle \mu^+ \mathbf{p} m; \nu, -\mathbf{p}, -m \mid C^{-1} \mathrm{H} C \mid \pi^+, \mathbf{0}, \mathbf{0} \rangle \\ &= \langle \mu^- \mathbf{p} m; \nu, -\mathbf{p}, -m \mid \mathrm{H} \mid \pi^+, \mathbf{0}, 0 \rangle \, \eta_\mu \eta_\nu \eta_\pi \,. \end{aligned} \tag{5.6}$$

Since $|\eta_\Pi \eta_\mu \eta_\nu| = 1$, equation (5.6) implies amongst its other consequences that the μ^+ and μ^- must have the same polarisation. But experiment shows that, on the contrary, the μ^+ and μ^- are strongly polarised in opposite directions.

G parity

The fact that by equation (5.3) C anticommutes with the third component of isospin occasionally makes it inconvenient to exploit. We shall

* Translator's note: phrases like 'charge conjugation is conserved' have passed into common usage in connection with discrete symmetry operations. Correct grammar would force one into 'invariance under charge conjugation is conserved...' or 'the...interactions are invariant under charge conjugation.'

introduce a new operator G, called *G parity*, whose conservation by the strong interactions is equivalent to that of C. Since G depends on the isospin operators, it is not conserved by the electromagnetic interactions.

Recall the action of C on the doublet (p, n). One has

$$C\,|\,p\rangle = |\,\bar{p}\rangle \qquad C\,|\,n\rangle = |\,\bar{n}\rangle \tag{5.7}$$

The operator $e^{i\pi I_2} = e^{i\pi\tau_2/2} = i\tau_2$ acts on this doublet according to

$$e^{i\pi I_2}\,|\,p\rangle = -\,|\,n\rangle \qquad e^{i\pi I_2}\rangle = |\,p\rangle \tag{5.8}$$

If we now define $G = Ce^{i\pi I_2}$, we have

$$G\,|\,\bar{p}\rangle = -|\,n\rangle \qquad G\,|\,\bar{n}\rangle = |\,p\rangle \tag{5.9}$$

so that G conserves I_3 in view of the fact that p and $\bar{n}$ both have $I_3 = +\frac{1}{2}$. Moreover we have the commutation relation

$$[G, \mathbf{I}] = 0. \tag{5.10}$$

Proof:

Since C changes the sign of I_3 we have: $CI_+ = I_-C$ and $CI_- = I_+C$, whence $[C, I_1] = 0$ and $\{C, I_2\} = 0$. Then equation (5.10) follows at once.

The action of G on a single-pion state is particularly interesting. To investigate this it is convenient to use components π_1, π_2, π_3, such that

$$|\,\pi^+\rangle = \frac{|\,\pi_1\rangle + i\,|\,\pi_2\rangle}{\sqrt{2}}, \; |\,\pi^-\rangle = +\frac{|\,\pi_1\rangle - i\,|\,\pi_2\rangle}{\sqrt{2}}, \; |\,\pi^0\rangle = |\,\pi_3\rangle \tag{5.11}$$

Since π^0 decays electromagnetically into two photons, we need assume only that the photon corresponds to a well defined value of C in order to have

$$C\,|\,\pi^0\rangle = |\,\pi^0\rangle \tag{5.12}$$

whence, using isospin,

$$C\,|\,\pi^+\rangle = |\,\pi^-\rangle, \qquad C\,|\,\pi^-\rangle = |\,\pi^+\rangle. \tag{5.13}$$

In view of (5.11) this gives

$$C\,|\,\pi_1\rangle = |\,\pi_1\rangle \qquad C\,|\,\pi_2\rangle = -|\,\pi_2\rangle \qquad C\,|\,\pi_3\rangle = |\,\pi_3\rangle \tag{5.14}$$

On the other hand, since $e^{i\pi I_2}$ is a rotation through π about the 2 axis, one finds

$$e^{i\pi I_2}\,|\,\pi_1\rangle = -|\,\pi_1\rangle, \; e^{i\pi I_2}\,|\,\pi_2\rangle = |\,\pi_2\rangle, \; e^{i\pi I_2}\,|\,\pi_3\rangle = -|\,\pi_3\rangle \tag{5.15}$$

whence, combining (5.14) and (5.15),

$$G\,|\,\pi\rangle = -|\,\pi\rangle \tag{5.16}$$

G conservation implies that *in a reaction with nothing but pions in both initial and final states, the difference between initial and final pion numbers is even.*

Note that this law does not apply when particles other than pions enter the reaction, as for instance in

$$\pi + N \rightarrow \pi + \pi + N$$

It is of great theoretical interest but cannot be applied directly to observable reactions because no pion targets are available.

TABLE: PARITIES AND G-PARITIES

Particle	γ		ν	e^-	e^+	μ^-	μ^+	π^-	π^+	π^0
Parity	−		ND	+	−	+	−	−	−	−
G parity (G)	ND		ND	ND	ND	ND	ND	−	−	−
Particle	K^+	K^-	K^0	$\bar{K}^0$	p	n	$\bar{p}$	$\bar{n}$	Λ^0	$\bar{\Lambda}^0$
Parity	−	−	−	−	+	+	−	−	+	−
G parity	ND	ND	ND	ND	ND	ND	ND	ND	ND	ND
Particle	Σ	$\bar{\Sigma}$	Ξ	$\bar{\Xi}$	Ω					
Parity	+	+	?+(th)	?−(th)	?+(th)					
G parity	ND	ND	ND	ND	ND					

ND stands for 'not defined'; ? for 'unknown'; th for 'predicted theoretically' (see the chapter on $SU(3)$).

CHAPTER 7

NONRELATIVISTIC COLLISION THEORY

We give an elementary introduction to the principal concepts that are useful in describing collisions between particles.

Now that we have become acquainted with the essential properties of individual particles, our next task is to study their collisions in order to discover from them the laws of dynamics.

1. Description of a collision

Consider an experiment where particles collide nonrelativistically: for instance, protons of momentum $\mathbf{k}$ collide with a target consisting of protons at rest. In actual fact the accelerator producing the beam delivers protons whose momenta vary about a mean value $\mathbf{k}$, and whose positions in space are more or less well defined. This packet moves in space, and changes its shape in the course of time because the particles do not all have exactly the same velocity; finally it hits the target. Some of the particles collide with a target nucleus (the target in this case being liquid hydrogen), while others go through the target without interacting. Particles in the former group are scattered in all directions and emerge rather like a cloud diverging from the target, while particles in the latter group continue to propagate in the direction of the incident packet.

In some cases one deals with a single particle rather than with a beam containing many. This would be the case for instance when a particle is produced inside a bubble chamber. But because the description of such an experiment by quantum mechanics has an essentially statistical character, it does not differ basically from our description of the evolution of a whole group of particles. In such cases one says that the wavefunction $\varphi(\mathbf{x}, t)$ of the particle obeys the Schroedinger equation for a free particle:

$$i \frac{\partial \varphi}{\partial t} = - \frac{1}{2m} \nabla^2 \varphi(\mathbf{x}, t) \tag{1.1}$$

In fact, $\varphi(\mathbf{x}, t)$ is a superposition of plane waves whose momenta vary about a mean value $\mathbf{k}$, and it is localised like a wavepacket having finite extension. This packet moves, changing its shape, until it hits the target. At that time one must take into account the interaction between the initial particle and the target (a proton in our case) by introducing a local potential; thus the Schroedinger equation is written as

$$i\frac{\partial\varphi}{\partial t} = -\frac{1}{2m}\nabla^2\varphi(\mathbf{x}, t) + V(\mathbf{x})\,\varphi(\mathbf{x}, t). \tag{1.2}$$

After a long enough time lapse, the probability density for the particle inside the target falls practically to zero, and once again $\varphi(\mathbf{x}, t)$ obeys the free Schroedinger equation (1.1); but now it is divided into two parts. The first part, analogous to the initial packet, is a superposition of plane waves whose momenta vary around $\mathbf{k}$; however, its total probability is less than that of the initial packet. The second part consists of diverging (outgoing) spherical waves whose momenta are close to $|\mathbf{k}|$ in magnitude; it looks like an expanding cloud (Figure 7.1.1). The total probability in the two parts taken together equals that in the incident packet. Since the collision occurs at a more or less well defined instant of time, the diverging cloud which was emitted at that time is localised in space and has a radius which grows at the rate k/m. Because its total probability

$$\int |\,\varphi^2\,|_{\mathbf{cloud}}\; d^3\mathbf{x}$$

remains constant after its emission, its wavefunction must decrease inversely with the radius as the latter increases.

It is possible to give a complete mathematical description of the time evolution of this wavepacket. But it is simpler mathematically to replace

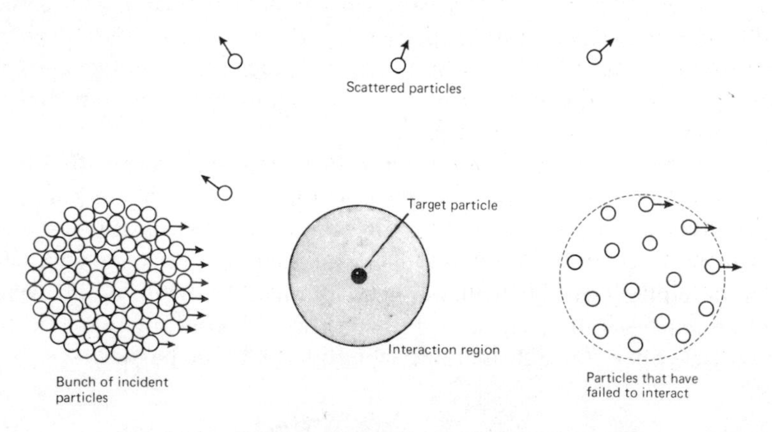

FIG. 7.1.1. Collision of a bunch of particles with a target.

the initial wavepacket by a single plane wave $e^{i\mathbf{k}\cdot\mathbf{x}}$. Such simplicity must be paid for by renouncing a description in time; indeed, since the energy and momentum are now sharply defined, the uncertainty principle prevents one from following the particle's history in detail. In this way it is possible to use the time-independent Schroedinger equation. Then the overall wavefunction obeys

$$\frac{\nabla^2}{2m}\psi(\mathbf{x}) + \left[\frac{k^2}{2m} - V(\mathbf{x})\right]\psi(\mathbf{x}) = 0. \tag{1.3}$$

For values of $\mathbf{x}$ corresponding to points outside the target, where the potential is therefore zero, $\psi(\mathbf{x})$ obeys the free-particle Schroedinger equation

$$\frac{\nabla^2}{2m}\psi(\mathbf{x}) + \frac{k^2}{2m}\psi(\mathbf{x}) = 0. \tag{1.4}$$

Moreover, we have seen that in this exterior region the wavefunction can be divided into three parts:

(a) a plane wave corresponding to the incident packet,
(b) another plane wave with the same momentum, corresponding to that part of the initial wavepacket which has not undergone collisions,
(c) a diverging wave corresponding to the scattered particles.

In a time-independent theory the two plane waves cannot be distinguished, and must both be described by one and the same expression $e^{i\mathbf{k}\cdot\mathbf{x}}$. As regards the diverging wave, we know that its momentum has magnitude k, and that it decreases like $1/r$; hence for large enough r it necessarily takes the form

$$f(\theta, \varphi)\frac{e^{ikr}}{r} \qquad r = |\mathbf{x}| \tag{1.5}$$

Here we have adopted spherical polar coordinates with the z axis along the incident momentum (Figure 7.1.2).

In most cases the potential depends only on r, and the particles are scattered with equal probability in any two directions differing only in azimuthal angle φ; thus $f(\theta, \varphi)$ depends only on the scattering angle θ. This function $f(\theta, \varphi)$ or $f(\theta)$ is called the *scattering amplitude*.

It is easy to find the connection between the scattering amplitude and the scattering cross-section. The incident wavefunction $e^{i\mathbf{k}\cdot\mathbf{x}}$ corresponds to a density of one particle per volume V:

$$\int_V |e^{i\mathbf{k}\cdot\mathbf{x}}|^2\, d^3x = V. \tag{1.6}$$

The flux Φ_i of incident particles is the number of particles per unit time crossing a unit surface Σ normal to $\mathbf{k}$. Therefore it is equal to the number

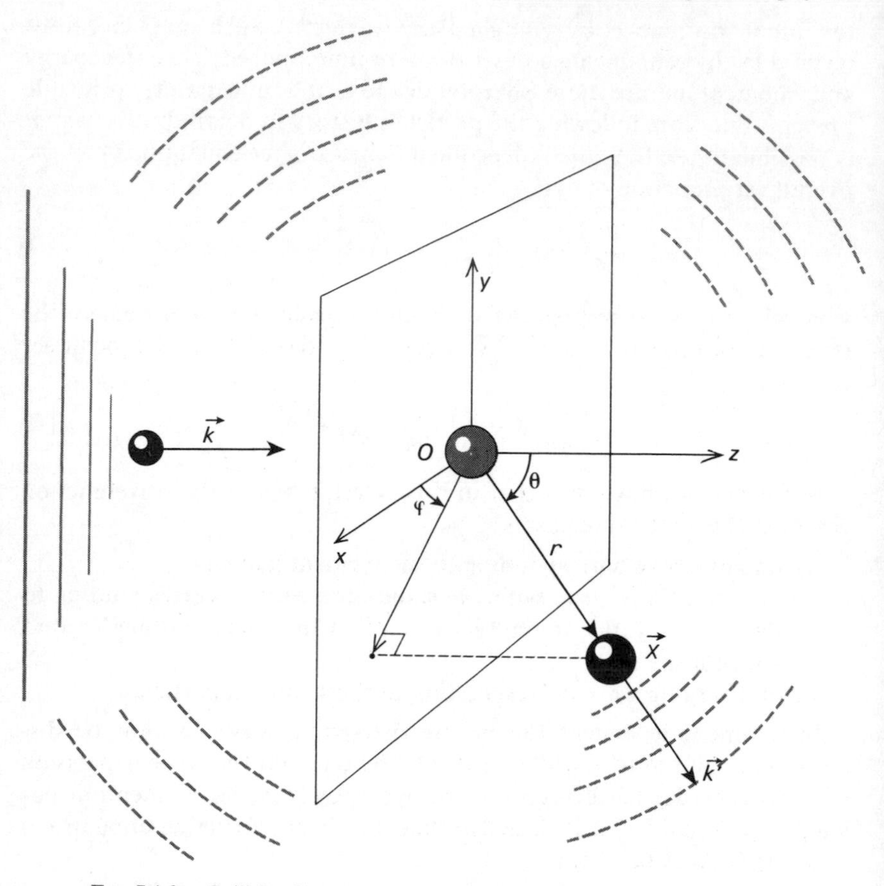

FIG. 7.1.2. Collision between two particles: time-independent description.

of particles which at a given instant are within a cylinder of base Σ and of height $L = \mathbf{k}/m = \mathbf{v}$; in other words $\Phi = v$. To obtain the number N of particles scattered per unit time into an element of solid angle $d\Omega$ around the direction (θ, φ), we note that the speed of such particles is v, and that their numerical density is $|f(\theta, \varphi)|^2/r^2$ per unit volume. Hence the number N of particles per unit time which cross a surface element subtending a solid angle $d\Omega$, and situated at a large distance r, is given by

$$N = v \times \text{density of particles} \times \text{area}$$

But the area is $r^2\, d\Omega$, whence $N = |f(\theta, \varphi)|^2 v\, d\Omega$. Finally, the differential scattering cross-section $d\sigma = N/\Phi$ is given by

$$\boxed{\frac{d\sigma}{d\Omega} = |f(\theta, \varphi)|^2} \quad . \tag{1.7}$$

Let us now see how to formulate a collision problem in nonrelativistic quantum mechanics. We need to find a solution $\psi(\mathbf{x})$ of the Schroedinger equation (1.3), which for large enough $\mathbf{x}$ behaves like

$$\psi(\mathbf{x}) = e^{i\mathbf{k}\cdot\mathbf{x}} + f(\theta)\frac{e^{ikr}}{r} + 0\left(\frac{1}{r^2}\right) \tag{1.8}$$

The symbol $0(1/r^2)$ represents a quantity which vanishes at least as fast as r^{-2} when r tends to infinity. The ultimate aim of scattering theory is to discover, from ψ, the exact form of the scattering amplitude $f(\theta)$; it is this which is interesting, since it determines the scattering cross-section through (1.7).

2. Integral form of the Schroedinger equation

From the mathematical point of view, a scattering problem consists in solving the partial differential Schroedinger equation (1.3), subject to the boundary condition (1.8). We shall now replace this problem by the solution of an integral equation which embodies both (1.3) and (1.8) simultaneously.

Rewrite the Schroedinger equation in the form

$$\left[H_0 - \frac{k^2}{2m} + V(\mathbf{x})\right]\psi(\mathbf{x}) = 0 \tag{2.1}$$

where the operator H_0 is defined by

$$H_0 = -\nabla^2/2m.$$

The initial wavefunction $\varphi(\mathbf{x}) = e^{i\mathbf{k}\cdot\mathbf{x}}$ obeys the free Schroedinger equation

$$\left(H_0 - \frac{k^2}{2m}\right)\varphi(\mathbf{x}) = 0. \tag{2.2}$$

Subtracting (2.2) from (2.1), we obtain an equation for $(\varphi(\mathbf{x}) - e^{i\mathbf{k}\cdot\mathbf{x}})$, i.e. for the nontrivial part of the wavefunction which we shall call the 'scattered wave.' This equation reads

$$\left(H_0 - \frac{k^2}{2m}\right)[\psi(\mathbf{x}) - \varphi(\mathbf{x})] = -V(\mathbf{x})\,\psi(\mathbf{x}). \tag{2.3}$$

Let us now assume that it is possible to define, formally, an operator inverse to the differential operator $H_0 - k^2/2m$, and write it as $(H_0 - k^2/2m)^{-1}$. Multiplying both sides of (2.3) by this operator, and writing $E = k^2/2m$, we obtain

$$\psi(\mathbf{x}) = e^{i\mathbf{k}\cdot\mathbf{x}} - \frac{1}{H_0 - E}V(\mathbf{x})\,\psi(\mathbf{x}). \tag{2.4}$$

If we want to safeguard ourselves against writing nonsense, we must now give a proper definition of the operator $(H_0 - E)^{-1}$. Adopting a more abstract formulation, we begin by considering a state vector written in Dirac notation as $|\alpha\rangle$. Thus the operator $(1/i)\boldsymbol{\nabla}$ in $\mathbf{x}$-space is a realisation of the abstract momentum operator $\mathbf{P}$, and $(-\nabla^2 - k^2)$ is a realisation of the abstract operator $(-k^2 + \mathbf{P}^2)$. What we have just introduced is essentially the abstract operator $H_0 - E = (\mathbf{P}^2 - k^2)/2m$. Our problem then is the following: how can we express $(k^2 - \mathbf{P}^2)^{-1}$ as an explicit operator acting on the wavefunction $\varphi_\alpha(\mathbf{x})$, where $|\alpha\rangle$ is an arbitrary state vector?

To answer this question, it is useful to represent the states $|\alpha\rangle$ not only by their wavefunctions $\varphi_\alpha(\mathbf{x})$ in $\mathbf{x}$-space, but also by their wavefunctions $\varphi_\alpha(\mathbf{p})$ in momentum space. That the latter may prove useful is suggested by the simple connection between our operator and the operator for momentum. Let us therefore introduce the set of eigenstates $|\mathbf{p}\rangle$ of the momentum operator $\mathbf{P}$, that is to say the set of states which satisfy

$$\mathbf{P} \mid \mathbf{p}\rangle = \mathbf{p} \mid \mathbf{p}\rangle.$$

In $\mathbf{x}$-space, $|\mathbf{p}\rangle$ is represented by the wavefunction $e^{i\mathbf{p}\cdot\mathbf{x}}$. Acting on $|\mathbf{p}\rangle$, the operator $(\mathbf{P}^2 - k^2)$ gives

$$(\mathbf{P}^2 - k^2)^{-1} \mid \mathbf{p}\rangle = (\mathbf{p}^2 - k^2)^{-1} \mid \mathbf{p}\rangle. \tag{2.5}$$

This equation shows that $(\mathbf{P}^2 - k^2)^{-1}$ is well defined, except when it acts on states corresponding to values of $\mathbf{p}$ such that $\mathbf{p}^2 = k^2$. Ignore this lack of definition for the moment, and express (2.5) in $\mathbf{x}$-space. To this end we use the selfevident formula

$$\sum_{\mathbf{p}} \mid \mathbf{p}\rangle\langle\mathbf{p} \mid = I$$

where the sum runs over all values of $\mathbf{p}$, and I is the unit operator. Then one has

$$\begin{aligned}(\mathbf{P}^2 - k^2)^{-1} \mid \alpha\rangle &= (\mathbf{P}^2 - k^2)^{-1} \sum_{\mathbf{p}} \mid \mathbf{p}\rangle\langle\mathbf{p} \mid \alpha\rangle \\ &= \sum_{\mathbf{p}} \frac{1}{p^2 - k^2} \langle\mathbf{p} \mid \alpha\rangle \mid \mathbf{p}\rangle \end{aligned} \tag{2.6}$$

To express (2.6) in $\mathbf{x}$-space, we replace $|\mathbf{p}\rangle$ by the corresponding wavefunction $e^{i\mathbf{p}\cdot\mathbf{x}}$, and $\langle\mathbf{p} \mid \alpha\rangle$ by its explicit form

$$\langle\mathbf{p} \mid \alpha\rangle = \int e^{-i\mathbf{p}\cdot\mathbf{y}}\, \varphi_\alpha(\mathbf{y})\, d^3y$$

which leads to

$$(H_0 - E)^{-1}\, \varphi_\alpha(\mathbf{x}) = 2m \int \frac{d^3\mathbf{p}}{(2\pi)^3}\, \frac{1}{\mathbf{p}^2 - k^2}\, e^{i\mathbf{p}\cdot\mathbf{x}} \int e^{-i\mathbf{p}\cdot\mathbf{y}}\, \varphi_\alpha(\mathbf{y})\, d^3y. \tag{2.7}$$

In writing these equations we have taken into account the fact that the wavefunctions $e^{i\mathbf{k}\cdot\mathbf{x}}$ correspond to a density of one particle per unit volume (quantisation volume = 1), and that the number of states is $Vd^3p/(2\pi)^3 = d^3p/(2\pi)^3$. Thus we can write

$$(H_0 - E)\,\varphi_\alpha(\mathbf{x}) = \int G(\mathbf{x} - \mathbf{y})\,\varphi_\alpha(\mathbf{y})\,d^3y \tag{2.8}$$

where the function $G(\mathbf{x})$, called the Green's function, is defined by

$$G(\mathbf{x}) = \frac{2m}{(2\pi)^3}\int e^{i\mathbf{p}\cdot\mathbf{x}}\,\frac{d^3p}{p^2 - k^2}\,. \tag{2.9}$$

Evaluation of the Green's function

$G(\mathbf{x})$ cannot be evaluated as it stands because the integrand diverges at $p^2 = k^2$. However, so far we have used Schroedinger's equation (2.1) without paying heed to the boundary condition (1.8). On comparing (2.4) and (1.8) one sees that for large $|\mathbf{x}|$, the scattered wave, $\psi(\mathbf{x}) - e^{i\mathbf{k}\cdot\mathbf{x}}$, must behave like a diverging spherical wave. We shall have to exploit this physical requirement in order to give a precise definition of $(H_0 - E)^{-1}$, i.e. of the integral (2.9).

Let us first eliminate the angle variables. It is clear from (2.9) that $G(\mathbf{x})$ is invariant under rotations, and therefore depends only on $r = |\mathbf{x}|$. Hence we can chose $\mathbf{x}$ to lie along any predetermined direction, which we shall take to be the z-axis; then $e^{i\mathbf{p}\cdot\mathbf{x}}$ is replaced by $e^{ipr\cos\alpha}$, where α is the angle between $\mathbf{p}$ and the z-axis. We can then integrate over the azimuthal angle which together with α fixes the direction of $\mathbf{p}$; this integration amounts to a simple multiplication by 2π, whence

$$G(r) = \frac{2m}{(2\pi)^2}\int \frac{1}{p^2 - k^2}\,e^{ipr\cos\alpha}\,p^2\,dp\,d\cos\alpha \tag{2.10}$$

The integration over $\cos\alpha$ yields

$$G(r) = \frac{4m}{(2\pi)^2}\int_0^\infty \frac{\sin pr}{r}\,\frac{pdp}{p^2 - k^2} \tag{2.11}$$

Exploiting the symmetry of the integrand under the replacement $p \to -p$, this can be written as

$$G(r) = \frac{2m}{4\pi^2}\int_{-\infty}^\infty \frac{\sin pr}{r}\,\frac{pdp}{p^2 - k^2}\,. \tag{2.12}$$

In view of the physical requirement that $G(r)$ lead to an outgoing scattered wave, it is natural to separate the integrand into its diverging and converging components, i.e. to rewrite $\sin pr$ as $(e^{ipr} - e^{-ipr})/2i$.

Further, we use the formula

$$\frac{p}{p^2 - k^2} = \frac{1}{2}\left[\frac{1}{p-k} + \frac{1}{p+k}\right]$$

to write

$$G(r) = \frac{m}{4\pi^2}\int_{-\infty}^{\infty} \frac{e^{ipr}\,dp}{2ir}\left[\frac{1}{p-k} + \frac{1}{p+k}\right]$$

$$-\frac{m}{4\pi^2}\int_{-\infty}^{\infty} \frac{e^{-ipr}\,dp}{2ir}\left[\frac{1}{p-k} + \frac{1}{p+k}\right]. \qquad (2.13)$$

Consider the first integral. It is not meaningful as it stands because the path of integration, along the real axis, includes the two poles of the integrand at $p = \pm k$. But it could be defined properly by interpreting it as an integral in the complex plane, and taking the integration path not along the real axis, but along one of the contours L_1, L_2, L_3, or L_4 shown in Figure (7.2.1). These contours differ through the way in which they encircle the poles. In every case, the integral can be evaluated by the calculus of residues. Indeed it is clear that e^{ipr} tends to zero exponentially if p is a complex number $p_1 + ip_2$ and if p_2 tends to $+\infty$, since then the modulus of e^{ipr} is $e^{-p_2 r}$. It follows that the first integral in (2.13), taken along the semicircle C in the upper half of the complex plane, tends to zero as the

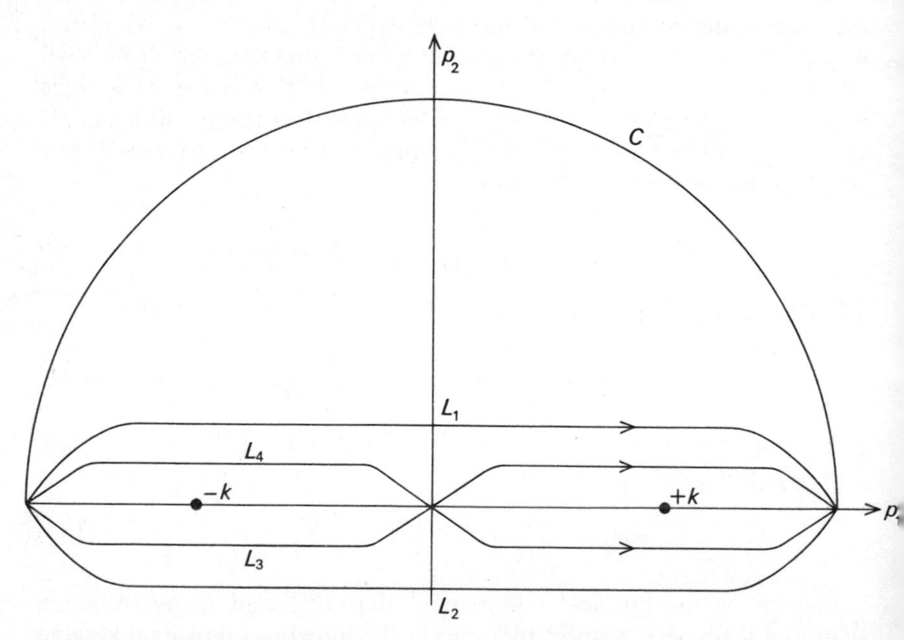

FIG. 7.2.1. The various integration contours for evaluating the Green's functions used in the first integral of equation (2.13).

radius of C increases. Hence we can add this integral along C to the integrals along L_1, L_2, L_3, or L_4, without changing their value. The resulting integrals are then equal to $2\pi i$ times the residues at the poles encircled by the contour consisting of C and L.

In this way, one obtains the following values for the first integral taken along the various contours:

$$\begin{aligned} L_1 &: 0 \\ L_2 &: \frac{m\pi}{4\pi^2 r}(e^{ikr} + e^{-ikr}) \\ L_3 &: \frac{m\pi}{4\pi^2 r} e^{-ikr} \\ L_4 &: \frac{m\pi}{4\pi^2 r} e^{ikr} \end{aligned} \tag{2.14}$$

The second integral can be found similarly. The only difference is that e^{-ipr} vanishes at infinity in the *lower* half plane, and that the semicircle C must now be drawn in this half plane. The contour $L + C$ is then traversed in the negative direction and the integral will equal $-2\pi i$ times the sum of the residues, which leads to the following:

$$\begin{aligned} L_1 &: \frac{m\pi}{4\pi^2 r}(e^{ikr} + e^{-ikr}) \\ L_2 &: 0 \\ L_3 &: \frac{m\pi}{4\pi^2 r} e^{-ikr} \\ L_4 &: \frac{m\pi}{4\pi^2 r} e^{ikr}. \end{aligned} \tag{2.15}$$

We see from this that there is only one way of defining $G(r)$ so that it behaves like an outgoing wave for large r; for this, we must replace (2.12) by

$$G(r) = \frac{2m}{4\pi^2} \int_{L_4} \frac{\sin pr}{r} \frac{p\,dp}{p^2 - k^2} \tag{2.16}$$

which, in view of (2.14) and (2.15), gives

$$G(r) = \frac{2m}{4\pi r} e^{ikr} \tag{2.17}$$

Physical interpretation of (2.16) *and* (2.17)

Equation (2.16) has been obtained by deforming the contour around the two poles at $p = \pm k$. It would have been equally possible to move the pole at $p = -k$ to just below the real axis and the pole at $p = +k$ to just above the axis, and to retain the axis itself as the integration contour (Figure 7.2.2). The poles would then lie at $p = \pm(k + i\epsilon)$, which amounts to adding to k an infinitesimally small positive imaginary part. Then, from (2.9),

$$G(r) = \frac{2m}{(2\pi)^3} \int e^{i\mathbf{p}\cdot\mathbf{x}} \frac{d^3p}{p^2 - (k + i\epsilon)^2} \tag{2.18}$$

The addition of such an imaginary part to k, in (2.17), causes $G(r)$ to decrease like $e^{-\epsilon r}$ for large r, which has no effect on the relevant properties of G when G multiplies a wavefunction. By contrast, had we replaced k by $k - i\epsilon$, we would have obtained an exponentially increasing factor $e^{\epsilon r}$; this would lead to a physically meaningless expression when multiplied into a wavefunction.

As regards the explicit form of (2.17), its structure can be explained by the following comments.

1. Its phase is given by e^{ikr}, which ensures that after the scattering event the free particles diverge like spherical waves with momenta of magnitude k.
2. For large r, the factor $1/r$ ensures that the number of particles per unit time crossing the surface of a sphere of radius r is independent of the radius. In other words the number of particles is conserved.

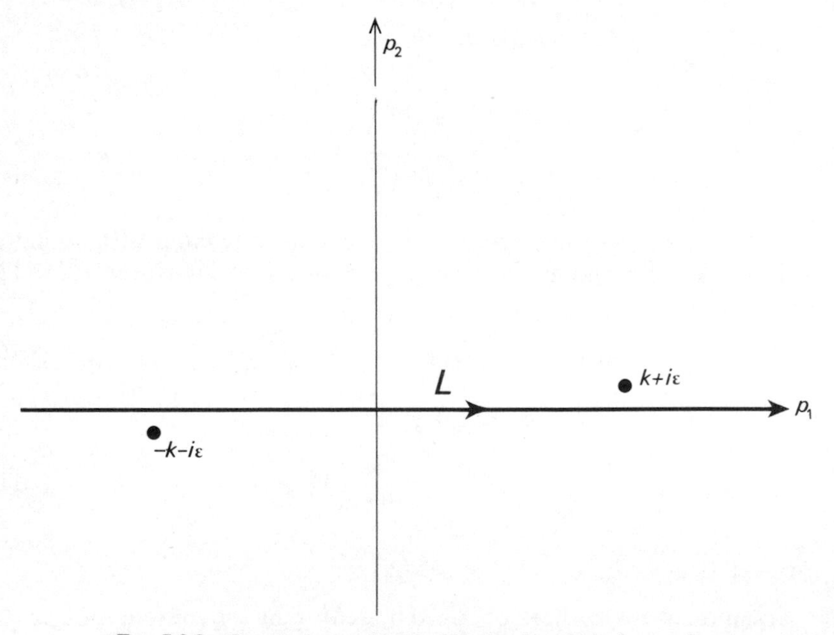

FIG. 7.2.2. Integration path with shifted poles, equivalent to L_4.

3. For small r, $G(r)$ behaves like $2m/4\pi r$. Indeed, recalling that $G(r)$ has been defined by

$$\frac{1}{2m}(P^2 - k^2)^{-1},$$

we see that one must have

$$(\nabla^2 + k^2)\, G(\mathbf{x}) = -2m\, \delta(\mathbf{x}) \tag{2.19}$$

in view of the fact that $G(\mathbf{x} - \mathbf{y})$ acted on by $-(\nabla_x{}^2 + k^2)$ must give the unit operator, which in $\mathbf{x}$-space is $\delta(\mathbf{x} - \mathbf{y})$. It follows that for $k^2 = 0$, $G(\mathbf{x})$ must tend to the potential of an electrostatic point charge $-2m$ situated at the origin, i.e. to $-2m/4\pi r$; to see this, note that (2.19) now tends to Poisson's equation for this potential

$$\nabla^2 V(\mathbf{x}) = \delta(\mathbf{x}).$$

Finally we rewrite equation (2.4) as the integral equation

$$\psi(\mathbf{x}) = e^{i\mathbf{k}\cdot\mathbf{x}} - \frac{2m}{4\pi}\int \frac{e^{ik|\mathbf{x}-\mathbf{y}|}}{|\mathbf{x}-\mathbf{y}|}\, V(\mathbf{y})\, \psi(\mathbf{y})\, d^3y \tag{2.20}$$

At first sight this seems more complicated than Schroedinger's equation, but it has the enormous advantage of incorporating automatically the boundary conditions for the scattering process.

3. Calculation of the scattering amplitude

According to equation (2.18) the scattering amplitude is determined by the asymptotic behaviour of the scattered wave $\psi(\mathbf{x}) - e^{i\mathbf{k}\cdot\mathbf{x}}$, which appears explicitly in equation (2.20). To discover this asymptotic behaviour, we must study the asymptotics of the integral

$$\int \frac{e^{ik|\mathbf{x}-\mathbf{y}|}}{|\mathbf{x}-\mathbf{y}|}\, V(\mathbf{y})\, \psi(\mathbf{y})\, d^3y. \tag{3.1}$$

If the interaction is local (or if $V(\mathbf{y})$ decreases fast enough as $|\mathbf{y}|$ tends to infinity), one can consider the integration region as essentially finite, so that in the denominator $|\mathbf{x} - \mathbf{y}|$, $\mathbf{y}$ is negligible compared to $\mathbf{x}$. By contrast, the phase factor $e^{ik|\mathbf{x}-\mathbf{y}|}$ depends crucially on $\mathbf{y}$ even when $\mathbf{x}$ is very large. The strongest dependence on $\mathbf{x}$ can be isolated by writing

$$|\mathbf{x} - \mathbf{y}| = \sqrt{(\mathbf{x} - \mathbf{y})^2} = \sqrt{x^2 - 2\mathbf{x}\cdot\mathbf{y} + y^2}$$

$$= |\mathbf{x}|\sqrt{1 - 2\frac{\mathbf{x}\cdot\mathbf{y}}{|\mathbf{x}|^2} + \frac{y^2}{|\mathbf{x}|^2}}$$

The term $\mathbf{y}^2/\mathbf{x}^2$ is negligible for large enough $|\mathbf{x}|$, whence

$$|\mathbf{x}-\mathbf{y}| \simeq |\mathbf{x}| \sqrt{1 - 2\frac{\mathbf{x}\cdot\mathbf{y}}{|\mathbf{x}|^2}} \simeq |\mathbf{x}| - \frac{\mathbf{x}\cdot\mathbf{y}}{|\mathbf{x}|}.$$

We now introduce a new notation. When a scattered particle is observed at a point $\mathbf{x}$ far from the target, we shall define $\mathbf{k}' = k(\mathbf{x}/|\mathbf{x}|)$. The vector $\mathbf{k}'$ is the momentum of the particle observed at $\mathbf{x}$, and its absolute value is clearly k. Writing $|\mathbf{x}| = r$, we have

$$e^{ik|\mathbf{x}-\mathbf{y}|} = e^{ikr - \mathbf{k}'\cdot\mathbf{y}}$$

whence

$$\psi(\mathbf{x}) - e^{i\mathbf{k}\cdot\mathbf{x}} \simeq -\frac{2m}{4\pi}\frac{e^{ikr}}{r}\int e^{-i\mathbf{k}'\cdot\mathbf{y}}\, V(\mathbf{y})\,\psi(\mathbf{y})\, d^3y. \tag{3.2}$$

For greater clarity we shall write the scattering amplitude as $f(\mathbf{k}, \mathbf{k}')$, since θ is simply the angle between $\mathbf{k}$ and $\mathbf{k}'$. Comparison of (3.2) and (1.1) yields an explicit expression for the scattering amplitude in terms of the wavefunction:

$$f(\mathbf{k}, \mathbf{k}') = -\frac{2m}{4\pi}\int e^{-i\mathbf{k}'\cdot\mathbf{y}}\, V(\mathbf{y})\,\psi(\mathbf{y})\, d^3y. \tag{3.3}$$

4. The Born series

If the integral equation (2.4) is rewritten in the form

$$\psi = \varphi_i - \frac{1}{H_0 - E - i\epsilon} V\psi \tag{4.1}$$

where φ_i is the initial plane wave, it can be solved by the *Born series:*

$$\psi = \varphi_i - \frac{1}{H_0 - E - i\epsilon}\varphi_i + \frac{1}{H_0 - E - i\epsilon} V \frac{1}{H_0 - E - i\epsilon} V\varphi_i + \cdots \tag{4.2}$$

This expansion can be interpreted physically by saying that the overall wavefunction consists of a series of terms describing the evolution of the scattering process in detail. The first term φ_i describes simultaneously both the incident wave and the transmitted plane wave corresponding to particles which have not been scattered. The second term $-(1/H_0 - E - i\epsilon)V\varphi_i$ describes particles which have interacted with the potential once: the initial wavefunction φ_i is multiplied by the value $V(\mathbf{y})$ of the potential at the point where the interaction of the particle with the target has taken place. This interaction results in a diverging spherical wave represented in our expression by the factor $-(H_0 - E - i\epsilon)^{-1}$, i.e. by $G(\mathbf{x} - \mathbf{y})$. The third term represents particles which have undergone two successive interactions, and so on.

The Born series can thus be considered to describe the overall scattering process as multiple scattering at individual points where the interaction has taken effect. It is clear that such a description will be useful only if the importance of successive scatterings decreases, i.e. only if the potential is weak. In particular the Born series cannot account adequately for the existence and properties of bound states and resonances. This is especially obvious for resonances: a scattering resonance can be thought of as the formation of a metastable state of the particle and the target, as in Bohr's model of the compound nucleus. Such a metastable state survives for much longer than the typical interaction time, which is the time a particle takes to cross the target once. But in order for the particle to be captured for such a long period, it must necessarily undergo many individual scattering interactions meanwhile. Thus a resonance corresponds precisely to those circumstances in which the successive terms of the Born series do not decrease. For a bound state the situation is evidently much worse. From a mathematical point of view it can be shown that it is precisely the possibility of bound states and resonances which limits the convergence of the Born series.

The form (3.3) of the scattering amplitude can be expressed as the matrix element

$$f_{fi} = -\frac{2m}{4\pi}(\varphi_f, V\psi_i) \tag{4.3}$$

where φ_f represents the plane wave $e^{i\mathbf{k}'\cdot\mathbf{y}}$ for the observed final state. Substituting the Born series for ψ, one obtains

$$f_{fi} = -\frac{2m}{4\pi}\left[(\varphi_f, V\varphi_i) - \left(\varphi_f V\frac{1}{H_0 - E - i\epsilon}V\varphi_i\right)\right.$$
$$\left. + \left(\varphi_f V\frac{1}{H_0 - E - i\epsilon}V\frac{1}{H_0 - E - i\epsilon}V\varphi_i\right)\cdots\right] \tag{4.4}$$

By appeal to

$$\frac{1}{H_0 - E - i\epsilon} = \int\frac{d^3\mathbf{p}}{(2\pi)^3}\,|\,\mathbf{p}\rangle\frac{1}{\dfrac{p^2}{2m} - \dfrac{(k + i\epsilon)^2}{2m}}\langle\mathbf{p}\,| \tag{4.5}$$

this can be written as

$$-\frac{4\pi}{2m}f(\mathbf{k}', \mathbf{k})$$
$$= \langle\mathbf{k}'\,|\,V\,|\,\mathbf{k}\rangle - \frac{1}{(2\pi)^3}\int\langle\mathbf{k}'\,|\,V\,|\,\mathbf{p}\rangle\frac{2m\,d^3p}{p^2 - (k + i\epsilon)^2}\langle\mathbf{p}\,|\,V\,|\,\mathbf{k}\rangle$$
$$+ \int\langle\mathbf{k}'\,|\,V\,|\,\mathbf{p}\rangle\frac{2m\,d^3p}{(2\pi)^3[p^2 - (k + i\epsilon)^2]}\langle\mathbf{p}\,|\,V\,|\,\mathbf{p}'\rangle$$
$$\times\frac{2m\,d^3p'}{(2\pi)^3[p'^2 - (k + i\epsilon)^2]}\langle\mathbf{p}'\,|\,V\,|\,\mathbf{k}\rangle - \cdots \tag{4.6}$$

where φ_i and φ_f represent the state vectors $|\mathbf{k}\rangle$ and $|\mathbf{k}'\rangle$, and where we have defined

$$\langle \mathbf{k}' \mid V \mid \mathbf{k} \rangle = \int e^{i(\mathbf{k}-\mathbf{k}')\cdot\mathbf{x}}\, V(\mathbf{x})\, d^3x. \tag{4.7}$$

If one keeps only the first term of the series (4.6), then according to (1.7) one obtains, for the scattering cross-section,

$$\frac{d\sigma}{d\Omega} = \left(\frac{2m}{4\pi}\right)^2 |\langle \mathbf{k}' \mid V \mid \mathbf{k} \rangle|^2 \tag{4.8}$$

which is simply Fermi's Golden Rule applied to the case in hand.

5. Partial wave expansion of the Schroedinger equation

(a) *Radial equation*

We shall now examine in greater detail those properties of the Schroedinger equation and of the scattering amplitude which result from invariance under rotations, i.e. from the fact that the potential $V(r)$ depends only on $r = |\mathbf{x}|$. In view of this fact we adopt spherical polar co-ordinates (r, θ, φ), with the polar axis taken along the momentum $\mathbf{k}$ of the incident beam. Then the wavefunction $\psi(\mathbf{x})$ depends only on r and on θ. To eliminate θ we expand $\psi(\mathbf{x})$ in a series of eigenfunctions of the angular momentum $L^2 = l(l + 1)$:

$$\psi(x) = \sum_{l=0}^{\infty} \frac{y_\ell(r)}{r} P_\ell(\cos\theta) \tag{5.1}$$

Here, the $P_\ell(\cos\theta)$ are Legendre polynomials. Then the three-dimensional Schroedinger equation separates into a set of one-dimensional equations. This is best seen by noting the classical vector equation

$$\mathbf{p}^2 = p_r{}^2 + \frac{(\mathbf{p}\times\mathbf{r})^2}{r^2} = p_r{}^2 + \frac{\mathbf{L}^2}{r^2} \tag{5.2}$$

where p_r is the component of $\mathbf{p}$ along r. It is also the momentum canonically conjugate to r, and as such must under quantisation be replaced by $(1/i)(d/dr)$, in the convention which assigns to y_ℓ the norm

$$\int |\, y(r)|^2\, dr$$

The factor $1/r$ in (5.1) is needed because in three dimensional space the scalar product involves integration with respect to $d^3r = r^2\, dr\, d\Omega$. According to (5.2), the Schroedinger equation is now given by

$$\sum_{\ell=0}^{\infty} \left[-\frac{1}{2m}\left(\frac{d}{dr}\right)^2 + \frac{L^2}{2mr^2} - k^2 + V(r)\right] y_\ell(r)\, P_\ell(\cos\theta) = 0 \tag{5.3}$$

For the function $y_\ell(r)$ this leads to

$$\left[\frac{d^2}{dr^2} + k^2 - \frac{\ell(\ell+1)}{r^2} - U(r)\right] y_\ell(r) = 0 \tag{5.4}$$

where account has been taken of $L^2 P_\ell(\cos\theta) = \ell(\ell+1)P_\ell(\cos\theta)$, and where we have defined $U(r) = 2mV(r)$. Equation (5.4) is called the radial Schroedinger equation.

(b) *Free particle*

When there is no interaction, $V(r) = 0$, and $\psi(\mathbf{x})$ reduces to the plane wave $e^{i\mathbf{k}\cdot\mathbf{x}}$, which can be expanded in terms of spherical waves according to the standard formula

$$e^{ikr\cos\theta} = \sum_{\ell=0}^{\infty} (-i)^\ell (2\ell+1) \left(\frac{r}{k}\right)^\ell \left(\frac{1}{r}\frac{d}{dr}\right)^\ell \frac{\sin kr}{kr} P_\ell(\cos\theta) \tag{5.5}$$

This shows that for free particles $y_\ell(r)$ is simply

$$y_\ell^0(r) = \frac{r^{\ell+1}}{k^\ell}\left(\frac{1}{r}\frac{d}{dr}\right)^\ell \frac{\sin kr}{kr}(-i)^\ell(2\ell+1). \tag{5.6}$$

Later on we shall need the asymptotic behaviour of $y_\ell^{(0)}(r)$ as r tends to infinity. It is easy to see that the operator $(1/r)(d/dr)$ acting on the factor $1/r$ in $\sin kr/r$ leads to terms of higher order in $1/r$ than it gives when acting on the factor $\sin kr$. Therefore,

$$\frac{1}{r}\frac{d}{dr}\frac{\sin kr}{kr} \simeq \frac{1}{kr}\frac{1}{r}\frac{d}{dr}(\sin kr) = \frac{1}{r^2}\sin\left(kr + \frac{\pi}{2}\right), \text{ etc., whence}$$

$$y_\ell^0(r) \simeq \frac{\sin\left(kr + \ell\frac{\pi}{2}\right)}{k}(-i)^\ell(2\ell+1). \tag{5.7}$$

(c) *Asymptotic behaviour of* $y_\ell(r)$

If the potential decreases sufficiently fast, then for very large r we can neglect the last two terms in (5.4), and the Schrodinger equation becomes

$$\left(\frac{d^2}{dr^2} + k^2\right) y_\ell(r) = 0 \tag{5.8}$$

showing that $y_\ell(r)$ is a linear combination of cos kr and sin kr. To facilitate comparison with (5.7) we write this combination in the form

$$y_\ell(r) \simeq C_\ell \sin\left(kr + \ell\frac{\pi}{2} + \delta_\ell\right) \tag{5.9}$$

where δ_ℓ vanishes if there is no potential. The angle δ_ℓ is called the elastic *phase shift* (or simply the phase shift) for the partial wave ℓ.

(d) *Asymptotic behaviour of* $\psi(\mathbf{x})$

By the definition of the scattering amplitude, the large r behaviour of $\psi(\mathbf{x})$ is

$$\psi(\mathbf{x}) \simeq e^{ikz} + f(\theta)\frac{e^{ikr}}{r} \tag{5.10}$$

$f(\theta)$ can be expanded in terms of Legendre polynomials by writing

$$f(\theta) = \sum_{\ell=0}^{\infty} (2\ell+1)\, a_\ell(k^2)\, P_\ell(\cos\theta). \tag{5.11}$$

The inverse is given by

$$a_\ell(k^2) = \frac{1}{2}\int_{-1}^{+1} f(k^2, \cos\theta)\, P_l(\cos\theta)\, d\cos\theta. \tag{5.12}$$

which is obvious from the orthogonality relation for the Legendre polynomials:

$$\frac{2\ell+1}{2}\int_{-1}^{+1} P_\ell(\cos\theta)\, P_n(\cos\theta)\, d(\cos\theta) = \delta_{\ell n}. \tag{5.13}$$

By using the expansions (5.5) of e^{ikz} and (5.11) of $f(\theta)$, we obtain the partial wave expansion of $\psi(\mathbf{x})$ applicable for large r:

$$\psi(\mathbf{x}) \simeq \sum (2\ell+1)\left[(-i)^\ell \frac{\sin\left(kr + \frac{\ell\pi}{2}\right)}{kr} + \frac{a_\ell(k^2)}{r}\, e^{ikr}\right] P_\ell(\cos\theta) \tag{5.14}$$

Moreover, equation (5.9) shows that under the same conditions this asymptotic form can be written as

$$\psi(\mathbf{x}) \simeq \sum C_\ell \frac{\sin\left(kr + \frac{\ell\pi}{2} + \delta_\ell\right)}{r} P_\ell(\cos\theta); \tag{5.15}$$

Equating the coefficients of $e^{\pm ikr}$ in expressions (5.14) and (5.15) one obtains

$$(2\ell+1)(-i)^\ell \frac{e^{+i\ell\pi/2}}{k} + (2\ell+1)\, a_\ell(k^2) = C_\ell\, e^{+il\pi/2+i\delta_\ell}$$

$$(2\ell+1)(-i)^\ell \frac{e^{-i\ell\pi/2}}{k} = C_\ell\, e^{-il\pi/2-i\delta_\ell}$$

which yield at once an expression for the coefficient $a_\ell(k^2)$ in terms of the phase shift δ_ℓ:

$$a_\ell(k^2) = \frac{1}{k} e^{i\delta_\ell}\sin\delta_\ell \quad \text{with} \quad C_\ell = \frac{2\ell+1}{k} e^{i\delta_\ell - i\ell\pi/2}. \tag{5.16}$$

The function $a_\ell(k^2)$ is called the *partial wave amplitude*, and equations

(5.11) and (5.16) express the scattering amplitude in terms of the phase shifts:

$$f(k^2, \cos\theta) = \sum_{\ell=0}^{\infty} (2\ell+1) \frac{e^{i\delta_\ell} \sin\delta_\ell}{k} P_\ell(\cos\theta) \tag{5.17}$$

It is important to note that the entire argument of this paragraph, and equation (5.17) in particular, are in all essentials independent of the Schroedinger equation.

6. Absorption

The above formalism can be interpreted in greater detail as follows. Equations (5.5) and (5.7) express the fact that the plane wave e^{ikz} is in each of its partial waves the sum of a converging (incoming) spherical wave

$$(2\ell+1) \frac{e^{-i\pi\ell}}{2ikr} e^{-ikr} \tag{6.1}$$

and of a diverging (outgoing) one

$$(2\ell+1) \frac{e^{ikr}}{2ikr}. \tag{6.2}$$

There is added to these the spherical scattered wave

$$(2\ell+1) \frac{e^{ikr}}{kr} e^{+i\delta_\ell} \sin\delta_\ell . \tag{6.3}$$

The total outgoing spherical wave in $\psi(\mathbf{x})$ includes both the scattered wave and the outgoing part of the plane wave, being given by the sum of (6.2) and (6.3):

$$(2\ell+1) \frac{e^{ikr}}{2ikr} (1 + 2i\, e^{i\delta_\ell} \sin\delta_\ell) = (2\ell+1) \frac{e^{ikr}}{2ikr} e^{2i\delta_\ell}. \tag{6.4}$$

Comparison of (6.1) and (6.4) shows the amplitude ratio of the outgoing to the incoming spherical wave to be

$$(-1)^\ell e^{2i\delta_\ell} = (-1)^\ell (1 + 2ika_\ell). \tag{6.5}$$

The fact that its modulus is 1 expresses the conservation of the number of particles in an elastic collision.

If there is inelasticity, there ought to be fewer particles of the incident kind in the outgoing spherical wave than in the incoming one. Thus, if one writes

$$\eta_\ell = 1 + 2ika_\ell \tag{6.6}$$

then one should expect

$$|\eta_\ell| < 1 \tag{6.7}$$

The elastic cross-section is always given by

$$\frac{d\sigma}{d\Omega} = \left|\sum (2\ell + 1)\, a_\ell(k^2)\, P_\ell(\cos\theta)\right|^2 \tag{6.8}$$

so that according to (5.13) the total elastic cross-section is

$$\sigma_{\text{el}} = 2\pi \int \frac{d\sigma}{d\Omega}\, d\cos\theta = 4\pi \sum_\ell (2\ell + 1)\, |a_\ell(k^2)|^2. \tag{6.9}$$

The inelastic cross-section is clearly proportional to the obliterated part of the incoming wave, i.e. to

$$1 - |\eta_\ell|^2$$

for a given incoming amplitude. The ratio of the number of particles scattered to the number absorbed is given by

$$\frac{|\eta_\ell|^2}{1 - |\eta_\ell|^2} = \frac{4\,|a_\ell(k^2)|^2\, k^2}{1 - |\eta_\ell|^2} \tag{6.10}$$

whence the total inelastic cross-section is

$$\sigma_{\text{inel}} = \pi\Sigma(2\ell + 1)\,\frac{(1 - |\eta_\ell|^2)}{k^2} \tag{6.11}$$

Thus the total cross-section is

$$\sigma_T = \sigma_{\text{el}} + \sigma_{\text{inel}}. \tag{6.12}$$

whence

$$\sigma_T = \pi \sum_\ell \frac{(2\ell + 1)}{k^2}\,[|1 - \eta_\ell|^2 + 1 - |\eta_\ell|^2] = \frac{2\pi}{k^2} \sum (2\ell + 1)(1 - \text{Re}\,\eta_\ell) \tag{6.13}$$

It is worth noting that the imaginary part of the scattering amplitude is given by

$$\text{Im} f(k, \theta) = \sum (2\ell + 1)\,\frac{(1 - \text{Re}\,\eta_\ell)}{2k}\, P_\ell(\cos\theta). \tag{6.14}$$

For forward scattering, $\theta = 0$, $P_\ell(\cos\theta) = 1$, and one gets

$$\text{Im} f(k, 0) = \sum (2\ell + 1)\,\frac{(1 - \text{Re}\,\eta_\ell)}{2k}. \tag{6.15}$$

Comparison of (6.13) and (6.15) gives the *optical theorem* which relates the total cross-section to the imaginary part of the forward scattering amplitude:

$$\boxed{\text{Im} f(k, 0) = \frac{k\sigma_T}{4\pi}} \tag{6.16}$$

7. Particles with spin

We shall give a quick survey of the changes in the principal results obtained so far, when one is interested in collisions of spin 0 particles (e.g. π^+) with spin $\frac{1}{2}$ particles (e.g. protons).

Denoting the spin state of the initial proton by $|\alpha\rangle$, the asymptotic form of the wavefunction becomes

$$e^{ikz}\,|\,\alpha\rangle + F(\theta, \varphi)\,\frac{e^{ikr}}{r}\,|\,\alpha\rangle \tag{7.1}$$

where F is now a 2×2 matrix. The number of particles observed in the spin state $|\beta\rangle$ and in the direction (θ, φ) is now proportional to $|\langle\beta|\, F\, |\alpha\rangle|^2$ instead of $|f(\theta, \varphi)|^2$, so that for initial spin state α and final spin state β the cross-section is

$$\frac{d\sigma_{\alpha\to\beta}}{d\Omega} = |\langle\beta\,|\,F(\theta, \varphi)\,|\,\alpha\rangle|^2. \tag{7.2}$$

In particular, if the target is unpolarised and if the final polarisation is not observed, one has

$$\frac{d\sigma}{d\Omega} = \frac{1}{2}\sum_{\substack{\alpha=\pm 1/2\\ \beta=\pm 1/2}} |\langle\beta\,|\,F(\theta, \varphi)|\,\alpha\rangle|^2 = \frac{1}{2}\,\mathrm{Trace}[F^{+}(\theta, \varphi)\,F(\theta, \varphi)] \tag{7.3}$$

The partial wave expansion of the wavefunction must be modified. Indeed, it is no longer the orbital angular momentum $\mathbf{L}$ which is conserved, but the total angular momentum $\mathbf{J} = \mathbf{L} + \mathbf{S}$, where $\mathbf{S}$ is the spin of the proton. For a given value of ℓ one can have the two values $j = \ell \pm \frac{1}{2}$. For given j there are two states, differing in ℓ by 1, and therefore having opposite parities. Instead of (5.1) we shall write

$$\psi(\mathbf{x}) = \sum \frac{y_\ell(r)}{r}\,|\,\ell jm\rangle \tag{7.4}$$

where $|\ell jm\alpha\rangle$ is a state with total angular momentum j and orbital angular momentum ℓ, so that

$$|\,\ell jm\rangle = C(\ell, m - \alpha, \tfrac{1}{2}, \alpha\,|\,j, m)\; Y_l^{m-\alpha}(\theta, \varphi)\,|\,\alpha\rangle. \tag{7.5}$$

Projection operators

The expression (7.5) can be considered as a 2×2 operator acting on the spinor $Y_\ell^{m-\alpha}(\theta, \varphi)\,|\alpha\rangle$, so that

$$|\,\ell mj\rangle = 0(\ell, j)\,|\,\ell m\alpha\rangle. \tag{7.6}$$

The matrix $0(\ell, j)$ could be written down explicitly by calculating the

Clebsch-Gordan coefficients appearing in (7.5). But it can also be found directly, as follows.

The operator $0(\ell, j)$ projects the two states $|\ell m\alpha\rangle$, $\alpha = \pm\frac{1}{2}$, onto the state $|\ell jm\rangle$, $j = \ell \pm \frac{1}{2}$. Now for given j and ℓ, the value of the operator $\boldsymbol{\sigma}\cdot\mathbf{L}$ is uniquely determined by the relation

$$\mathbf{J}^2 = j(j+1) = \left(\mathbf{L} + \frac{\boldsymbol{\sigma}}{2}\right)^2 = \ell(\ell+1) + \mathbf{L}\cdot\boldsymbol{\sigma} + \frac{3}{4} \tag{7.7}$$

Thus we have, for

$$\begin{aligned} j &= \ell + \tfrac{1}{2} \qquad \boldsymbol{\sigma}\cdot\mathbf{L} = \ell \\ j &= \ell - \tfrac{1}{2} \qquad \boldsymbol{\sigma}\cdot\mathbf{L} = -\ell - 1. \end{aligned} \tag{7.8}$$

Therefore the projection operators $0(\ell, j)$ are given as follows: for

$$j = \ell + \tfrac{1}{2} \quad \text{by} \quad 0^+ = \frac{\boldsymbol{\sigma}\cdot\mathbf{L} + (\ell+1)}{\ell + (\ell+1)} = \frac{\ell + 1 + \boldsymbol{\sigma}\cdot\mathbf{L}}{2\ell+1}$$

for

$$j = \ell - \tfrac{1}{2} \quad \text{by} \quad 0^- = \frac{\boldsymbol{\sigma}\cdot\mathbf{L} - \ell}{-(\ell+1) - \ell} = \frac{\ell - \boldsymbol{\sigma}\cdot\mathbf{L}}{2\ell+1}. \tag{7.9}$$

We have simply used (7.8) to construct an operator 0^+ which takes the value 1 (0) for $j = \ell + \frac{1}{2}$ $(\ell - \frac{1}{2})$, and conversely for 0^-.

Thus equation (7.4) takes the form

$$\psi(\mathbf{x}) = \sum \frac{y_\ell^j(r)}{r} 0(\ell, j)\,|\,\ell m\alpha\rangle\,. \tag{7.10}$$

Similarly, we shall denote by $y_{\ell\pm}(r)$ the quantity $y_\ell{}^j$ where $j = \ell \pm \frac{1}{2}$, and write $0(\ell_\pm)$ for $0(\ell, j)$.

Phase shifts

By complete analogy with (5.9), the asymptotic behaviour of $y_{\ell\pm}(r)$ is given by

$$y_{\ell\pm}(r) = C_{\ell\pm} \sin\left(kr + \frac{\ell\pi}{2} + \delta_{\ell\pm}\right). \tag{7.11}$$

The incident plane wave is given by

$$e^{ikz}\,|\,\alpha\rangle \simeq \sum_\ell (-i)^\ell (2\ell+1) \frac{\sin\left(kr + \dfrac{\ell\pi}{2}\right)}{kr} P_\ell(\cos\theta)\,|\,\alpha\rangle.$$

But

$$P_\ell(\cos\theta) = \sqrt{\frac{4\pi}{2\ell+1}}\, Y_\ell{}^0(\theta, \varphi)$$

whence

$$e^{ikz}\,|\,\alpha\rangle \simeq \sum_\ell (-i)^\ell \sqrt{\frac{2\ell+1}{4\pi}} \frac{\sin\left(kr + \dfrac{\ell\pi}{2}\right)}{kr}\,|\,\ell 0\alpha\rangle. \tag{7.12}$$

Comparing the asymptotic behaviours of (7.1) and of the partial waves, one obtains

$$\sum (-i)^\ell \sqrt{\frac{2\ell+1}{4\pi}} \frac{\sin\left(kr + \frac{\ell\pi}{2}\right)}{r} |\ell 0\alpha\rangle + F(\theta,\varphi)\frac{e^{ikr}}{r} |\alpha\rangle$$

$$= \sum_{\ell,j} C_{\ell\pm} \frac{\sin\left(kr + \frac{\ell\pi}{2} + \delta_{\ell\pm}\right)}{r} 0_\ell^\pm |\ell m\alpha\rangle.$$

Equating coefficients of e^{-ikr} gives

$$(-i)^\ell \sqrt{\frac{2\ell+1}{4\pi}} \frac{1}{k} I = \sum_\pm C_{\ell\pm} e^{-i\delta_{\ell\pm}} 0^\pm$$

whence, in view of (7.9),

$$C_{\ell\pm} = (-i)^\ell \frac{\sqrt{2\ell+1}}{k\sqrt{4\pi}} e^{i\delta_{\ell\pm}}.$$

Equating coefficients of e^{ikr} gives

$$F(\theta,\varphi) = \sum_\ell \sqrt{\frac{2\ell+1}{4\pi}} \frac{1}{k} [e^{i\delta_{\ell+}} \sin\delta_{\ell+} 0_\ell^+ + e^{i\delta_{\ell-}} \sin\delta_{\ell-} 0_\ell^-] \,|\ell 0\alpha\rangle$$

Thus

$$F(\theta) = \sum_\ell (2\ell+1)[a_{\ell+}(k^2)\, 0_\ell^+ + a_{\ell-}(k^2)\, 0_\ell^-]\, P_\ell(\cos\theta) \tag{7.13}$$

where

$$a_{\ell\pm}(k^2) = \frac{e^{i\delta_{\ell\pm}} \sin\delta_{\ell\pm}}{k}. \tag{7.14}$$

The expression (7.13) is a natural generalisation of (5.17).

Alternative form of the scattering amplitude

Recall $\mathbf{L} = \mathbf{x} \times \mathbf{p} = -\mathbf{p} \times \mathbf{x}$, and let $\mathbf{k}(\mathbf{k}')$ be the momenta of the initial (final) pions. Then one has

$$\mathbf{L}P_\ell(\cos\theta) = -\mathbf{k} \times \mathbf{x}P_\ell(k\cdot k') = +i\mathbf{k} \times \nabla_\mathbf{k} P_\ell(k\cdot k') = \hat{\mathbf{k}} \times \hat{\mathbf{k}}' P_\ell'(\hat{\mathbf{k}}\cdot\hat{\mathbf{k}}') \tag{7.15}$$

where $\hat{\mathbf{k}}$ is the unit vector along $\mathbf{k}$. Under these conditions the amplitude assumes the form

$$F(\theta) = f(\theta) + i\boldsymbol{\sigma}\cdot\mathbf{n}g(\theta) \tag{7.16}$$

where

$$\begin{aligned} f(\theta) &= \sum_\ell [(\ell+1)\, a_{\ell+} + \ell a_{\ell-}]\, P_\ell(\cos\theta) \\ g(\theta) &= \sum_\ell [a_{\ell+} - a_{\ell-}]\, P_\ell'(\cos\theta). \end{aligned} \tag{7.17}$$

$\mathbf{n} = \mathbf{k} \times \mathbf{k}'$ is a vector normal to the scattering plane and has the modulus $\sin \theta$.

Cross-sections and polarisations

According to (7.3), and in view of $\mathbf{n}^2 = \sin^2 \theta$, the differential cross-section is

$$\frac{d\sigma}{d\Omega} = |f(\theta)|^2 + |g(\theta)|^2 \sin^2 \theta \tag{7.18}$$

The polarisation of a recoil proton for scattering in the direction θ is given by

$$P = \frac{\text{number of protons with spin up} - \text{number of protons with spin down}}{\text{total number of protons}},$$

whence

$$P = \frac{\sum_\alpha \{|\langle \frac{1}{2} | F(\theta) | \alpha \rangle|^2 - |\langle -\frac{1}{2} | F(\theta) | \alpha \rangle|^2\}}{\sum_\alpha \{|\langle \frac{1}{2} | F(\theta) | \alpha \rangle|^2 + |\langle -\frac{1}{2} | F(\theta) | \alpha \rangle|^2\}}.$$

We have seen already that polarisation is possible only in a direction normal to the scattering plane, so that it is natural to take the spin quantisation axis along $\mathbf{n}$, whence

$$\mathbf{P} = \frac{|f(\theta) + g(\theta) \sin \theta|^2 - |f(\theta) - g(\theta) \sin \theta|^2}{|f(\theta) + g(\theta) \sin \theta|^2 + |f(\theta) - g(\theta) \sin \theta|^2} \hat{\mathbf{n}}$$

$$= \frac{2 \operatorname{Re} f(\theta) g^*(\theta) \sin \theta}{|f(\theta)|^2 + |g(\theta)|^2 \sin^2 \theta} \hat{\mathbf{n}} \tag{7.19}$$

$g(\theta)$ is commonly known as the spin-flip amplitude, and $f(\theta)$ as the coherent amplitude.

The effects of absorption

If there is absorption, the cross-sections are given by

$$\sigma_{\text{el}} = \int [|f(\theta)|^2 + |g(\theta)|^2 \sin^2 \theta] \, d\Omega.$$

In view of the orthogonality relations

$$\int_{-1}^{+1} P_\ell(\cos \theta) \, P_n(\cos \theta) \, d \cos \theta = \delta_{\ell n} \frac{2\ell(\ell + 1)}{2\ell + 1}$$

$$\int P_\ell'(\cos \theta) \, P_n'(\cos \theta) \, d \cos \theta \sin^2 \theta = \delta_{\ell n} \frac{2\ell(\ell + 1)}{2\ell + 1}$$

one finds

$$\sigma_{\text{el}} = 4\pi \sum_{\ell} [(\ell + 1) \mid a_{\ell+} \mid^2 + \ell \mid a_{\ell-} \mid^2] = 4\pi \sum_{\ell j} (j + \tfrac{1}{2}) \mid a_{\ell j} \mid^2$$

$$= \frac{\pi}{k^2} \sum_{\ell j} (j + \tfrac{1}{2}) \mid \eta_{\ell j} - 1 \mid^2 \quad (7.20)$$

where

$$a_{\ell j} = \frac{\eta_{\ell j} - 1}{2ik} \qquad \mid \eta_{\ell j} \mid \leqslant 1. \quad (7.21)$$

The absorption cross-section is given by

$$\sigma_{\text{inel}} = \frac{\pi}{k^2} \sum_{\ell j} (j + \tfrac{1}{2})(1 - \mid \eta_{\ell j} \mid^2). \quad (7.22)$$

The total cross-section is

$$\sigma_T = \sigma_{\text{el}} + \sigma_{\text{inel}} = \frac{\pi}{k^2} \sum_{\ell j} (j + \tfrac{1}{2})\, 2(1 - \text{Re}\, \eta_{\ell j}) \quad (7.23)$$

so that the optical theorem takes the form

$$\text{Im}\, f(\theta) = \frac{k\sigma_T}{4\pi} \quad (7.24)$$

where $f(\theta)$ is now the coherent amplitude.

Problems

1. By using the expression

$$\nabla^2\psi = \frac{1}{r^2}\frac{\partial}{\partial r}\left(r^2 \frac{\partial\psi}{\partial r}\right) + \frac{1}{r^2 \sin\theta}\frac{\partial}{\partial\theta}\left(\sin\theta \frac{\partial\psi}{\partial\theta}\right) + \frac{1}{r^2 \sin^2\theta}\frac{\partial^2\varphi}{\partial\varphi^2}$$

show that $f(\theta, \varphi)e^{\pm ikr}/r$ solves the Schroedinger equation up to terms of order $1/r^2$.

2. Evaluate the differential cross-section for π^+ on completely polarised protons in terms of the functions $f(\theta)$ and $g(\theta)$.
3. Calculate the differential cross-section and the polarisation of the particle B in the reaction

$$a + A \rightarrow b + B$$

where all particles are nonrelativistic, a and b have spin 0, A and B have spin $\frac{1}{2}$, and the product of the intrinsic parities of the initial particles is opposite to that of the final particles.

CHAPTER 8

BOUND STATES AND RESONANCES

We analyse the solutions of the radial Schroedinger equation in order to study the most prominent features of phase shift and of partial wave amplitudes $a_\ell(k^2)$. In such an analysis it is convenient and instructive to consider complex values of the energy $k^2/2m$: One can then show that $a_\ell(k^2)$ is an analytic function of k^2 in the complex k^2 plane, with two branch cuts. One of these cuts is of dynamical origin, in the sense that its position and the discontinuity across it depend on the potential. The other cut is fixed, and is necessary to ensure unitarity; it underlies the validity of the effective range approximation. Apart from these cuts $a_\ell(k^2)$ also has poles. Some of the poles are due to the existence of bound states, while others are interpreted as resonances. Once we have studied in some detail the notion of a resonance, we shall show in what sense it is similar both mathematically and physically to a bound state.

1. The radial equation

Consider again the scattering problem for two nonrelativistic spinless particles interacting through a central potential. We have seen that the wavefunction $y_\ell(r)$ of a state with angular momentum ℓ obeys the radial Schroedinger equation

$$\frac{d^2y_\ell(r)}{dr^2} + \left[k^2 - U(r) - \frac{\ell(\ell+1)}{r^2}\right] y_\ell(r) = 0. \tag{1.1}$$

We now investigate what general properties of the scattering amplitude

$$a_\ell(k^2) = \frac{e^{i\delta_\ell} \sin \delta_\ell}{k} \tag{1.2}$$

can be deduced from equation (1.1). Later on we shall meet good reasons for believing that all potentials actually occurring in nature are superpositions of Yukawa potentials. A pure Yukawa potential has the form

$$V(r) = 2mU(r) = g\,\frac{e^{-\mu r}}{r}. \tag{1.3}$$

It is singular at the origin and decreases exponentially as r tends to infinity. $r_0 = \mu^{-1}$ is called the range of the potential. Neglecting all spin effects, it can be shown from field theory that a π^0 of mass μ exchanged between two protons produces in first approximation a Yukawa potential of the form (1.3) with μ given by the pion mass. A superposition of Yukawa potentials is of the form

$$V(r) = \int_{\mu_0}^{\infty} g(\mu) \frac{e^{-\mu r}}{r} \, d\mu \tag{1.4}$$

where μ_0^{-1} is now called the range.

Properties of the wavefunction

We shall study the behaviour of $y_\ell(r)$ as r tends to 0 or to infinity.

As r tends to zero, the most singular term is the centrifugal barrier, k^2 and $U(r)$ being negligible by comparison; thus the equation becomes

$$\frac{d^2 y_\ell}{dr^2} - \frac{\ell(\ell+1)}{r^2} y_\ell(r) = 0.$$

It is easy to find two independent solutions of this equation, namely

$$y_\ell^{(1)}(r) = r^{\ell+1} \qquad y_\ell^{(2)}(r) = r^{-\ell}.$$

The second solution must be rejected because it leads to infinite values of the wavefunction at $r = 0$. For the physical solution we chose a normalisation different from that of the last chapter, and write

$$\lim_{r \to 0} \frac{y_\ell(r)}{r^{\ell+1}} = 1. \tag{1.5}$$

As r tends to infinity, the potential $U(r)$ and the centrifugal term $\ell(\ell + 1)/r^2$ are negligible compared to k^2, and the equation becomes

$$\frac{d^2 y_\ell(r)}{dr^2} + k^2 y_\ell(r) = 0,$$

which gives the asymptotic form of $y_\ell(r)$:

$$y_\ell(r) \simeq \varphi_\ell^-(k^2)\, e^{ikr} + \varphi_\ell^+(k^2)\, e^{-ikr}. \tag{1.6}$$

This equation defines the coefficients φ^+ and φ^-, which are called the Jost functions. Since $y_\ell(r)$ is a real function, φ^+ and φ^- are complex conjugates.

Connection with the phase shift

In the last chapter we saw that the wavefunction behaves like

$$y_\ell(r) \simeq \text{const} \times \sin\left(kr - \frac{\ell\pi}{2} + \delta_\ell\right) \qquad \text{as} \qquad r \to \infty$$

so that

$$y_\ell(r) \simeq \frac{\text{const}}{i} [e^{-i(\ell\pi/2)+i\delta_\ell} e^{ikr} + e^{i(\ell\pi/2)-i\delta_\ell} e^{-ikr}]. \tag{1.7}$$

Comparing equations (1.6) and (1.7), we can express the phase shift in terms of the Jost functions

$$e^{2i\delta_\ell} = (-1)^{\ell+1} \frac{\varphi_\ell^-(k^2)}{\varphi_\ell^+(k^2)}. \tag{1.8}$$

Hence the phase shift and the Jost functions can be studied interchangeably.

Bound states

Recall that for negative energies k^2, i.e. for $k = iK$, the asymptotic behaviour (1.6) still applies, and can be written as

$$y_\ell(r) \simeq \varphi_\ell^-(E)\, e^{-Kr} + \varphi_\ell^+(E)\, e^{Kr}.$$

A bound state with energy E_0 exists whenever this function is square integrable, which will be the case if

$$\varphi_\ell^+(E_0) = 0. \tag{1.9}$$

Note that the expression (1.8), whose physical meaning is no longer self-evident, will now have a pole.

2. Analytic properties of the scattering amplitude

The dependence of the radial equation on k^2 is particularly simple. We know that in order to extract maximum information about the solution of a differential equation, it should be considered as a function of the parameters with the latter assuming complex values. By a theorem due to Poincaré, $y_\ell(r, k^2)$ is an analytic function of k^2 if it is defined as the solution of a differential equation whose coefficients are analytic in k^2, and subject to boundary conditions which are also analytic in k^2. This theorem holds when all the coefficients are regular; hence it is not directly applicable to equation (1.1) which is singular at the origin $r = 0$. But the origin is actually a singularity of a rather simple type (technically speaking, it is a regular singularity), which makes it possible to generalise Poincaré's theorem to show that $y_\ell(r, k^2)$ is indeed an analytic function of k^2.

We shall not derive these mathematical properties in detail; neither shall we analyse rigorously their consequences regarding the scattering amplitude. The reason is that from the point of view of particle physics very little interest attaches to such techniques for analysing differential equations, important though they are in many other domains. In general therefore we shall make do with heuristic arguments. Our final object is to generalise the scattering amplitude

$$a_\ell(k^2) = \frac{e^{i\delta_\ell}\sin\delta_\ell}{k} = \frac{e^{2i\delta_\ell}-1}{2ik} \tag{2.1}$$

so that it can be regarded as a function of a complex variable k^2; to study its singularities (poles, branch points, etc.); and to analyse the connection between these singularities and the properties of the potential.

Therefore we face the problem in the following form. Define $y_\ell(r, k^2)$ as the solution of the radial equation subject to the boundary condition (1.5). It is an analytic function of k^2. Study its asymptotic behaviour and from this deduce the Jost functions by appeal to (1.6). Finally, define $a_\ell(k^2)$ by (1.8) and (2.1). The only nontrivial part of this programme is the definition of the Jost functions. We shall see that there exist two sets of values of k^2 for which this definition is not straightforward, and they will lead us to the singularities of $\varphi_\ell^{\pm}(k^2)$.

3. Left-hand cut

In order to discover the asymptotic behaviour of (1.6) we have neglected $U(r)$ and $\ell(\ell + 1)/r^2$ compared to k^2. This is certainly justified for real values of k but must be reconsidered in more detail when k is complex. When the approximation does apply it continues to validate the asymptotic formula (1.6).

However, at this point we must define more closely just what we mean by k, since it is only k^2 which enters directly. We chose to define k as that square root $\sqrt{k^2}$ whose imaginary part is positive. Actually this choice is suggested by the fact that the physical region corresponds to $k + i\epsilon$, as we saw in connection with the Green's function.

There is one case where analyticity in k^2 of the wavefunction $y_\ell(r, k^2)$ does not imply analyticity for the Jost functions, because of the failure of the asymptotic formula (1.6). When k^2 is real and negative, the first exponential in (1.6) is a decreasing one since $e^{ikr} = e^{-|k|r}$; but the second is exponentially increasing, $e^{-ikr} = e^{|k|r}$.

In view of the fact that we are dealing with Yukawa potentials which decrease exponentially with r like $e^{-\mu_0 r}$, we are justified in isolating the first term of (1.6) only if the decreasing exponential e^{ikr} does not decrease too fast. If $|k|$ exceeds $\mu_0/2$, then it does not make sense to neglect in the radial equation the term where the exponential $e^{-\mu_0 r}$ multiplies the increasing exponential $e^{|k|r}$, resulting in $e^{(|k|-\mu_0)r}$, while retaining the term where the coefficient k^2 multiplies $e^{-|k|r}$.

We must conclude that equation (1.6) does not make sense when k^2 is real and less then $-\mu_0{}^2/4$. For such values of k^2, the Jost function $\varphi_\ell^{(-)}(k^2)$ is not defined; by contrast, the coefficient $\varphi_\ell^{(+)}$ of the exponentially increasing term is well defined, and remains analytic. Indeed, one can prove rigorously that $\varphi_\ell^{(-)}(k^2)$ has a singularity on the negative real axis at $k^2 = -\mu_0{}^2/4$. The usual convention is to make a branch cut of $\varphi_\ell^{(-)}(k^2)$ from $-\infty$ to $-\mu_0{}^2/4$ along the real axis. It is called the *left-hand cut*. Note that its position depends on the *range* of the potential (Figure 8.3.1).

The left-hand cut and the Born series

The presence of the left-hand cut in the Jost functions evidently implies its presence also in $a_\ell(k^2)$. One can learn more about it by using the formula

$$f(k^2, \cos\theta) = \sum_\ell (2\ell + 1)\, a_\ell(k^2)\, P_\ell(\cos\theta) \tag{3.1}$$

and the Born series

$$f_{fi} = -\frac{2m}{4\pi}\left[(\varphi_f, V\varphi_i) - \left(\varphi_f, V\frac{1}{H_0 - E - i\epsilon}V\varphi_i\right) + \cdots\right] \tag{3.2}$$

To this end, consider a pure Yukawa potential

$$V(r) = g\,\frac{e^{-\mu r}}{r}. \tag{3.3}$$

The first Born approximation is

$$\langle \mathbf{k}' \mid V \mid \mathbf{k}\rangle = g\int e^{i(\mathbf{k}-\mathbf{k}')\cdot\mathbf{x}}\,\frac{e^{-\mu r}}{r}\,d^3x \tag{3.4}$$

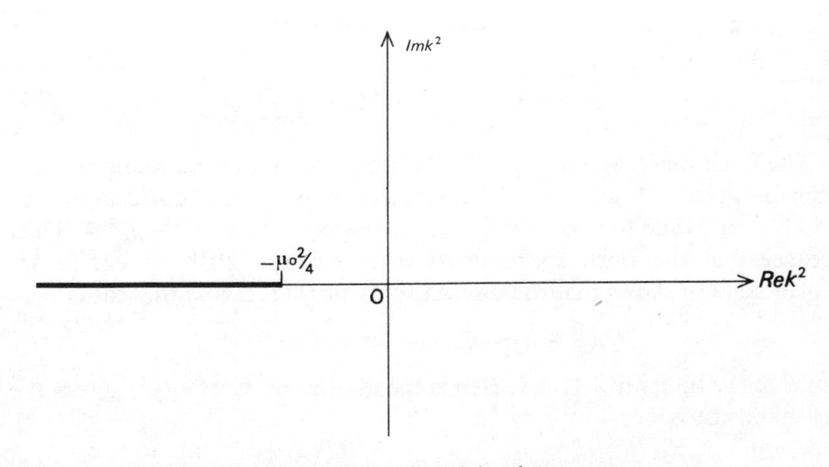

FIG. 8.3.1. The left-hand cut.

We evaluate the integral in spherical polar co-ordinates, in which it appears as

$$\langle \mathbf{k}' \mid V \mid \mathbf{k} \rangle = g \int e^{i\Delta r \cos\theta - \mu r}\, r^2\, dr\, d\cos\theta\, d\varphi \qquad \text{with} \qquad \mathbf{\Delta} = \mathbf{k}' - \mathbf{k}$$

The vector $\mathbf{\Delta}$ is called the *momentum transfer*. Integration over φ gives a factor 2π; the integration over $\cos\theta$ is straightforward and yields

$$\langle \mathbf{k}' \mid V \mid \mathbf{k} \rangle = 2\pi g \int_0^\infty \frac{e^{i\Delta r} - e^{-i\Delta r}}{i\Delta r}\, e^{-\mu r}\, r\, dr.$$

The remaining integral is also straightforward, and leads to

$$\langle \mathbf{k}' \mid V \mid \mathbf{k} \rangle = \frac{2\pi g}{i\Delta}\left\{\frac{1}{\mu - i\Delta} - \frac{1}{\mu + i\Delta}\right\} = \frac{4\pi g}{\Delta^2 + \mu^2}$$

whence finally we obtain, in first Born approximation,

$$f(k^2, \cos\theta) = -\frac{2mg}{\Delta^2 + \mu^2}. \tag{3.5}$$

To deduce $a_\ell(k^2)$ from (3.5), we need the expression for Δ^2 in terms of $\cos\theta$:

$$\Delta^2 = 2k^2(1 - \cos\theta)$$

and the expansion (3.2). It follows from these, by appeal to the orthogonality of the Legendre polynomials, that

$$a_\ell(k^2) = -mg \int_{-1}^{+1} \frac{P_\ell(\cos\theta)\, d\cos\theta}{2k^2(1 - \cos\theta) + \mu^2}.$$

The integral representation of the Legendre functions of the second kind,

$$Q_\ell(x) = +\frac{1}{2}\int_{-1}^{+1} \frac{P_\ell(\cos\theta)\, d\cos\theta}{x - \cos\theta} \tag{3.6}$$

gives immediately

$$a_\ell(k^2) = -\frac{mg}{k^2} Q_\ell\left(1 + \frac{\mu^2}{2k^2}\right). \tag{3.7}$$

The function $Q_\ell(z)$ is analytic in the complex z plane cut along the real axis between -1 and $+1$. The singular point $z = 1$ corresponds to $k^2 = -\infty$, while the point $z = -1$ corresponds to $k^2 = -\mu_0{}^2/4$. Thus we see that the Born approximation (3.7) has a left-hand cut in k^2. Formula (3.6) shows that the discontinuity of $Q_\ell(z)$ across this cut is

$$Q_\ell(x + i\epsilon) - Q_\ell(x - i\epsilon) = -i\pi P_\ell(x).$$

for x in the interval $-1, +1$. Hence the discontinuity of $a_\ell(k^2)$ across the left hand cut is

$$a_\ell(k^2 + i\epsilon) - a_\ell(k^2 - i\epsilon) = -\frac{img\pi}{k^2} P_\ell\left(1 + \frac{\mu^2}{2k^2}\right). \tag{3.8}$$

Higher terms in the Born approximation lead to similar results; we see that the position of the cut depends on the range of the potential, while the discontinuity depends on the strength of the potential. For this reason the cut is sometimes called the dynamical cut.

4. Right-hand cut

Another case where the asymptotic formula (1.6) fails to apply arises for $k^2 = 0$, where $U(r) + \ell(\ell + 1)/r^2$ cannot be neglected compared to k^2.

Actually the nature of this singularity is readily visible from (1.6). If we start from a given value of $k^2 = k_0{}^2 + i\epsilon$, where k_0 is small, and encircle the origin once, then k becomes $-k$, and the roles of the functions $\varphi_\ell^{(+)}(k^2)$ and $\varphi_\ell^{(-)}(k^2)$ are interchanged; in other words

$$\varphi_\ell^{(+)}(k) = \varphi_\ell^{(-)}(-k)$$

Hence the point $k^2 = 0$ is a branch point of order 2 (Figure 8.4.1).

In order to understand the situation more clearly, we shall consider $\varphi_\ell^{(-)}(k^2)$ as a function of k rather than of k^2, and write simply

$$\varphi_\ell^{(-)}(k^2) = \varphi_\ell(k) \tag{4.1}$$

We know that $\varphi_\ell^{(-)}(k^2)$ is an analytic function of k^2 in the complex k^2 plane cut along the real axis from $-\infty$ to $-\mu^2/4$. This implies that it is an analytic function of k in the upper half complex k plane ($\text{Im } k > 0$) cut from $+i\infty$ to $i\mu/2$ (Figure 8.4.2).

There is a remarkable symmetry between the Jost functions under the interchange of k and $-k$. Indeed, this interchange affects neither the radial Schroedinger equation which depends only on k^2, nor the boundary condition defining $y_\ell(k^2, r)$ which is independent of k^2. Therefore it leaves $y_\ell(k^2, r)$ unchanged. But it interchanges the two exponentials $e^{\pm ikr}$ in (1.6), whence

$$\varphi_\ell(k) = \varphi_\ell(-k) \tag{4.2}$$

The connection (4.2) shows that the analytic continuation of $\varphi_\ell(k)$ into the lower half complex plane ($\text{Im } k < 0$) is simply $\varphi_\ell^{(+)}(k^2)$; we have seen already that the latter has no singularities in this half plane.

Riemann sheets

In order to visualise the Jost functions as functions of k^2, it is useful to consider the k^2 plane as consisting actually of two superimposed planes, or sheets, corresponding to the two possible choices of k. To specify this choice, we shall say that for $\text{Im } k > 0$ we are on the *physical sheet*, while

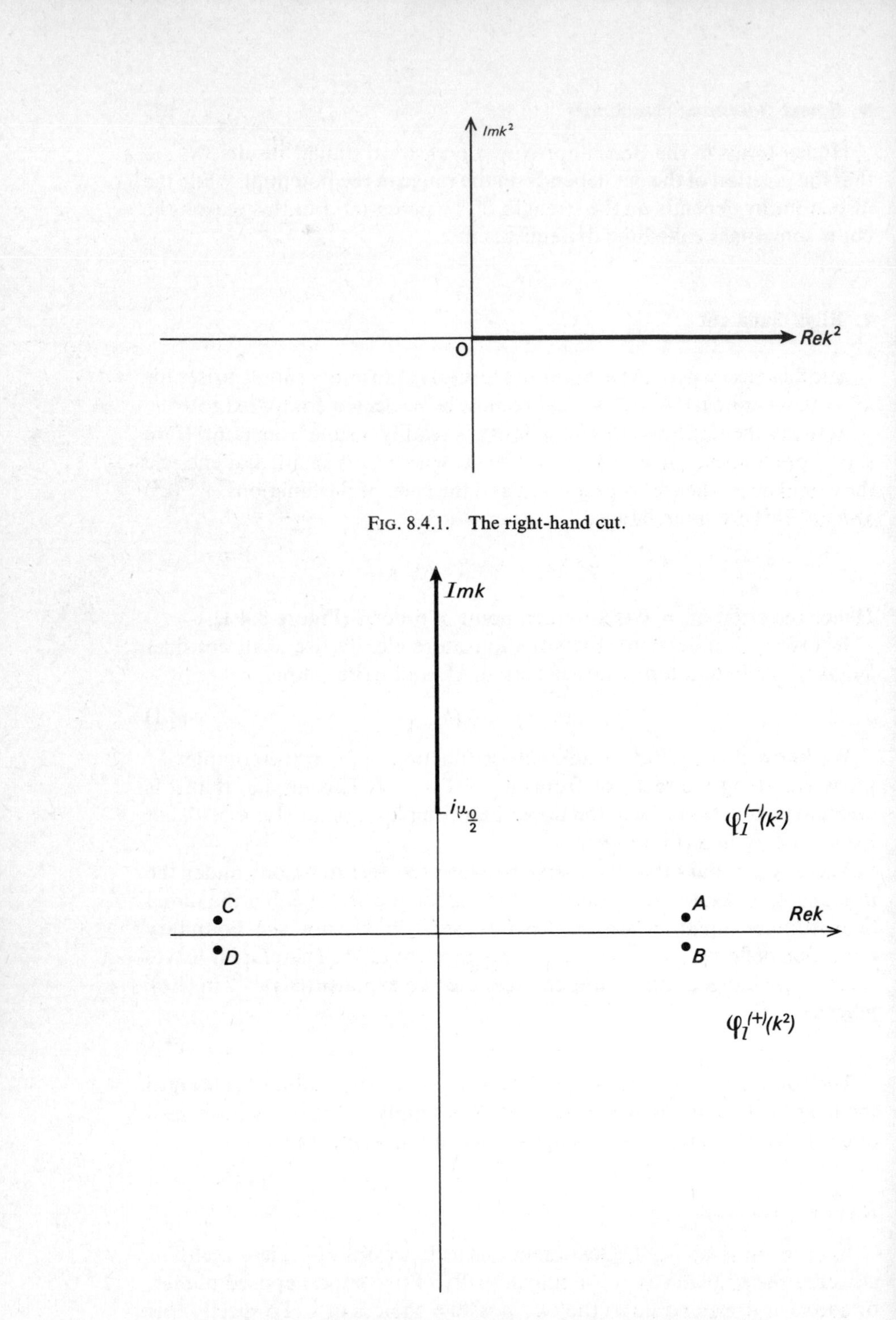

FIG. 8.4.1. The right-hand cut.

FIG. 8.4.2. The domain of analyticity in k of the Jost function $\phi_\ell(k)$.

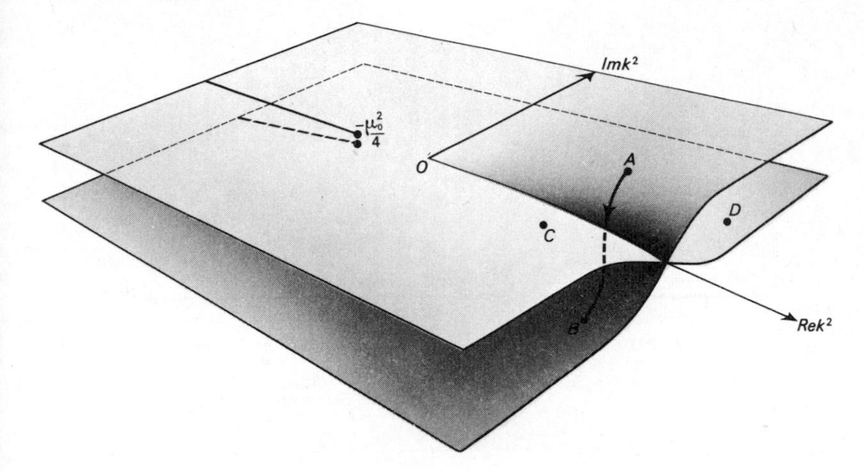

FIG. 8.4.3. Riemann surface of a scattering amplitude.

for Im $k < 0$ we are on the *unphysical sheet*. This nomenclature arises from the fact that physically measurable values of $a_\ell(k^2)$ correspond to real positive values of k with an infinitesimal positive imaginary part. The two sheets have a common boundary along Im $k = 0$, i.e. along the real positive k^2 axis (Figure 8.4.3). Their connection across this boundary is a cross-over; in other words $A(k^2 + i\epsilon)$ (on the physical sheet) adjoins $B(k^2 - i\epsilon)$ (on the unphysical sheet), while $C(k^2 - i\epsilon)$ (on the physical sheet) is considered to be remote both from A and from B, since $k_c \sim -k_A$, $-k_B$. In case of doubt it is often simpler to work directly in the k plane rather than in the k^2 plane.

A set of such sheets defined over the complex plane constitutes, by definition, a Riemann surface. This surface is uniquely defined by the boundaries between sheets (in our case the real positive k^2 axis), and by their connections across these boundaries. A point which like the origin $k^2 = 0$ is common to several sheets is called a branch point.

We have now completed our analysis of the singularity of $\varphi_\ell^{(\pm)}(k^2)$ at the origin; it is simply a branch point of order 2, or in other words a branch point common to only two sheets. It is a kinematic singularity in the sense that its nature and its position do not depend on the potential.

5. Analytic properties of partial wave amplitudes

We shall now investigate the properties of the amplitude $a_\ell(k^2)$, given that the Jost functions $\varphi_\ell^{(\pm)}(k^2)$ are analytic on the Riemann surface described in the last paragraph.

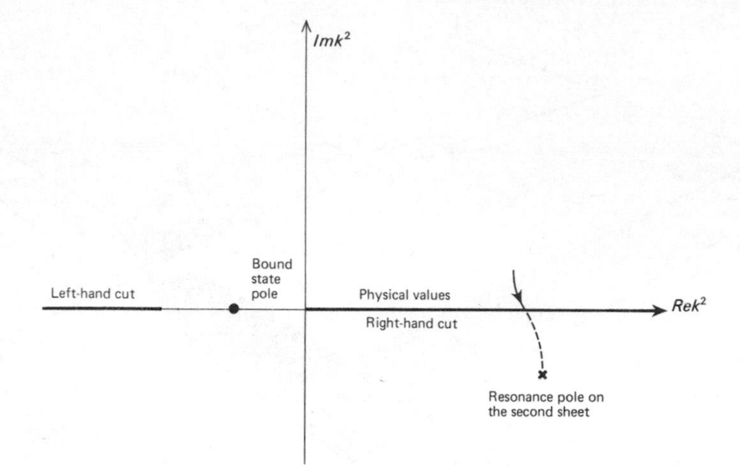

FIG. 8.5.1. Singularities of a scattering amplitude.

According to formulae (1.8) and (2.1), one has

$$a_\ell(k^2) = \frac{1}{2ik}\left[(-1)^{\ell+1}\frac{\varphi_\ell^-(k^2)}{\varphi_\ell^+(k^2)} - 1\right]. \tag{5.1}$$

Hence $a_\ell(k^2)$ is an analytic function of k^2 having the same singularities as the Jost functions, and is defined on a two-sheeted Riemann surface. Moreover, $a_\ell(k^2)$ can have poles wherever $\varphi_\ell^{(+)}(k^2)$ has a zero (Figure 8.5.1).

(a) *Connections between the two sheets*

We shall try to clarify the relation between the values assumed by $a_\ell(k^2)$ on the two sheets. To this end it is more convenient to consider the function $S_\ell(k^2)$ defined by

$$S_\ell(k^2) = e^{2i\delta_\ell} = (-1)^{\ell+1}\frac{\varphi_\ell^-(k^2)}{\varphi_\ell^+(k^2)}. \tag{5.2}$$

Consider once more the two points B and C of the Riemann surface which lie on top of each other and correspond to the same value of k^2. Our problem is to determine S_ℓ at the point B from its known value at C, thus reducing the study of the entire amplitude to the first sheet only. To avoid confusion, we shall continue to use $S_\ell(k^2)$ for the value of the function on the physical sheet, and shall write $S_\ell{}^I(k^2)$ for its value when k^2 is on the unphysical sheet. The points B and C correspond to opposite values of k, whence (4.2) at once leads to

$$S^I(k^2) = (-1)^{\ell+1}\frac{\varphi_\ell^{(-)}(-k)}{\varphi_\ell^{(+)}(-k)} = (-1)^{\ell+1}\frac{\varphi^{(+)}(k)}{\varphi^{(-)}(k)} = \frac{1}{S(k^2)} \tag{5.3}$$

From this we can find the connection between $a_\ell(k^2)$ on the physical sheet and its continuation onto the unphysical sheet. To this end we note that on the physical sheet

$$a_\ell(k^2) = \frac{S_\ell(k^2) - 1}{2ik}$$

while on the unphysical sheet

$$a_\ell^I(k^2) = \frac{S_\ell^I(k^2) - 1}{2i(-k)}$$

in view of the fact that the analytic continuation of k is $k^I = -k$. Therefore

$$a_\ell^I(k^2) = \frac{a_\ell(k^2)}{1 + 2ika_\ell(k^2)}. \tag{5.4}$$

The implications of equation (5.4) can be summed up by saying that everything that happens on the second sheet is completely determined by what happens on the physical sheet. In particular, $a_\ell(k^2)$ has a left-hand cut on the second sheet, for k^2 real negative and less than $-\mu_0{}^2/4$. Consequently, we are not allowed to cross the real positive k^2 axis without taking special precautions.

(b) *Bound states*

Consider again the poles of $a_\ell(k^2)$, i.e. the zeros of $\varphi_\ell^{(+)}(k^2)$. When $\varphi_\ell^{(+)}(k^2)$ has a zero, the asymptotic form of $y_\ell(r)$ is

$$y_\ell(r) \underset{r\to\infty}{\sim} \varphi^-(\ell, k^2)\, e^{ikr}. \tag{5.5}$$

If this occurs on the physical sheet ($\operatorname{Im} k > 0$), then $y_\ell(r)$ decreases exponentially with r, so that the wavefunction is normalisable; this, corresponding as it does to a physical state, is necessarily a bound state. But in that case k^2 must be real and negative. Thus on the physical sheet there can occur poles corresponding to bound states. If the binding energy of such a state is B, then its energy is $E = k^2/2m = -B$, whence $k^2 = -2mB$. No other poles can occur on the physical sheet.*

(c) *Resonances*

Consider now the poles which can appear on the second sheet. Since $\operatorname{Im} k < 0$, the function (5.5) grows exponentially with r. Hence it is not the wavefunction of a physically realisable state, and the pole can occur anywhere in the complex plane.

* This statement is almost true, and applies to all the potentials of interest in nuclear and particle physics. However, mathematically speaking there do exist potentials giving poles of $S_\ell(k)$ on the positive imaginary k axis, due to poles of $\phi_\ell^{(+)}(k)$ rather than to zeros of $\phi_\ell^{(-)}(k)$. Such poles are called 'redundant poles'; their physical significance is the same as that of the left-hand cut singularities, and they have nothing to do with bound states. The exponential potential is an example.

The presence of a pole on the unphysical sheet, for instance at the point B, entails several consequences as regards the values assumed by the function $S_\ell(k)$ at the points ACD; we shall now investigate these in detail.

Since the coefficients in Schroedinger's equation are real, the wave-functions $y_\ell(k^2, r)$ and $y_\ell(k^{2*}, r)$ corresponding to the complex conjugate values k^2 and k^{2*} are themselves complex conjugates; in other words

$$y_\ell(k^2, r) = [y_\ell(k^{2*}, r)]^*. \tag{5.6}$$

Evidently, one has

$$y_\ell(k^2, r) \underset{r\to\infty}{\sim} \varphi_\ell^{(-)}(k^2)\, e^{ikr} + \varphi_\ell^{(+)}(k^2)\, e^{-ikr}.$$

To k^{2*} there corresponds the value $-k^*$ of $\sqrt{k^{2*}}$, by virtue of our convention for chosing the imaginary parts of k; consequently

$$y_\ell(k^{2*}, r) \underset{r\to\infty}{\sim} \varphi_\ell^{(-)}(k^{2*})\, e^{-ik*r} + \varphi_\ell^{(+)}(k^{2*})\, e^{+ik*r}.$$

By virtue of (5.6), this yields the remarkable symmetry

$$\varphi_\ell^{+}(k^{2*}) = [\varphi_\ell^{+}(k^2)]^* \qquad \varphi_\ell^{-}(k^{2*}) = [\varphi_\ell^{-}(k^2)]^*. \tag{5.7}$$

By (5.2), this entails

$$S_\ell(k^{2*}) = [S_\ell(k^2)]^*. \tag{5.8}$$

The last equation shows that, on both first and the second sheets, the function $S_\ell(k^2)$ assumes complex conjugate values at two points related by reflection in the real axis.

Suppose now that the function $S_\ell{}^I(k^2)$ has a pole at B fairly close to the real positive axis, where

$$k^2 = k_0{}^2 - i\frac{\Gamma}{2}. \tag{5.9}$$

Then according to (5.8), $S_\ell{}^I(k^2)$ will have a second pole on the unphysical sheet at the point D symmetric to C, where

$$k^2 = k_0{}^2 + i\frac{\Gamma}{2}.$$

The relation (5.3) between the values assumed by $S_\ell(k^2)$ on the two sheets then shows that on the first sheet $S_\ell(k^2)$ has two zeros at $k^2 = k_0{}^2 \pm i\Gamma/2$, i.e. at the points A and C.

Next, we recall that the physical phase shift is defined as the limit

$$e^{2i\delta_\ell(k^2)} = \lim_{\epsilon\to 0} S_\ell(k^2 + i\epsilon) \tag{5.10}$$

taken on the first sheet. In other words the physical matrix element $e^{2i\delta_\ell(k^2)}$ is the limit of the analytic function $S_\ell(k^2)$ as k^2 approaches the upper edge of the right-hand cut on the physical sheet, i.e. the point M of the Riemann surface. For small enough Γ, the points A and B are close to M, and the variation of $S_\ell(k^2)$ with the energy k^2 is governed chiefly by the pole at B and the zero at A; thus one can write

$$S_\ell(k^2) = \frac{k^2 - k_0^2 - i\dfrac{\Gamma}{2}}{k^2 - k_0^2 + i\dfrac{\Gamma}{2}} S_P(k^2), \tag{5.11}$$

where $S_P(k^2)$ is a slowly varying function of k^2. Since the quotient shown explicitly in (5.11) has unit modulus for real k^2, the equation yields the following representation for the phase shift:

$$\delta_\ell(k^2) = \delta_R(k^2) + \delta_P(k^2) \tag{5.12}$$

where $S_P(k^2) = e^{2i\delta_P(k^2)}$. One often calls $\delta_P(k^2)$ the background phase shift, while $\delta_R(k^2)$ is given by

$$e^{2i\delta_R(k^2)} = \frac{k^2 - k_0^2 - i\dfrac{\Gamma}{2}}{k^2 - k_0^2 + i\dfrac{\Gamma}{2}}. \tag{5.13}$$

An alternative form of (5.12) is

$$\delta_\ell(k^2) = \text{Arc tg}\,\frac{\Gamma}{2(k_0^2 - k^2)} + \delta_P(k^2). \tag{5.14}$$

In the presence of such a pole close to the real axis on the unphysical sheet, one says that one is dealing with a *resonance*. The phase shift then behaves as shown in Figure (8.5.2). We see that it passes through $\pi/2$ at energy k_0^2 (called the energy of the resonance) if the background phase shift is negligible. If $\delta_P(k^2)$ is not negligible then the phase shift can pass through $\pi/2$ at a value of k^2 different from k_0^2.

Neglecting the background, the relation between $a_\ell(k^2)$ and $S_\ell(k^2)$ shows

$$a_\ell(k^2) = \frac{S_\ell(k^2) - 1}{2ik} = -\frac{\dfrac{\Gamma}{2}}{k^2 - k_0^2 + i\dfrac{\Gamma}{2}} \tag{5.15}$$

We saw in the last chapter that a given partial wave contributes to the total cross-section an amount proportional to $|a_\ell(k^2)|^2$, i.e. proportional to the function

$$\frac{\dfrac{\Gamma^2}{4}}{(k^2 - k_0^2)^2 + \dfrac{\Gamma^2}{4}} \tag{5.16}$$

which is shown in figure (8.5.3). A curve of this general appearance is said to have a Breit–Wigner (or Lorentz) shape, while the formulae (5.13),

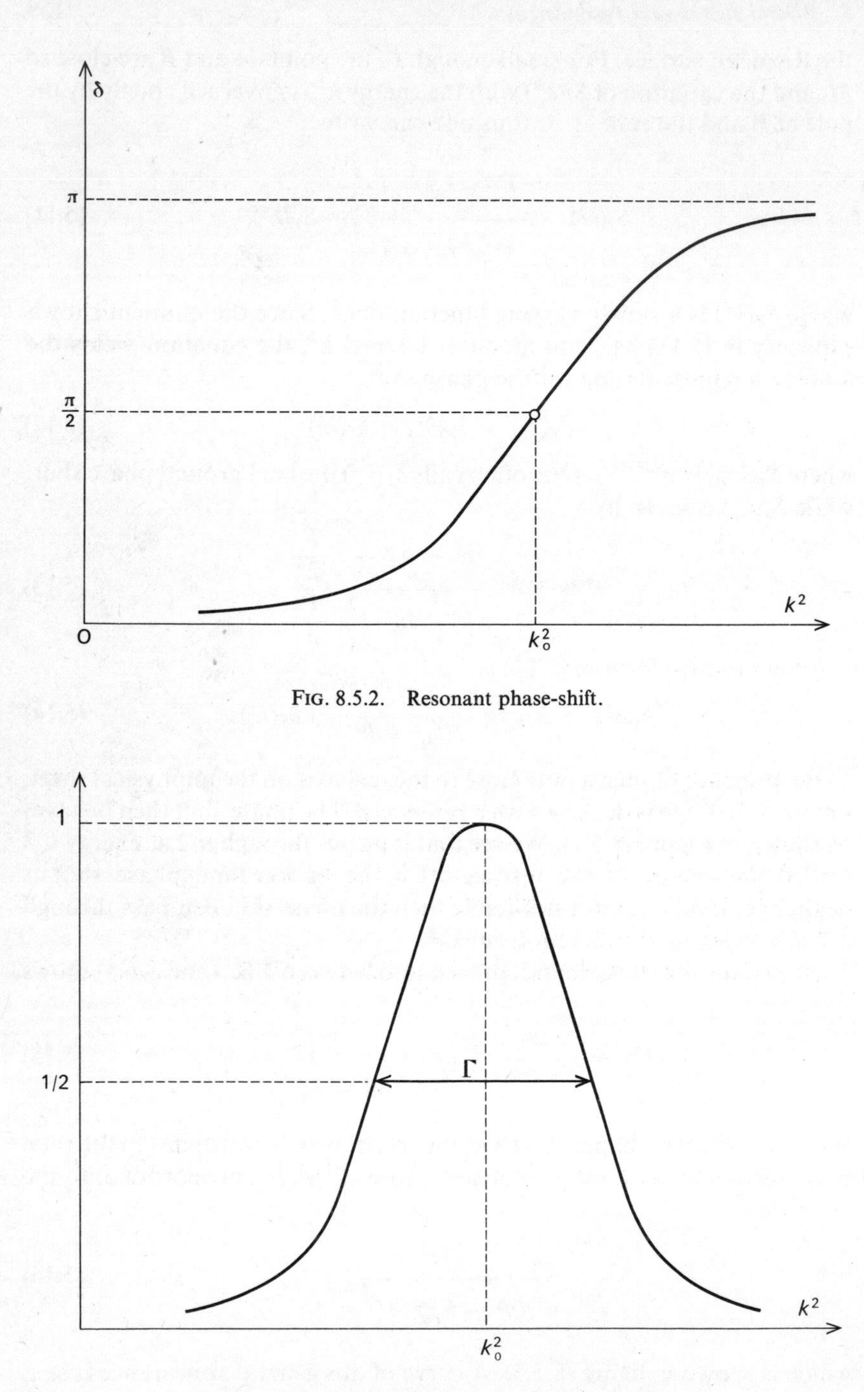

FIG. 8.5.2. Resonant phase-shift.

FIG. 8.5.3. Breit–Wigner curve (equation 5.16).

(5.14), (5.15), and (5.16) are, indiscriminately, called Breit–Wigner formulae.

Note from the figure that the resonance appears like a characteristically shaped peak in the cross-section, which has a maximum at $k^2 = k_0{}^2$. (Once again the background can shift the curve slightly.) The width of the peak at half-maximum is Γ; hence Γ is called the width of the resonance.

To sum up, a resonance with energy $k_0{}^2$ and width Γ is due to a pole on the unphysical sheet at $k^2 = k_0{}^2 - i\Gamma/2$.

It is worth noting the analogy between the scattering amplitude close to a resonance, and the impedance of a resonant circuit as a function of the frequency.

(d) *Effective range formulae*

We wish to establish the so-called 'effective range formula', which is much used in practice; we shall need to exploit the following result:

As k^2 tends to 0, $a_\ell(k^2)$ tends to 0 like $k^{2\ell}$.

Proof

The simplest proof relies on the fact, to be discussed in chapter 12, that the scattering amplitude is an analytic function of k^2 and of $\Delta^2 = 2k^2(1 - \cos\theta)$. Hence, for small k^2, we can expand $f(k^2, \Delta^2)$ as a Taylor series in Δ^2:

$$f(k^2, \Delta^2) = f(k^2, 0) + \Delta^2 f(k^2, 0) + \frac{\Delta^2}{2!} f(k^2, 0) + \ldots \tag{5.17}$$

From this, $a_\ell(k^2)$ is evaluated by appeal to

$$a_\ell(k^2) = \frac{1}{2}\int_{-1}^{+1} f(k^2, \cos\theta)\, P_\ell(\cos\theta)\, d\cos\theta\,. \tag{5.18}$$

Since Δ^2 is linear in $\cos\theta$, the first $(\ell - 1)$ terms of (5.17) form a polynomial in $\cos\theta$ of degree $(\ell - 1)$, which does not contribute to the integral. The higher terms of (5.17) all contain a factor $k^{2\ell}$.

Next we note that the scattering amplitude can be written in the form

$$a_\ell(k^2) = \frac{e^{i\delta_\ell}\sin\delta_\ell}{k} = \frac{\sin\delta_\ell}{ke^{-i\delta_\ell}} = \frac{\sin\delta_\ell}{k(\cos\delta_\ell - i\sin\delta_\ell)} = \frac{1}{k\operatorname{cotg}\delta_\ell - ik}\,. \tag{5.19}$$

The function

$$R(k) = k\operatorname{cotg}\delta_\ell(k)$$

has some interesting properties. Equation (5.19) allows one to write

$$R(k) = \frac{1}{a_\ell(k)} + ik$$

which shows at once that $R(k)$ has the same singularities as $a_\ell(k)$ except for the latter's poles and except for the fact that $R(k)$ has poles where $a_\ell(k)$ vanishes. Actually the unitarity condition on the amplitude implies that $R(k)$ *has no right-hand cut*. Evidently one has

$$a_\ell(k) - a_\ell^*(k) = 2ika_\ell(k)\, a_\ell^*(k),$$

and equivalently

$$\frac{1}{a_\ell^*(k)} - \frac{1}{a_\ell(k)} = 2ik$$

or

$$R^*(k) - R(k) = 0$$

which shows that $R(k)$ has no right-hand cut.

The threshold behaviour of $a_\ell(k^2)$ shows that $R(k)$ tends to infinity like $k^{-2\ell}$ as k tends to zero. Collecting all this information, we see that $k^{2\ell} R(k^2) = k^{2\ell+1} \operatorname{cotg} \delta_\ell(k^2)$ is an analytic function of k^2 (and not only of k), and that in the neighbourhood of the origin it can be expanded into the Taylor series

$$k^{2\ell+1} \operatorname{cotg} \delta_\ell(k^2) = -\frac{1}{a} + \frac{1}{2} r k^2 + \cdots \tag{5.20}$$

This is known as the effective range formula; a is called the scattering length (for an S wave it has the dimension of length), and r the effective range.

The Born approximation (3.5) suggest that for small k^2 a negative (attractive) potential leads to a positive phase shift, i.e. to a positive value of the cotangent, and a negative scattering length (same sign as the potential), and vice versa. A large value of a signals a large value of δ_ℓ, and therefore a marked interaction at low energy.

The formula (5.20) is always well obeyed at low energy.

(e) *Low energy resonance*

When a resonance occurs at low energy (small k_0^2), the Breit–Wigner formula (5.11) may need to be modified in order to ensure the correct threshold behaviour of $a_\ell(k^2)$ for small k^2. In that case one writes

$$S_\ell(k^2) = \frac{k^2 - k_0^2 - i\dfrac{\Gamma}{2}\left(\dfrac{k}{k_0}\right)^{2\ell+1}}{k^2 - k_0^2 + i\dfrac{\Gamma}{2}\left(\dfrac{k}{k_0}\right)^{2\ell+1}} S_P(k^2) \tag{5.21}$$

which evidently leaves unchanged both the position of the pole and the residue. Neglecting the deviation of $S_P(k^2)$ from unity, one finds

$$a_\ell(k^2) \simeq \frac{\dfrac{\Gamma}{2}\left(\dfrac{k}{k_0}\right)^{2\ell+1}}{(k^2 - k_0) + i\dfrac{\Gamma}{2}\left(\dfrac{k}{k_0}\right)^{2\ell+1}} = \frac{1}{\dfrac{2}{\Gamma}\left(\dfrac{k_0}{k}\right)^{2\ell+1}(k_0{}^2 - k^2) - 1} \tag{5.22}$$

which gives

$$k^{2\ell+1} \operatorname{cotg} \delta_\ell \simeq \frac{2}{\Gamma} k_0^{2\ell+1}(k_0{}^2 - k^2) \tag{5.22}$$

whence

$$a = -\frac{\Gamma}{2k_0^{2\ell+3}}, \qquad r_0 = \frac{2k_0^{2\ell+1}}{\Gamma}.$$

(f) *Other poles*

We have interpreted as a resonance a pole on the unphysical sheet, provided it is close enough to the positive real axis. As the pole recedes, it becomes more and more difficult to detect; experimentally, the Breit–Wigner curve flattens out until it becomes invisible, and the effects of $\delta_P(k^2)$ can become dominant.

There also exist noteworthy poles of a different kind, giving rise to so-called *virtual states*. These are situated on the negative real axis on the second sheet. Their effect on the effective range formula is very similar to that of a bound state.

6. Example of a resonance: N^*_{33}

Consider low energy π^+-p scattering, which occurs in a state of pure isotopic spin $\frac{3}{2}$. For nonrelativistic protons the formulae of the last chapter yield

$$\frac{d\sigma_+}{d\Omega} = |f(\cos\theta)|^2 + |g(\cos\theta)|^2 \sin^2\theta \tag{6.1}$$

$$f(E, \cos\theta) = \sum_\ell [(\ell + 1)\, a_{\ell+} + \ell a_{\ell-}]\, P_\ell(\cos\theta)$$

$$g(E, \cos\theta) = \sum_\ell (a_{\ell+}(E) - a_{\ell-}(E))\, P_\ell'(\cos\theta) \tag{6.2}$$

with

$$a_{\ell j} = \frac{e^{2i\delta_{\ell j}} - 1}{2ik} = \frac{1}{k} e^{i\delta_{\ell j}} \sin \delta_{\ell j}\,. \tag{6.3}$$

We have seen that $a_\ell(k^2)$ vanishes like $k^{2\ell}$ as k tends to 0. Hence for small enough energies we need retain only S and P waves. Under these

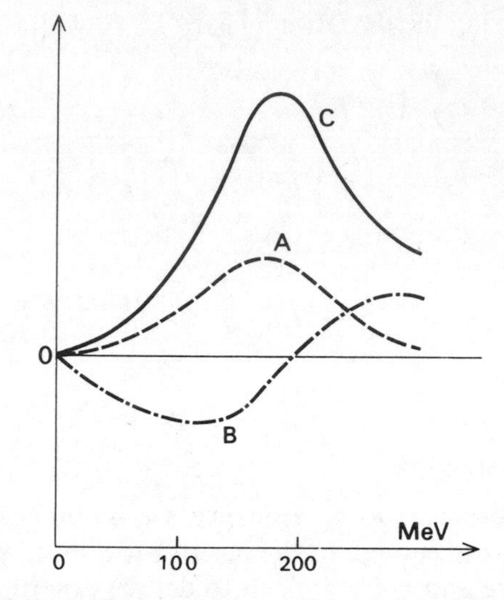

FIG. 8.6.1. Variation with energy of the coefficients A, B, C in equation (6.6)

conditions equation (6.2) becomes

$$\begin{aligned} f(E, \cos\theta) &= a_{0\frac{1}{2}} + (2a_{1\frac{3}{2}} + a_{1\frac{1}{2}})\cos\theta \\ g(E, \cos\theta) &= a_{1\frac{3}{2}} - a_{1\frac{1}{2}} \end{aligned} \tag{6.4}$$

whence, by (6.1),

$$\frac{d\sigma_+}{d\Omega} = \frac{1}{k^2}(A + B\cos\theta + C\cos^2\theta) \tag{6.5}$$

Here,

$$\begin{aligned} A &= \sin^2\delta_{0\frac{1}{2}} + \sin^2\delta_{1\frac{1}{2}} + \sin^2\delta_{1\frac{3}{2}} - 2\sin\delta_{1\frac{3}{2}}\sin\delta_{1\frac{1}{2}}\cos(\delta_{1\frac{3}{2}} - \delta_{1\frac{1}{2}}) \\ B &= 4\cos(\delta_{1\frac{3}{2}} - \delta_{0\frac{1}{2}})\sin\delta_{0\frac{1}{2}}\sin\delta_{1\frac{3}{2}} + 2\cos(\delta_{1\frac{1}{2}} - \delta_{0\frac{1}{2}})\sin\delta_{0\frac{1}{2}}\sin\delta_{1\frac{1}{2}} \\ C &= 3\sin^2\delta_{1\frac{3}{2}} + 6\sin\delta_{1\frac{3}{2}}\sin\delta_{1\frac{1}{2}}\cos(\delta_{1\frac{3}{2}} - \delta_{1\frac{1}{2}}) \end{aligned} \tag{6.6}$$

and the total cross-section is given by

$$\sigma_+ = \frac{4\pi}{k^2}\left(A + \frac{C}{3}\right) = \frac{4\pi}{k^2}(\sin^2\delta_{0\frac{1}{2}} + \sin^2\delta_{1\frac{1}{2}} + 2\sin^2\delta_{1\frac{3}{2}}). \tag{6.7}$$

The measured total cross-sections yield a typical Breit–Wigner curve, centered at 190 MeV, and very slightly distorted by its proximity to threshold, as shown in figure (1.5.2).

Moreover, at 190 MeV the cross-sections for $\pi^+p \to \pi^+p$, $\pi^-p \to \pi^-p$, $\pi^-p \to \pi^0n$ are practically in the ratios $\sigma_+ : \sigma_- : \sigma_0 \approx 9:2:1$, which are the values predicted for a state of pure isotopic spin $\frac{3}{2}$. Hence at 190 MeV there is a single dominant phase shift; it corresponds to isotopic spin $\frac{3}{2}$, and passes through a resonance.

When one of the phase shifts passes through $\pi/2$, as happens at a resonance, equation (6.7) shows that σ_+ reaches a maximum of $4\pi/k^2$ if one is dealing with a phase shift for $j = \frac{1}{2}$, and a maximum of $8\pi/k^2$ for $j = \frac{3}{2}$. But at 190 MeV, $8\pi/k^2$ is 75 mb, showing by comparison with the curve that it is the $P_{3/2}$ phase shift which resonates. This is confirmed by observing that at the resonance energy, as suggested by the formulae (6.6), C and A have maxima with $C \approx 3A$, while B has a zero (figure 8.6.1).

7. Example of a bound state: the deuteron

The deuteron is a proton–neutron bound state with isotopic spin 0 and spin 1. It is a mixture of S and D states, with S dominant (7% D state). Its binding energy is $B = 2$ MeV. Note that the mass m entering the Schroedinger equation is the reduced mass $m = m_p m_n/(m_p + m_n) = m_p/2$. Denoting by k the relative momentum of the proton and neutron in the centre of mass frame, one has $E = k^2/2m = k^2/m_p$.

The S wave amplitude $a_0(k^2)$ has a pole at $k^2 = -k_0{}^2 = -2mB$, close to the origin. In the vicinity of this pole one has

$$a_0(k^2) = \frac{g^2}{k^2 + k_0{}^2}\,.$$

In the vicinity of $k^2 = 0$ one has $a_0(k^2) = g^2/k_0{}^2 \approx 1/k \operatorname{cotg} \delta$, whence the $p - n$ triplet S-wave scattering length is given by

$$a = \frac{g^2}{2mB}\,. \tag{7.1}$$

Example of a virtual state:

The triplet proton–proton S wave amplitude has a real pole on the unphysical sheet very close to the origin ($B \sim 1$ MeV). The above argument is applicable if we change k into $-k$, conformably with the change from one sheet to the other; this changes the sign of the scattering length, whence

$$a = -\frac{g^2}{2mB}\,. \tag{7.2}$$

a is very large, corresponding to a large triplet S-wave phase shift at low energy (note the sign).

The fact that bound states and resonances are both associated with poles of the partial wave amplitudes suggests that there is a deepgoing analogy between them. Indeed, one model of narrow resonances, due to Niels Bohr, treats nuclear resonances as metastable bound states.

Width and mean life

The Breit–Wigner formula as written in (5.16) could be interpreted by saying that a resonances is a 'wavepacket' with energy roughly between ${k_0}^2 - \Gamma/2$ and ${k_0}^2 + \Gamma/2$; or more precisely, with an energy distribution whose probability amplitude is proportional to

$$\frac{\dfrac{\Gamma}{2}}{({k_0}^2 - {k_0}^2) + i\dfrac{\Gamma}{2}}\, dk^2 \tag{7.3}$$

The variation of such a packet with time is given by

$$\int_{-\infty}^{+\infty} \frac{dE\, e^{-iEt}}{(E - E_0) + i\dfrac{\Gamma}{2}} \sim e^{-iE_0 - \Gamma t/2}. \tag{7.4}$$

In other words the survival probability of the packet decreases with time like $e^{-\Gamma t}$. Therefore a resonance is basically an *unstable particle* whose mean life is connected with the width by $\tau = 1/\Gamma$.

Conversely, an unstable particle can be thought of as a resonance. Thus the Λ can be considered as a pion-nucleon resonance of width $1/2.51 \times 10^{-10}$ sec $= 2.6 \times 10^{-6}$ eV. By adopting this point of view one commits oneself to treating resonances and particles on an equal footing.

8. Multichannel resonance

In many problems one must use several wavefunctions simultaneously, because inelastic reactions are important. Thus a π-Λ collision can result in elastic scattering where the final state is also $\Lambda\pi$; or in the production of a $\Sigma\pi$ system or even in a KN system if the energy is high enough.

The Schroedinger equation can describe all these processes if we label the different *channels* (in our case, the channels are $\pi^-\Lambda$, $\pi^-\Sigma^0$, $\pi^0\Sigma^-$, K^-n) by an index i ranging from 1 to n (here, $n = 4$), and if we introduce n wavefunctions $\psi_i(\mathbf{x}_i)$ obeying a system of coupled Schroedinger equations

$$\frac{1}{2m_i} \nabla_i^2 \psi_i(\mathbf{x}) + E\psi_i(\mathbf{x}) - \sum_j V_{ij}(x)\, \psi_j(\mathbf{x}) = 0 \tag{8.1}$$

Note that the Hermitecity of the Hamiltonian restricts us to Hermitean potentials, so that

$$V_{ij}(\mathbf{x}) = V_{ji}^*(\mathbf{x}). \tag{8.2}$$

The analysis of such a system of equations is closely analogous to that of the ordinary Schroedinger equation. Neglecting spin for clarity, the wavefunction for an initial state with two particles of relative momentum k_i in channel i has the asymptotic form

$$\psi_j(\mathbf{x}) \simeq \delta_{ij}\, e^{i\mathbf{k}_i\mathbf{x}} + F_{ij}(\theta)\frac{e^{ik_j r}}{r}. \tag{8.3}$$

Then the differential cross-section for the reaction $i \to j$ (e.g. $\pi^-\Lambda \to \Sigma^0\pi^-$) is given by

$$\frac{d\sigma_{ij}}{d\Omega} = |F_{ij}(\theta)|^2. \tag{8.4}$$

This is analysed into partial waves by the expansion

$$\psi_i(\mathbf{x}) = \sum_\ell \frac{y_{\ell i}(r)}{r} P_\ell(\cos\theta) \tag{8.5}$$

whence one obtains the system of Schroedinger equations for a given angular momentum

$$\frac{d^2 y_{\ell j}(r)}{dr^2} + \left[E - \frac{\ell(\ell+1)}{r^2}\right] \times y_{\ell j}(r) - \sum_k V_{jk}(r)\, y_{\ell k}(r) = 0. \tag{8.6}$$

It follows that for large r the asymptotic form of the wavefunction is

$$y_{\ell j}(r) \simeq D_j^{(\ell)}\, e^{ik_j r} - C_j^{(\ell)}\, e^{-ik_j r}. \tag{8.7}$$

In this formula k_j denotes the momentum of the particles in channel j, defined by $E = k_j^2/2m_j$. The incoming and outgoing waves have been separated in the formula (8.7) for the following reason: since the initial state is a plane wave in channel i, an incoming spherical wave exists only in this channel, whence

$$C_j^{(\ell)} = C^{(\ell)}\delta_{ij}. \tag{8.8}$$

On the other hand, the total flux of incoming particles is given by $v_i\,|C|^2$, v_i being the relative velocity of the two initial particles; the total flux of outgoing particles in all channels is, according to (8.7), given by

$$\sum_j v_j\,|\,D_j^{(\ell)}\,|^2. \tag{8.9}$$

Since every incoming particle gives rise to an outgoing particle in some channel, and since we are using a time-independent formalism, the total number of incoming particles per second must equal the total number of outgoing ones; hence

$$\sum_j v_j\,|\,D_j^{(\ell)}\,|^2 = v_i\,|\,C\,|^2. \tag{8.10}$$

This conservation of flux is made explicit by writing

$$D_j^{(\ell)} = C^{(\ell)} \sqrt{\frac{v_i}{v_j}}\, S_{ij}^{(\ell)}. \tag{8.11}$$

Since equation (8.10) is obeyed for any entrance channel i, it shows that the matrix S is unitary:

$$\sum_j S_{ij}^{(\ell)} S_{jk}^{(\ell)\dagger} = \delta_{jk}. \tag{8.12}$$

The amplitude $F_{ij}(\theta)$ and the matrix S are connected through the expansion of $F_{ij}(\theta)$ in terms of Legendre polynomials

$$F_{ij}(\theta) = \sum_\ell (2\ell + 1)\, a_{ij}^\ell(E)\, P_\ell(\cos\theta). \tag{8.13}$$

Equating the coefficients of $(e^{\pm ikr}/r)\, P_\ell(\cos\theta)$ in the asymptotic expressions (8.3) and (8.5), one obtains

$$\left(-\frac{1}{2i}\right)\frac{(-i)^\ell}{k_i} = \delta_{ij} C^{(\ell)}$$

$$\left(\frac{1}{2i}\right)\frac{(-i)^\ell}{k_i}\,\delta_{ij} + a_{ij}^\ell(E) = -C^{(\ell)} S_{ij}^{(\ell)} \sqrt{\frac{v_i}{v_j}}$$

whence

$$a_{ij}^\ell(E) = \frac{(-i)^\ell}{2ik_i}\sqrt{\frac{v_i}{v_j}}\,[S_{ij}^\ell(E) - \delta_{ij}]. \tag{8.14}$$

Since the matrix $S^{(\ell)}(E)$ is unitary it can be diagonalised and written as

$$S^{(\ell)} = A\Sigma^{(\ell)}A^{-1} \tag{8.15}$$

where A is an orthogonal matrix (unitary and real) and the diagonal matrix $\Sigma^{(\ell)}$ has the form

$$\Sigma^{(\ell)} = \begin{bmatrix} e^{2i\delta_1} & 0 & \cdots & 0 \\ 0 & e^{2i\delta_2} & & 0 \\ \cdots & & \ddots & \\ 0 & 0 & & e^{2i\delta_n} \end{bmatrix} \tag{8.16}$$

Consider for simplicity a case with two channels ($n = 2$). Then one has

$$A = \begin{bmatrix} \cos\alpha & -\sin\alpha \\ \sin\alpha & \cos\alpha \end{bmatrix} \qquad \Sigma^{(\ell)} = \begin{bmatrix} e^{2i\delta_1} & 0 \\ 0 & e^{2i\delta_2} \end{bmatrix} = \begin{bmatrix} \Sigma_1(E) & 0 \\ 0 & \Sigma_2(E) \end{bmatrix}. \tag{8.17}$$

The mixing angle α and the eigenphase shifts δ_1 and δ_2 are functions of the energy.

In the light of (8.14) the elastic scattering amplitude for the reaction $1 \to 1$ is given by

$$a_{11}^{(\ell)}(E) = \frac{(-i)^\ell}{2ik_1}\,[\cos^2\alpha\, e^{2i\delta_1} + \sin^2\alpha\, e^{2i\delta_2} - 1]. \tag{8.18}$$

Thus we recover the fact that the collision matrix element $\eta e^{2i\delta}$ given by

$$\eta\, e^{2i\delta} = \cos^2\alpha\, e^{2i\delta_1} + \sin^2\alpha\, e^{2i\delta_2} \tag{8.19}$$

has a coefficient of inelasticity $\eta < 1$.

A resonance corresponds to the case where one of the eigenphase shifts, say δ_1, satisfies a Breit–Wigner formula. If we neglect δ_2 as a first approximation, the matrix assumes the form

$$\begin{bmatrix} \cos^2\alpha\Sigma_1 & \cos\alpha\sin\alpha\Sigma_1 \\ \cos\alpha\sin\alpha\Sigma_1 & \sin^2\alpha\Sigma_1 \end{bmatrix}$$

where

$$\Sigma_1 = \frac{E - E_0 - i\dfrac{\Gamma}{2}}{E - E_0 + i\dfrac{\Gamma}{2}} \tag{8.20}$$

Thus, if we ignore quantities varying slowly near E_0, the resonant part of the matrix $a_{ij}^{(l)}$ becomes

$$a^{(\ell)} = (-i)^\ell \begin{bmatrix} \cos^2\alpha \dfrac{\dfrac{\Gamma}{2}}{E - E_0 + i\dfrac{\Gamma}{2}}\dfrac{1}{k_1} & \cos\alpha\sin\alpha \dfrac{\dfrac{\Gamma}{2}}{E - E_0 + i\dfrac{\Gamma}{2}}\dfrac{1}{k_1}\sqrt{\dfrac{v_1}{v_2}} \\ \cos\alpha\sin\alpha \dfrac{\dfrac{\Gamma}{2}}{E - E_0 + i\dfrac{\Gamma}{2}}\dfrac{1}{k_2}\sqrt{\dfrac{v_2}{v_1}} & \sin^2\alpha \dfrac{\dfrac{\Gamma}{2}}{E - E_0 + i\dfrac{\Gamma}{2}}\dfrac{1}{k_2} \end{bmatrix} \tag{8.21}$$

Defining

$$\cos\alpha\sqrt{\Gamma} = \sqrt{\Gamma_1} \qquad \sin\alpha\sqrt{\Gamma} = \sqrt{\Gamma_2}$$

we have therefore

$$\boxed{a_{ij}^{(\ell)} = (-i)^\ell \frac{1}{k_i}\sqrt{\frac{v_i}{v_j}}\,\frac{\sqrt{\Gamma_i\Gamma_j}/2}{E - E_0 + i\dfrac{\Gamma}{2}}} \tag{8.22}$$

where

$$\Gamma_1 + \Gamma_2 = \Gamma.$$

The formula (8.22) is called the multichannel Breit–Wigner formula. It shows that the same resonance appears in the matrix elements for all the reactions which are possible at the resonance energy, with an intensity depending on the partial widths Γ_1, Γ_2, ..., Γ_n; these are related to the width Γ of the resonance curve by

$$\Gamma_1 + \Gamma_2 + \cdots + \Gamma_n = \Gamma \tag{8.23}$$

Equation (8.23) is often interpreted by identifying Γ with the total decay rate, i.e. with the reciprocal of the mean life, $\Gamma = \tau^{-1}$, and Γ_1, Γ_2, ... with the partial decay rates.

In this way one can regard a resonance of the K^--nucleon system as a very short lived unstable particle, decaying by the modes K^-N, $\Lambda\pi$, $\Sigma\pi$, $\Lambda\pi\pi$, $\Sigma\pi\pi$, etc. with relative probabilities Γ_1/Γ, Γ_2/Γ, ... as indicated by the formula (8.22).

CHAPTER 9

THE RESONANCES

In this chapter we show how resonances are discovered experimentally, and how one determines their quantum numbers (spin, parity, isotopic spin).

1. Direct observation

We saw in the last chapter how the N_{33}^* resonance shows up in the π^+p system, and how its spin and isotopic spin are determined. This direct method depends on a study of the scattering process, and is applicable to resonances formed by nucleons and beam particles (i.e. in practice, π, K, and nucleons).

(a) *Pion-nucleon resonances*

Their main properties are given in table 1. Most were discovered by observing a peak in the total or in the total elastic scattering cross-section.

Seeing that the pion-nucleon system has only two isospin states, $\frac{1}{2}$ and $\frac{3}{2}$, and that the π^+-proton system has pure isospin $\frac{3}{2}$, it is easy to determine the isotopic spin of a resonance: if the peak occurs in the π^+p total cross-section, the isotopic spin is $\frac{3}{2}$; otherwise it is $\frac{1}{2}$. This can be confirmed by the ratio of the peaks observed in π^+p and π^-p, which is, respectively, 0:1 or 3:1.

It is a more delicate matter to determine the spin, and one exploits several methods simultaneously.

(i) The height of the peak in the total cross-section is given theoretically by equation (7.23) of chapter 7:

$$\frac{(2j+1)\pi}{k^2}(1+|\eta_{\ell j}|) \tag{1.1}$$

If the inelasticity coefficient $|\eta_{\ell j}|$ is close to 1, then j can readily be found from this formula. But it is not always easy to determine the coefficient $|\eta_{\ell j}|$, and the height of the peak itself is poorly known, because it is difficult

TABLE 1

Pion–nucleon resonances

Symbol	Isotopic spin I	Spin and parity J^P	Mass (MeV)	Width (MeV)
p	1/2	$1/2^+$	938,3	
n	1/2	$1/2^+$	939,6	
N (1470)	1/2	$1/2^+$	1460	260
N (1518)	1/2	$3/2^-$	1515	115
N (1550)	1/2	$1/2^-$	1525	80
N (1680)	1/2	$5/2^-$	1675	145
N (1688)	1/2	$5/2^+$	1690	125
N (1710)	1/2	$1/2^-$	1715	280
N (1750)	1/2	$1/2^+$	1785	405
N (2190)	1/2	$7/2^-$	2190	300
N (2650)	1/2	?	2650	360
N (3030)	1/2	?	3030	400
Δ (1236)	3/2	$3/2^+$	1236,0	120
Δ (1640)	3/2	$1/2^-$	1630	160
Δ (1690)	3/2	$3/2^-$	1670	225
Δ (1910)	3/2	$5/2^+$	1880	250
Δ (1930)	3/2	$1/2^+$	1905	300
Δ (1950)	3/2	$7/2^+$	1940	210
Δ (2420)	3/2	$11/2^+$	2420	310
Δ (2850)	3/2	?	2850	400
Δ (3230)	3/2	?	3230	440

The designation of a particle derives from the first measurement of its mass. This explains how, with increasing accuracy, the entries in the fourth column have come to differ from those in the first. The symbol N identifies particles or resonances with isospin $\frac{1}{2}$; and the symbol Δ identifies resonances with isospin $\frac{3}{2}$.

to separate the peak, represented by the Breit–Wigner formula, from the background.

(ii) One can investigate the differential cross-section. In practice, the most efficient method is a systematic determination of the energy dependence of the coefficients $A_n(E)$ in the formula

$$\frac{d\sigma}{d\Omega} = \sum_{n=0}^{N} A_n(E)\, P_n(\cos\theta) \tag{1.2}$$

Using the formulae of Chapter 7, §7, it is easy to connect the $A_n(E)$ with the phase shifts. One can then display the contributions of the resonant phase shifts to the variation with energy of the coefficients $A_n(E)$.

Resonance parities can be determined by the same method. Actually in the pion-nucleon system the resonances are so closely spaced that the resonant phase shifts interfere over certain energy regions between two resonances. Consider two such resonances α and β with parities ϵ_α and ϵ_β respectively. If ϵ_α is positive (negative), then the orbital angular momentum of α is odd (even), remembering that the pion's intrinsic parity is negative. If the parities of α and β are the same, then the parities of their orbital angular momenta are also the same; accordingly, their contributions either both change sign, or they both remain unchanged, when one replaces θ by $\theta + \pi$ in the scattering amplitudes $f(\theta)$ and $g(\theta)$. Taking the squared moduli of $f(\theta)$ and $g(\theta)$ to obtain $d\sigma/d\Omega$, it follows that the interference terms between two resonances α and β of equal parity are unchanged when θ is changed to $\pi + \theta$; hence they appear only in the coefficients $A_n(E)$ for n even. Similarly, if α and β have opposite parities, they interfere only in the coefficients with n odd. Knowing the parity of the $(\frac{3}{2}, \frac{3}{2})$ resonance, those of the higher resonances can be determined successively.

(b) *Nucleon-nucleon resonances*

These are the easiest to observe and to investigate: there are none.

(c) *Antinucleon-nucleon resonances*

To date, none have been observed convincingly.

(d) *K-nucleon resonances*

Since the K and the nucleon both have isotopic spin $\frac{1}{2}$, the total isospin of a K-nucleon system can be 0 or 1.

The third component of the isospin of a K^+p system is 1, whence $I = 1$. There is no prominent structure, such as would indicate a resonance, in the K^+p total cross-section.

The system with isotopic spin 0 can be realised in the scattering of K^+

TABLE 2—TABLE OF PARTICLES AND RESONANCES WITH BARYON NUMBER 1 AND STRANGENESS −1.

Symbol	Isotopic spin	Spin and parity	Mass (MeV)	Width (MeV)	Decay	
					Mode	Branching ratio %
Λ	0	$1/2^+$	1115,6			
Λ (1405)	0	$1/2^-$	1405	40	$\Sigma\pi$	100
Λ (1520)	0	$3/2^-$	1518,8	16	$N\bar{K}$	45 ± 4
					$\Sigma\pi$	45 ± 4
					$\Lambda\pi\pi$	10 ± 1
Λ (1670)	0	$1/2^-$	1670	25	$N\bar{K}$	14
					$\Lambda\eta$	33
					$\Sigma\pi$	45
Λ (1700)	0	$3/2^-$	1690	40	$N\bar{K}$	25
					$\Sigma\Pi$	35
					$\Lambda\pi\pi$	≈ 20
					$\Sigma\pi\pi$	≈ 20
Λ (1815)	0	$5/2^+$	1815	75	$N\bar{K}$	65
					$\Sigma\pi$	11
					$\Sigma(1385)\pi$	9
					$\Lambda\eta$	1
Λ (1830)	0	$5/2^-$	1830	80	$N\bar{K}$	10
					$\Sigma\pi$	35
Λ (2100)	0	$7/2^-$	2100	140	$N\bar{K}$	30
					$\Sigma\pi$	4
					$\Lambda\eta$	<3
					ΞK	~ 1
					$\Lambda\omega$	<10
Λ (2350)	0	?	2350	210	$N\bar{K}$	
Σ	1	$1/2^+$	Σ^+ 1189,4			
			Σ^0 1192,5			
			Σ^- 1197,3			
Σ (1385)	1	$3/2^+$	1382	36	$\Lambda\pi$	90 ± 3
					$\Sigma\pi$	10 ± 3
Σ (1610)	1	?	1615	65	$N\bar{K}$	small
					$\Lambda\pi$	dominant
					$\Sigma(1385)\pi$	observed
					– $\Lambda(1405)\pi$	large
Σ (1660)	1	$3/2^-$	1660	50	– $N\bar{K}$	small
Σ (1700)	1	?	1700	110	– $\Lambda\pi$	large
					– $\Sigma\pi$	poorly known
Σ (1765)	1	$5/2^-$	1765	100	$N\bar{K}$	46
					$\Lambda\pi$	16
					$\Lambda(1520)\pi$	15
					$\Sigma(1385)\pi$	15
					$\Sigma\eta$	~ 1
					$\Sigma\pi$	~ 1
Σ (1915)	1	$5/2^+$	1905	60	$N\bar{K}$	10
					$\Lambda\pi$	5
Σ (2030)	1	$7/2^+$	2030	120	$N\bar{K}$	10
					$\Lambda\pi$	35
					$\Sigma\pi$	10
					ΞK	<2
Σ (2250)	1	?	2250	200		
Σ (2455)	1	?	2455	120		
Σ (2595)	1	?	2595	~ 140		

by neutrons, i.e. in practice in K^+-deuteron scattering. Once again there is an absence of structure, i.e. of resonances.

(e) $\bar{K}$-nucleon resonances

Here too the isotopic spin can be 0 or 1. But the K^--proton system is a mixture of isospins 0 and 1, as shown by the formula

$$|K^-p\rangle = \frac{1}{\sqrt{2}}[|I = 0, I_3 = 0\rangle + |I = 1, I_3 = 0\rangle] \tag{1.3}$$

The scattering amplitudes for the two isospin states can be found from a study of K^-n scattering. In practice this involves directing a K^- beam onto a deuterium target, and attempting to interpret the observations in terms of K^-p and K^-n interactions.

One important feature of K^--nucleon interactions is that the inelasticity is always large. The reason is that even at very low K^- energies the following reactions can always occur:

$$\bar{K}^- + N \to \Lambda + \pi \tag{1.4}$$

$$\bar{K} + N \to \Lambda + \pi + \pi \tag{1.5}$$

$$\bar{K} + N \to \Sigma + \pi \tag{1.6}$$

$$\bar{K} + N \to \Sigma + \pi + \pi \tag{1.7}$$

Hence there exist resonances that are common to the systems $\bar{K}N$, $\Lambda\pi$, $\Sigma\pi$, etc.

The reactions (1.4) and (1.6) are excellent discriminators for the isospins of the resonances. Thus a zero isospin K^-p resonance cannot show up in reaction (1.4).

Table 2 lists the K^--nucleon resonances, as well as $\Lambda\pi$ and $\Sigma\pi$ resonances below the K^--nucleon threshold, which correspond to the same quantum numbers. The latter have been found by a method outlined in the next paragraph.

2. Observation of resonances in inelastic processes

Many resonances cannot be observed directly in elastic scattering because their constituent particles are unstable, examples being $\pi\Lambda$ and $\pi\Sigma$ resonances with masses below the K-nucleon threshold, $\Xi\pi$, $\pi\pi$, $K\pi$ resonances etc. These must be detected by different techniques. To appreciate the value of such techniques, it seems best to apply them to observe a resonance that we know already, namely the N^*_{33}.

In our study of resonances we have seen already that there is certainly no basic difference between resonances and particles, apart from the very short mean lives of the former. This might suggest that a resonance be observed directly through its production in a reaction like

$$\pi^+ + p \to N^{*++} + \pi^0 \tag{2.1}$$

If the N^{*++} were a genuine particle, its mass would be sharply defined. But the fact that its actual mean life is so short spreads its mass over an interval of width Γ. It implies also that the N^* produced in reaction (1) decays before it has had time to travel an appreciable distance, so that what one actually observes is the reaction

$$\pi^+ + p \to \pi^+ + p + \pi^0 \tag{2.2}$$

The process (2.1) must be disentangled from (2.2), since the latter can proceed even when the π^+ and the proton are not produced in a resonant state.

If the N^* mass were very sharply defined then in the centre of mass frame of reaction (2.1) the π^0 would be produced with a well-defined momentum. The spread of the N^* mass entails a spread in the π^0 momentum. Similarly, if the N^* mass were well defined, then the π^+ resulting from its decay would have a well-defined momentum in the N^* rest frame. The N^* can be emitted in various directions; the Lorentz transformation between its rest frame and the overall centre of mass frame depends on its direction, and produces a spread in the π^+ momentum. This spread is further increased by the uncertainty in the N^* mass.

Figure (9.2.1) gives the momentum spectrum of the π^0 in the centre of mass frame, for an incident π^+ energy close to 1 GeV. It shows a marked peak centred on the momentum that a π^0 would have if it were emitted by an N^* with a mass equal to the mean mass. The broader hump is explained as follows. Besides reactions (2.1) and (2.2) one can also have the process

$$\pi^+ + p \to N^{*+} + \pi^+ \tag{2.3}$$

$$N^{*+} \to p + \pi^0 \tag{2.4}$$

The π^0 produced by the decaying N^{*+} is completely analogous to the π^+ in reaction (2.2) and its momentum distribution must be wider than it would be if it were produced by reaction (2.1). It is these π^0's which give rise to the wide low hump in the spectrum.

Instead of observing the momentum spectrum of the π^0, one could just as well consider the mass spectrum of the $p\pi^+$ system. Since there is a one-to-one correspondence between the momentum of the π^0 and the mass of the $p\pi^+$ system, the result would be a shape very similar to Fig. 9.2.1. One finds that the mass spectrum has a peak centred around 1237 MeV, with a width of about 100 MeV. On the other hand, if one considers the total cross-section for π^+-proton scattering as a function of

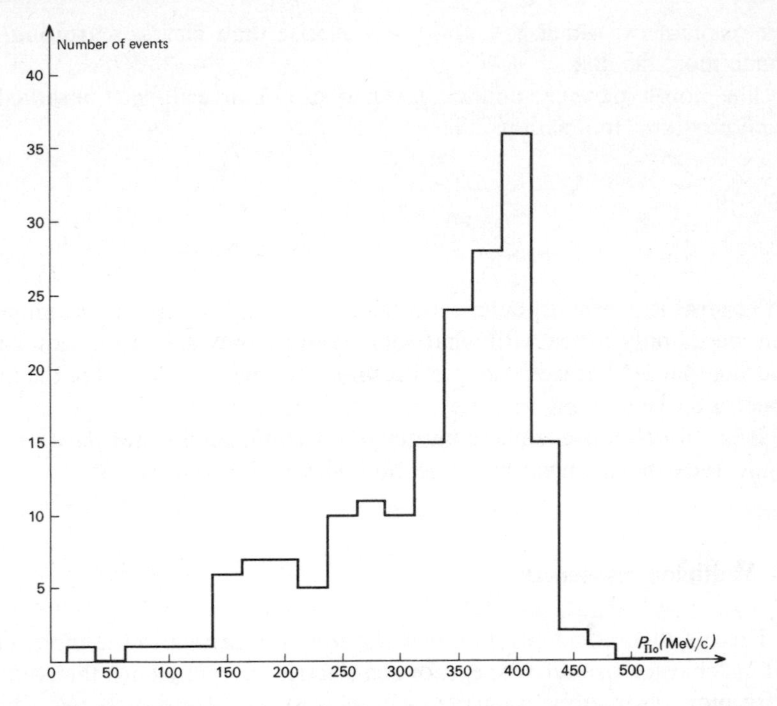

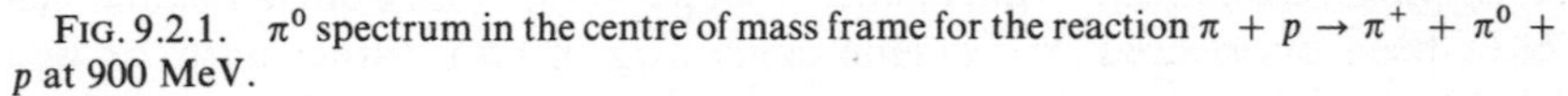
FIG. 9.2.1. π^0 spectrum in the centre of mass frame for the reaction $\pi + p \to \pi^+ + \pi^0 + p$ at 900 MeV.

the total mass of the system, then one finds a resonance peak which is also centred around this value of 1237 MeV, and has a width 94 $\pm$ 16 MeV. As a consequence, one can equally well find evidence for the N^* from its mass spectrum in an inelastic process, as from the same spectrum observed in elastic scattering.

In practice the N^* is observed in many reactions with a pion and a nucleon in the final state. Some examples are

$$\pi + N \to N + \pi + \pi + \pi$$
$$p + p \to p + p + \pi^0$$
$$p + p \to p + n + \pi^+$$
$$K^+ + p \to p + \pi^+ + K^0$$

In all these reactions the $N\pi$ mass spectrum has a peak similar to that in (2.1). The peak varies slightly both in position (by 20 to 30 MeV between extremes), and in width; this is generally explained as due to a variation with mass of the direct reaction analogous to (2.2).

To sum up, we see that inelastic processes provide a method of looking for resonances, which is slightly less precise than elastic scattering but much more flexible.

The pion-hyperon resonances given in table 2 have all been observed by such methods, in reactions like

$$\bar{K} + N \rightarrow Y^* + \pi \qquad Y^* \rightarrow \begin{cases} \Lambda \\ \Sigma \end{cases} + \pi \tag{2.5}$$

$$\bar{K} + N \rightarrow \Xi^* + K \qquad \Xi^* \rightarrow \Xi + \pi \tag{2.6}$$

In general it is easy to determine the isotopic spins of such resonances: one needs only to see with what total changes they are produced. Thus, the fact that N^* has isotopic spin $\frac{3}{2}$ results in its being observed in the mass spectra of the systems $p\pi^+$, $p\pi^0$, $n\pi^+$, $n\pi^0$, $p\pi^-$, $n\pi^-$.

It is a much more delicate matter to determine spins and parities. We shall study this problem in connection with multipion resonances.

3. Multipion resonances

From a theoretical point of view the simplest particle is the pion, since all its 'charges' are zero except for the electric charge. Hence the study of multipion resonances is particularly interesting. Many such resonances have been discovered, in systems consisting of two, three, four, or five pions. For instance, in the reaction

$$\pi^+ + d \rightarrow p + p + \pi^- + \pi^+ + \pi^0$$

the mass spectrum of the three pions shows two very prominent narrow peaks, one centred on 790 MeV and the other on 550 MeV.

Table 3 lists the multipion resonances known at present. We shall investigate in greater detail how to determine the quantum numbers of some of them. These quantum numbers are, isotopic spin, spin, parity, and G-parity.

4. The ω meson

In reactions like

$$\pi^+ + d \rightarrow p + p + \pi^+ + \pi^- + \pi^0 \tag{4.1}$$

$$\bar{p} + p \rightarrow \pi^+ + \pi^- + \pi^+ + \pi^- + \pi^0 \tag{4.2}$$

one observes a very prominent and very narrow peak in the $\pi^+\pi^-\pi^0$ mass spectrum, centred at 790 MeV, and with a width of 9 MeV when measured under the most favourable conditions (Figure 9.3.1). The fact that no such peak is seen in systems of nonzero charge (e.g. $\pi^+\pi^+\pi^-$) shows that one is dealing with a resonance, i.e. with a meson having a very short mean life, of isotopic spin 0. Its G-parity is -1, since it decays into three pions but not, to any noteworthy extent, into two or four. It is called the ω meson.

Since its isospin is 0, its wavefunction in isospace is given by the triple scalar product

$$\epsilon_{ijk} \mid \pi_i \pi_j \pi_k\rangle \tag{4.3}$$

where $i, j = 1, 2, 3$. This is antisymmetric; since the pions obey Bose statistics, the spatial wavefunction, and consequently the decay amplitude

$$\langle \pi^+(\mathbf{p}_+)\, \pi(\mathbf{p}_-)\, \pi^0(\mathbf{p}_0)|\, H \mid \omega\rangle = A(\mathbf{p}_+\,, \mathbf{p}_-\,, \mathbf{p}_0) \tag{4.4}$$

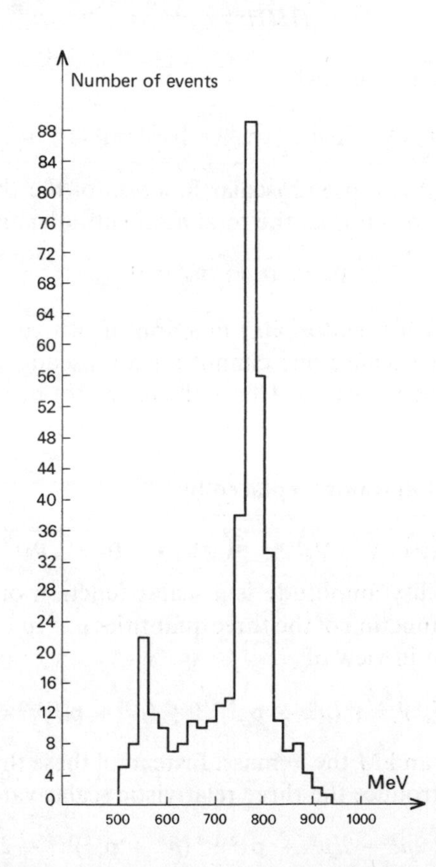

FIG. 9.3.1. Mass spectrum of the three-pion system in the reaction $\pi^+ d \to p + p + \pi^+ + \pi^- + \pi^0$. The high peak on the right represents the effects of the ω, the one on the left those of the η.

must be completely antisymmetric in the momenta p_+, p_-, p_0 of the three pions in the ω rest frame.

Next, we investigate the form assumed by the decay amplitude A for various values of the ω spin and parity.

(a) *Scalar* ω

Since the intrinsic parity of each pion is negative, one has for the parity Π,

$$\Pi \mid \omega\rangle = +\mid \omega\rangle \tag{4.5}$$

$$\Pi \mid \mathbf{p}_+, \mathbf{p}_-, \mathbf{p}_0\rangle = -\mid -\mathbf{p}_+, -\mathbf{p}_-, -\mathbf{p}_0\rangle \tag{4.6}$$

The decay of the ω is very fast and must therefore be due to the strong interactions, so that H conserves parity:

$$\Pi H \Pi^{-1} = H \tag{4.7}$$

This leads straightforwardly to

$$A(\mathbf{p}_+, \mathbf{p}_-, \mathbf{p}_0) = -A(-\mathbf{p}_+, -\mathbf{p}_-, -\mathbf{p}_0) \tag{4.8}$$

Hence $A(\mathbf{p}_+, \mathbf{p}_-, \mathbf{p}_0)$ is a pseudoscalar function of the three vectors $\mathbf{p}_+$, $\mathbf{p}_-$, $\mathbf{p}_0$. But in the ω rest frame, the total momentum vanishes so that one has

$$\mathbf{p}_+ + \mathbf{p}_- + \mathbf{p}_0 = 0 \tag{4.9}$$

Hence A is actually a pseudoscalar function of *two* vectors $\mathbf{p}_+$ and $\mathbf{p}_-$, which is impossible because one cannot form a pseudoscalar with fewer than three independent vectors. Hence the ω cannot be a scalar particle.

(b) *Pseudoscalar* ω

The expression (4.8) is now replaced by

$$A(\mathbf{p}_+, \mathbf{p}_-, \mathbf{p}_0) = A(-\mathbf{p}_+, -\mathbf{p}_-, -\mathbf{p}_0) \tag{4.10}$$

so that the probability amplitude is a scalar function of $\mathbf{p}_+$, $\mathbf{p}_-$, $\mathbf{p}_0$. By virtue of (4.9) it is a function of the three quantities $\mathbf{p}_+{}^2$, $\mathbf{p}_-{}^2$, and $\mathbf{p}_0{}^2$ which are not independent in view of

$$(\mu^2 + \mathbf{p}_+{}^2)^{1/2} + (\mu^2 + \mathbf{p}_-{}^2)^{1/2} + (\mu^2 + \mathbf{p}_0{}^2)^{1/2} = M \tag{4.11}$$

where μ is the pion and M the ω mass. Instead of these three variables we can equally well introduce the three relativistic scalar variables

$$\begin{aligned} s &= (p_+ + p_-)^2 = 2\mu^2 + 2(\mu^2 + \mathbf{p}_+{}^2)^{1/2}(\mu^2 + \mathbf{p}_-{}^2)^{1/2} - 2\mathbf{p}_+ \cdot \mathbf{p}_- \\ t &= (p_+ + p_0)^2 \\ u &= (p_- + p_0)^2 \end{aligned} \tag{4.12}$$

Note that a permutation of the three mesons is equivalent to a permutation of the three variables s, t, u. It is easy to check that the interdependence of $\mathbf{p}_+^2, \mathbf{p}_-^2$, and $\mathbf{p}_0^2$ leads to

$$s + t + u = M^2 + 3\mu^2 \tag{4.13}$$

The function A, being a scalar antisymmetric in the three mesons, must have the form

$$A(s, t, u) = \alpha(s, t, u)(s - u)(t - s)(u - t) \tag{4.14}$$

where $\alpha(s, t, u)$ is a symmetric function of s, t, u.

(c) *Vector* ω

In this case the state vector of the ω transforms under rotations like a vector, and has the form $|\omega j\rangle, j = 1, 2, 3$; hence the probability amplitude must also transform lide a vector under rotations:

$$A_j(\mathbf{p}_+, \mathbf{p}_-, \mathbf{p}_0) = \langle \mathbf{p}_+, \mathbf{p}_-, \mathbf{p}_0 \mid H \mid \omega j\rangle \qquad j = 1, 2, 3 \tag{4.15}$$

The parity of the ω now being $-$, the same argument as before shows that $A_j(\mathbf{p}_+, \mathbf{p}_-, \mathbf{p}_0)$ is invariant under reflection of the coordinate axes and is therefore an axial vector. Thus one has

$$A_j(\mathbf{p}_+, \mathbf{p}_-, \mathbf{p}_0) = \alpha(s, t, u)(\mathbf{p}_+ \wedge \mathbf{p}_-) \tag{4.16}$$

The linear dependence (4.9) entails that the vector product is totally antisymmetric in $\mathbf{p}_+, \mathbf{p}_-, \mathbf{p}_0$, since

$$\mathbf{p}_+ \wedge \mathbf{p}_- = -\mathbf{p}_+ \wedge \mathbf{p}_0 = \mathbf{p}_- \wedge \mathbf{p}_0 \tag{4.17}$$

Thus the function $\alpha(s, t, u)$ must be totally symmetric in s, t, u.

(d) *Axial vector* ω

By contrast to the previous case, $A_j(\mathbf{p}_+, \mathbf{p}_-, \mathbf{p}_0)$ must now be a (polar) vector, and has the form

$$A_j(\mathbf{p}_+, \mathbf{p}_-, \mathbf{p}_0) = \alpha(s, t, u)[p_j^+(u - t) + p_j^-(s - u) + p_j^0(t - s)] \tag{4.18}$$

where $\alpha(s, t, u)$ is symmetric.

We shall not consider the case of higher spins ($J = 2, 3, \ldots$) which in any case are better treated by a different technique.

Decay probability: the three amplitudes (4.14), (4.16), and (4.18) evidently lead to different decay probabilities for the ω. For future reference, we calculate this probability for the general case where a resonance (or unstable particle) decays into three particles 1, 2, 3 with masses m_1, m_2, m_3, energies E_1, E_2, E_3, and momenta $\mathbf{p}_1, \mathbf{p}_2, \mathbf{p}_3$. We shall use invariant normalisation for the states, so that $d^3\mathbf{p}_1/(2\pi)^3E_1$ is the number of states with momenta in the range $d^3\mathbf{p}_1$. If the resonance has spin J, the decay amplitude has $(2J + 1)$ components and can be written in the form

$$A_m(\mathbf{p}_1, \mathbf{p}_2, \mathbf{p}_3) \qquad m = -j, -j+1, \ldots, j-1, j \tag{4.19}$$

If the resonance is not polarised, the decay probability is proportional to

$$\sum_m \int |A_m(\mathbf{p}_1, \mathbf{p}_2, \mathbf{p}_3)|^2 \frac{d^3\mathbf{p}_1}{E_1}\frac{d^3\mathbf{p}_2}{E_2}\frac{d^3\mathbf{p}}{E_3} \times \delta(M - E_1 - E_2 - E_3)\,\delta(\mathbf{p}_1 + \mathbf{p}_2 + \mathbf{p}_3) \tag{4.20}$$

First we integrate over $\mathbf{p}_3$; defining $x = \hat{\mathbf{p}}_1 \cdot \hat{\mathbf{p}}_2$ and expressing $d^3\mathbf{p}_2$ as $p_2{}^2\, dp_2\, d\Omega_2$, we get

$$\int \sum_m |A_m(\mathbf{p}_1, \mathbf{p}_2, \mathbf{p}_3)|^2 \frac{d^3p_1 p_2{}^2\, dp_2\, d\Omega_2}{E_1E_2E_3} \times \delta(M - \sqrt{p_1{}^2 + m_1{}^2} - \sqrt{p_2{}^2 + m_2{}^2} - \sqrt{p_1{}^2 + p_2{}^2 + 2p_1p_2x + m_3{}^2})$$

Further, $d\Omega_2 = dx\, d\varphi$ where φ is an azimuthal angle. Integrating over x by use of the δ function we obtain a factor $(p_1p_2/E_3)^{-1}$, whence

$$\int \sum_m |A_m(\mathbf{p}_1 \cdot \mathbf{p}_2, \mathbf{p}_3)|^2 \frac{p_1{}^2\, dp_1\, d\Omega_1\, p_2{}^2\, dp_2\, d\varphi}{E_1E_2E_3}\frac{E_3}{p_1p_2} = \int \sum_m |A_m(\mathbf{p}_1 \cdot \mathbf{p}_2, \mathbf{p}_3)|^2\, d\Omega_1\, d\varphi\, dE_1\, dE_2 \tag{4.21}$$

In the last step, we have used $p_1\, dp_1 = E_1\, dE_1$, and $p_2\, dp_2 = E_2\, dE_2$. Since the expression $\sum |A_m(\mathbf{p}_1, \mathbf{p}_2, \mathbf{p}_3)|^2$ is invariant under rotations, the integrations over $d\Omega_1$, and $d\varphi$ are trivial and give simply a factor $4\pi \times 2\pi$. Therefore the probability is proportional to

$$\int \sum_m |A_m(\mathbf{p}_1, \mathbf{p}_2, \mathbf{p}_3)|^2\, dE_1\, dE_2 \tag{4.22}$$

Dalitz plot

The formula (4.22) is extremely simple in that it shows the phase space factor to be proportional to $dE_1\, dE_2$. It follows that if we represent the observed decays on a plot where each event is logged as a point whose coordinates are the appropriate values of E_1 and E_2, then the density of points on the plot is proportional to the matrix element squared.

The coordinates can be either E_1 and E_2, or the kinetic energies T_1 and T_2, where

$$E_1 = m_1 + T_1 \qquad E_2 = m_2 + T_2$$

as shown in figure (9.4.1). All events must fall within the curve C, whose interior corresponds to $E_1 + E_2 + E_3 = M, E_3 \geqslant m_3$.

Instead of using the coordinates T_1 and T_2, one can introduce the scalar variables s, t, u, which are actually proportional to E_1, E_2, E_3. One has

$$\begin{aligned} s &= (p_2 + p_3)^2 = (P - p_1)^2 = (M - E_1)^2 - \mathbf{p}_1^{\,2} = M^2 + m_1^{\,2} - 2m_1E_1 \\ t &= (p_1 + p_3)^2 = M^2 + m_2^{\,2} - 2m_2E_2 \\ u &= (p_1 + p_2)^2 = M^2 = m_3^{\,2} - 2m_3E_3 \end{aligned} \tag{4.23}$$

The relation

$$s + t + u = M^2 + m_1^{\,2} + m_2^{\,2} + m_3^{\,2} \tag{4.24}$$

suggests that one should use triangular coordinates, which have the advantage of being symmetric in the three particles. Here again the accessible values of (s, t, u) correspond to the interior of a certain curve (C)

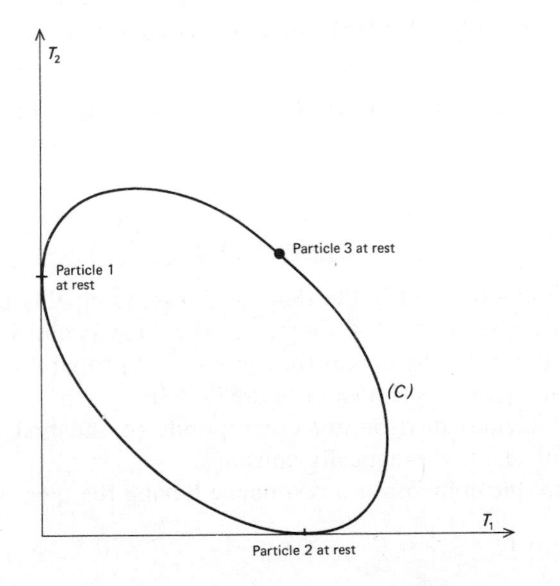

FIG. 9.4.1. The boundary of the Dalitz plot.

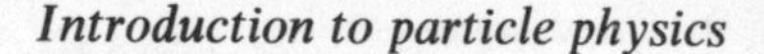

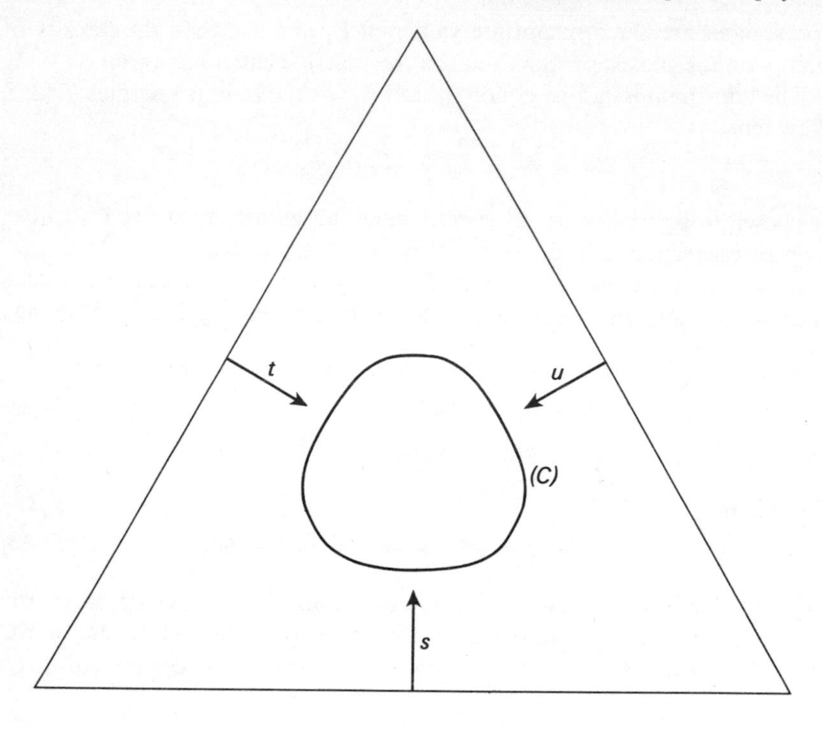

FIG. 9.4.2. The Dalitz plot in triangular coordinates (s, t, u).

(figure 9.4.2). The equation for (C) and some of its properties are investigated in chapter 13. In terms of these triangular variables the element of area is given by

$$\frac{2}{\sqrt{3}}\,ds\,dt = \frac{1}{2m_1m_2\sqrt{3}}\,dE_1\,dE_2 \tag{4.25}$$

whence we can still identify the density of events on this plot with the square of the matrix element A. For the ω, the three formulae (4.14), (4.16) and (4.18) differ markedly in that they predict the probability to vanish in very different regions, as shown in figure (9.4.3).

The experimental distribution corresponds to the first case, i.e. to $J^P = 1^-$, with $\alpha(s, t, u)$ practically constant.

To sum up, the ω meson is a resonance having the quantum numbers

$$J^P = 1^- \qquad I = 0 \qquad G = -1 \qquad B = 0 \qquad S = 0$$

It is also an eigenstate of charge conjugation with

$$C = G(-)^I = -1$$

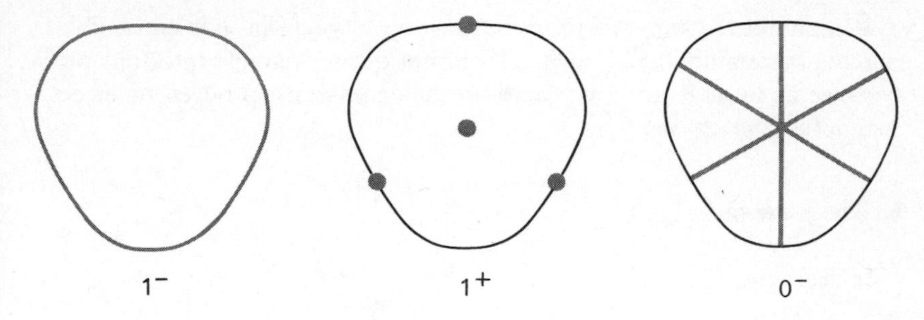

FIG. 9.4.3. Regions of zero probability on the Dalitz plot for total spin and parity 1^-, 1^+, and 0^-.

5. The η meson

The smaller peak observed beside the ω in the reaction

$$\pi^+ + d \to p + p + \pi^+ + \pi^- + \pi^0 \tag{5.1}$$

has a mass of 550 MeV and a width of 2.6 KeV. It is called the η meson. It is observed in many other reactions not only in the $\pi^+\pi^-\pi^0$ system but also in other systems, for which the following branching ratios have been measured:

$\pi^+\pi^-\pi^0$	$\pi^+\pi^-\gamma$	$\pi^0\pi^0\pi^0$	$\pi^0\gamma\gamma$	$\gamma\gamma$
$23.3 \pm 1.1\%$	$5.5 \pm 0.5\%$	$29.4 \pm 2.8\%$	$2.5 \pm 2.8\%$	$38.1 \pm 2.3\%$

Thus, as many η's decay with the emission of one or more photons as into 3π. This suggests that all the decays could be due to electromagnetic interactions.

Since all the observed decay modes are neutral, the isotopic spin is zero. But if the decay is electromagnetic, it does not conserve isospin, and we cannot rely on the isospin value to deduce the symmetry properties of the three pion decay amplitude.

Experimentally, the Dalitz plot for three pion decay is populated approximately uniformly, which corresponds to $A(\mathbf{p}_1, \mathbf{p}_2, \mathbf{p}_3) = $ constant. But the only function of these vectors which has constant modulus is actually independent of the vectors, and corresponds to

$$J^P = 0^- \tag{5.2}$$

While considering the ω we saw that an isospin 0 pseudoscalar meson cannot decay into three pions with conservation of isotopic spin. Hence it is natural to assume that G-parity is violated in η decay, so that for the η one has

$$G = +1 \tag{5.3}$$

It cannot decay into two pions because with two pions it is impossible to make a system having $J^P = 0^-$. It cannot decay strongly into four pions because its mass is too low; therefore its decay must proceed by electromagnetic interactions.

6. The ρ meson

In reactions like

$$\pi + N \to \pi + \pi + N$$
$$\pi + N \to \pi + \pi + \pi + N$$
$$\bar{p} + p \to n\pi$$

one finds a peak in the two-pion mass spectrum; it is centred on 750 MeV, has a width of approximately 100 MeV, and appears in systems with charge $+1(\pi^+\pi^0)$, $0(\pi^+\pi^-)$, and $-1(\pi^-\pi^0)$. It does not appear in systems with charge $+2(\pi^+\pi^+)$ and $-2(\pi^-\pi^-)$. Hence one is dealing with a two-pion resonance having isotopic spin 1; it is called the ρ meson.

Since each pion has isospin 1, their isospin 1 combination is the vector product $\boldsymbol{\pi}_1 \wedge \boldsymbol{\pi}_2$ in isospace. Therefore the wavefunction of the ρ is antisymmetric in isotopic spin, and must be antisymmetric in space, which means that the ρ spin is odd: $J = 1, 3, \ldots$.

To determine the spin one proceeds as follows. Consider for instance the reaction

$$\pi^- + p \to \pi^- + \pi^0 + p \tag{6.1}$$

At energies above 1 BeV, N^* production no longer dominates, and we consider those events (6.1) for which the $\pi^-\pi^0$ mass falls under the ρ peak. They are a mixture of events where the $\pi^-\pi^0$ mass is close to the ρ by accident, and of other events

$$\pi^- + p \to \rho^- + p \tag{6.2}$$

Since the production proceeds through strong interactions, the ρ is polarised perpendicularly to the production plane. Taking the normal to the production plane as the spin quantisation axis, the ρ^- is a statistical mixture of states $|jm\rangle$. We go over to its rest frame by a Lorentz transformation Λ along its momentum; this transforms the statistical mixture into another, having the density matrix D

$$D_{m'm}(\rho \text{ center-of-mass})$$
$$= \mathscr{D}^j_{m'm_1}(L^{-1}(\Lambda\mathbf{p})\Lambda L(\mathbf{p}))\, D^{(1)}_{m_1'm_1}(\text{total c.m.})\mathscr{D}^j_{m_1m}(L(\mathbf{p})\Lambda L^{-1}(\Lambda\mathbf{p})) \tag{6.3}$$

The angular distribution of the two pions in the ρ rest frame is given by

$$\frac{dN}{d\Omega} = \sum_{m'm} Y_j^{m'*}(\theta, \varphi)\, D_{m'm} Y_j^m(\theta, \varphi) \tag{6.4}$$

and will therefore be a superposition of Legendre polynomials

$$\frac{dN}{d\Omega} = \sum_{n=0}^{2j} \alpha_n P_n(\cos\theta) \tag{6.5}$$

In practice one finds a strong component with $n = 2$, whence $J = 1$.

Since the ρ is a two-pion system, its G-parity is $+1$. To sum up, the ρ has $J^P = 1^-, G = +1$. Note that the spins and parities of the ρ and the ω are the same. Table 3 lists some meson resonances with strangeness zero, and table 4 lists the strange ones.

TABLE 3

MESON RESONANCES WITH STRANGENESS 0

Symbol	Isotopic spin and G parity I^G	Spin and parity J^P	Mass (MeV)	Width (MeV)	Strong interaction decay moves: Mode	Branching ratio %
$\pi^\pm$	1^-	0^-	139,58		weak interactions	
π^0			134,97	~	electromagnetic interactions	
η	0^+	0^-	548,8	2,6 keV	electromagnetic interactions	
ρ	1^+	1^-	765	125 ± 20	$\pi\pi$	~100
ω	0^-	1^-	783	13	$\pi^+\pi^-\pi^0$	~90
η' or X^0	0^+	0^-	958	<4	$\eta\pi\pi$	71 ± 4
					$\pi^+\pi^-\gamma$	22 ± 3
					$\gamma\gamma$	6 ± 3
δ (962)	1^-	0^+ ?	962	<5	$\eta\pi$?	
ϕ	0^-	1^-	1019	3,7	K^+K^-	48
					$K_L K_S$	33
					$\pi^+\pi^-\pi^0$	20
A_1	0^+	0^+	1070	80	3π	existence still in doubt
B	1^+	1^+ ?	1221	123	$\omega\pi$	~100
f	0^+	2^+	1264	145	$\pi\pi$	
D	0^+		1285	31	$K\bar{K}\pi$	
A_{2L}	1^-	?	1269	26	$\rho\pi$	
A_{2H}	1^-	2^+ ?	1315	24	$\rho\pi$	
E	0^+	0^- ?	1424	71	$K\bar{K}\pi$	
f'	0^+	2^+	1514	73	$K\bar{K}$	72
					$K\bar{K}\pi$	10
					$\pi\pi$	<14
					$\eta\pi\pi$	?
					$\eta\eta$	?

TABLE 4

STRANGE MESON RESONANCES

Symbol	I	J^P	Mass (MeV)	Γ (MeV)	Decay mode	Branching ratio %
$K^*(890)$	1/2	1^-	891,4	49,7	$K\pi$	
$K^*(1320)$	1/2	1^+?	1330	70	$K\pi\pi$	
$K^*(1420)$	1/2	2^+?	1422	90	$K\pi$	51 ± 5
					$K^*\pi$	33 ± 3
					$K\rho$	11 ± 4
$K^*(1780)$	1/2	?	1775	72	$K^*\pi$	
					$K^*(1420)\pi$	
					$K\rho$	

Problems

1.* *Isobar model*

We wish to describe the reaction

$$\pi^+ + N \rightarrow N^{*++} + \pi^0,$$

under the following assumptions. (1) the N^* mass is distributed according to

$$\frac{dP}{dM} = \frac{(2\pi)^{-1}\Gamma}{(M - M_0)^2 + \dfrac{\Gamma^2}{4}}$$

(2) the N^* is produced isotropically in the centre of mass frame.

Calculate numerically the momentum spectrum of the π^0 in the centre of mass frame, given that the incident π^+ has energy 1 GeV in the laboratory frame.

2. From isotopic spin conservation calculate the branching ratio between the two reactions

$$\pi^+ + N \rightarrow N^{*++} + \pi^0 \qquad (1)$$

$$\pi^+ + N \rightarrow N^{*+} + \pi^+ \qquad (2)$$

3.* *Adair analysis*

Consider the reaction sequence

$$\pi^+ + p \rightarrow N^{*++} + \pi^0 \qquad (1)$$

$$N^{*++} \rightarrow p + \pi^+ \qquad (2)$$

Assume that the π^0 is emitted at zero angle relative to the direction of the incident π^+. Write down the density matrix for the N^{*++}, given an unpolarised proton target, by classifying the states according to their spin component along the incident π^+ direction. Taking the N^* as nonrelativistic, give the angular distribution of the final π^+ relative to the direction of the initial π^+.

How would these results change if the N^* spin were $\frac{1}{2}$?

This method can be used to determine the spin of a resonance by selecting those events where the resonance is emitted at very small angles to the forward (or backward) direction.

4. Show that invariance under charge conjugation forbids the decay $\eta \to \pi^0\pi^0\gamma$.
5. Show that because of parity conservation the ω cannot decay into $\eta + \pi$, not even electromagnetically.
6. Show that the ρ cannot decay into $\eta + \pi$.

CHAPTER 10

THE DIRAC EQUATION

In this chapter we show how one can describe spin $\frac{1}{2}$ particles in an explicitly covariant manner by using a formalism, namely the Dirac equation, equivalent to that of chapter 4. The new formalism will be used in Part 2 of this book.

1. General survey. Lorentz group and $SL(2C)$

(a) *General survey*

Having studied the most prominent qualitative properties of particles, we shall soon have to turn more closely to the dynamics of interactions. The fact that all stable particles (except the photon) have spin $\frac{1}{2}$, and that they naturally play an especially important part in experiments, will lead us to an analysis of systems featuring such particles. We saw in chapter 4 that a state of a single spin $\frac{1}{2}$ particle, with mass m, is specified by a vector $|\frac{1}{2}\alpha p\rangle$, where $\alpha = \pm\frac{1}{2}$ and $p^2 = m^2$ ($p_0 > 0$). A Lorentz transformation acts on this vector through a unitary operator $U\{a, \Lambda\}$ having the explicit form

$$U\{a, \Lambda\} \mid \tfrac{1}{2}\,\alpha, p\rangle = e^{ia\cdot\Lambda p} \sum_{\alpha'} \mathscr{D}^{(1/2)}_{\alpha'\alpha}(L^{-1}(\Lambda p)\,\Lambda L(p)) \mid \tfrac{1}{2}\,\alpha', \Lambda p\rangle \qquad (1.1)$$

In this chapter we shall omit the spin index $\frac{1}{2}$, it being understood that only spin $\frac{1}{2}$ particles are under discussion.

Our immediate aim is to give a description of the states which, though equivalent to (1.1), is different in form; to this end we use a manifestly covariant formalism known as the Dirac equation. Note that it is perfectly possible to treat interactions on the basis of (1.1); but such a treatment would be so different from the classical ones that at this point we prefer to introduce Dirac's formulation.

(b) *Correspondence between R_3 and $SU(2)$*

The Lorentz transformation $R = L^{-1}(\Lambda p)\Lambda L(p)$ appearing in (1.1) is a rotation, so that $\mathscr{D}^{(1/2)}(R)$ is a matrix representing R in the subspace with angular momentum $j = \frac{1}{2}$. We saw in chapter 3 that $A = \mathscr{D}^{(1/2)}(R)$ is a unitary unimodular matrix, or in other words that

$$AA^+ = A^+A = I \tag{1.2}$$

$$\det A = 1 \tag{1.3}$$

In chapter 4 we saw that if for every 3-dimensional vector $\mathbf{x}$ we define a Hermitean 2×2 matrix X by

$$X = \mathbf{x} \cdot \boldsymbol{\sigma} \tag{1.4}$$

then this matrix has the properties

$$\det X = -\mathbf{x}^2 \tag{1.5}$$

$$X = X^+ \tag{1.6}$$

The inverse relation is

$$\mathbf{x} = \tfrac{1}{2}\,\text{Trace}\,(X\boldsymbol{\sigma}). \tag{1.7}$$

The 3×3 matrix R and the 2×2 matrix A are related explicitly. If one writes the rotation R in the form

$$x_i' = R_{ij}x_j \tag{1.8}$$

and introduces the matrix X' associated with the transform $\mathbf{x}'$ of $\mathbf{x}$:

$$X' = \mathbf{x}' \cdot \boldsymbol{\sigma}$$

then one has

$$X' = AXA^+. \tag{1.9}$$

Comparing (1.8) and (1.9) we can easily express R in terms of A and vice versa, by equating the coefficients of the x_j in the expression for the x_i'.

(c) *Lorentz group and $SL(2C)$*

Next, we go over from three-dimensional to Minkowski space. To this end we note that a matrix like X is not the most general 2×2 Hermitean matrix, because the trace of X is zero. Indeed, Hermitean 2×2 matrices depend on 4 parameters and can always be written in the form

$$X = x_0\mathbb{1} - \mathbf{x} \cdot \boldsymbol{\sigma} \tag{1.10}$$

It is convenient to unify our notation by writing $\sigma_0 = \mathbb{1}$, so that by introducing a four-vector σ_μ:

$$\sigma_\mu = (\sigma_0\,, \sigma_1\,, \sigma_2\,, \sigma_3)$$

we can write (1.10) as

$$X = \sigma_\mu x_\mu\,. \tag{1.11}$$

Conversely, any four-vector $(x_0, \mathbf{x})$ can be associated in this way with a Hermitean 2×2 matrix, the vector being expressed in terms of the matrix by the inverse relation

$$x_\mu = \tfrac{1}{2}\,\text{Trace}\,(X\sigma_\mu) \tag{1.12}$$

Here we have used the relation *Trace* $\sigma_\mu \sigma_\nu = 2\delta_{\mu\nu}$, which should be checked.

Explicitly, the matrix X can be written as

$$X = \begin{bmatrix} x_0 + x_3 & x_1 - ix_2 \\ x_1 + ix_2 & x_0 - x_3 \end{bmatrix} \tag{1.13}$$

whence one has

$$\det X = x_0{}^2 - x_1{}^2 - x_2{}^2 - x_3{}^2 = g_{\mu\nu} x_\mu x_\nu = x^2. \tag{1.14}$$

We recall that the group of rotations conserving the three-dimensional metric (1.5) can be described by the transformations (1.9); in the same way it is natural to introduce the transformations conserving the relativistic metric (1.14) (i.e. basically, the Lorentz transformations) by writing

$$X' = AXA^+ \tag{1.15}$$

which ensures that the Hermitecity of X' is maintained.

In order to conserve the metric (1.14) one demands

$$\det A = 1 \tag{1.16}$$

but abandons the unitarity condition (1.6) which guaranteed the invariance of the trace of X.

Note that the matrices A satisfying (1.16) form the group $SL(2C)$ introduced in chapter 3. We saw that $SL(2C)$, like the Lorentz group, depends on 6 parameters. From the foregoing, to every matrix A there corresponds a Lorentz transformation Λ such that

$$x_\mu' = \Lambda_{\mu\nu} x_\nu\,. \tag{1.17}$$

Conversely, one can show that to every transformation of the proper orthochronous Lorentz group, i.e. to every Lorentz transformation satisfying

$$\det \Lambda = +1 \qquad \Lambda_{00} > 1 \tag{1.18}$$

there correspond two unimodular matrices $\pm A$. (The fact that $+A$ and $-A$ give rise to the same transformation Λ is obvious from the quadratic nature of (1.15).)

The matrices A clearly constitute a two-dimensional representation of the proper orthochronous Lorentz group. Note that this representation is *not* unitary. The unitary matrices A representing R_3 form a subgroup of the unimodular matrices.

(d) *Complex conjugation*

The complex conjugates of the Pauli matrices can be written as

$$\begin{aligned} \sigma_1{}^* &= \sigma_1 = \sigma_2(-\sigma_1)\,\sigma_2 \\ \sigma_2{}^* &= -\sigma_2 = \sigma_2(-\sigma_2)\,\sigma_2 \\ \sigma_3{}^* &= \sigma_3 = \sigma_2(-\sigma_3)\,\sigma_2 \end{aligned} \tag{1.19}$$

Accordingly, if one considers a unitary matrix of the form $e^{i\boldsymbol{\lambda}\cdot\boldsymbol{\sigma}} = A$, one has

$$A^* = e^{-i\lambda\sigma^*} = e^{i\boldsymbol{\lambda}\cdot(-\boldsymbol{\sigma}^*)} = \sigma_2 A \sigma_2 \,. \tag{1.20}$$

In view of $\sigma_2 = \sigma_2^{-1}$, this relation shows that a unitary matrix A is connected to A^* through a unitary change of basis in the two-dimensional (spinor) space. This result harmonises with the fact that the two, mutually complex conjugate, multiplication rules $A = BC$ and $A^* = B^*C^*$ amount to one and the same group composition law, while R_3 has only a single irreducible 2 dimensional representation (up to changes of basis). It follows that A and A^* must necessarily be unitary equivalent.

It is more convenient to use instead of σ_2 the real matrix

$$\zeta = i\sigma_2 \tag{1.21}$$

and to write, for unitary A,

$$A = \zeta A^* \zeta^{-1} \tag{1.22}$$

(e) *Infinitesimal generators*

In chapter 3, §3, we saw that the infinitesimal generators of $SL(2C)$ are the six traceless matrices

$$J_1 = \sigma_1/2 \quad J_2 = \sigma_2/2 \quad J_3 = \sigma_3/2 \quad K_1 = i\sigma_1/2 \quad K_2 = i\sigma_2/2 \quad K_3 = i\sigma_3/2 \tag{1.23}$$

One checks at once that the commutation relations of these generators,

$$\begin{aligned} &[J_1\,, J_2] = iJ_3 \qquad [J_2\,, J_3] = iJ_1 \qquad [J_3\,, J_1] = iJ_2 \\ &[J_1\,, K_2] = iK_3 \qquad [J_2\,, K_3] = iK_1 \qquad [J_3\,, K_1] = iK_2 \\ &[J_1\,, K_1] = [J_2\,, K_2] = [J_3\,, K_3] = 0 \\ &[K_1\,, K_2] = -iJ_3 \qquad [K_2\,, K_3] = -iJ_1 \qquad [K_3\,, K_1] = -iJ_2 \end{aligned} \tag{1.24}$$

coincide exactly with those of the infinitesimal generators of the Lorentz group. The same commutation relations (1.24) are equally satisfied by a different choice of representation, as in

$$\mathbf{J}' = \boldsymbol{\sigma}/2 \qquad \mathbf{K}' = -i\boldsymbol{\sigma}/2 \tag{1.23a}$$

Hence a priori there are two different ways of associating 2×2 matrices with the infinitesimal generators of Lorentz transformations. The choices (1.23) and (1.23a) lead to two distinct representations of the Lorentz group. They are evidently not equivalent, because a unitary transformation of the form

$$\boldsymbol{\sigma} \rightarrow S\boldsymbol{\sigma}S^{-1}$$

necessarily conserves **K** if it conserves **J**, and cannot therefore transform (**J**, **K**) into (**J**′, **K**′).

(f) *Parity*

A matrix belonging to the representation (1.23) of the Lorentz group can always be written in the form

$$A = e^{i\boldsymbol{\lambda}\cdot\boldsymbol{\sigma}+\boldsymbol{\mu}\cdot\boldsymbol{\sigma}} \tag{1.25}$$

where $\boldsymbol{\lambda}$ and $\boldsymbol{\mu}$ are real vectors. This is simply the definition of the Lie algebra given in chapter 3. Under the action of (1.22), which is not an equivalence relation (except for unitary matrices), the matrix A is according to (1.19) transformed into the new matrix

$$A' = e^{i\boldsymbol{\lambda}\cdot\boldsymbol{\sigma}-\boldsymbol{\mu}\cdot\boldsymbol{\sigma}} \tag{1.25a}$$

This shows that the transformation (1.23) does indeed realise the correspondence between the generators (**J**, **K**) and (**J**′, **K**′) of the last paragraph.

Note that this transformation has the same effect as the parity operation, because we know that **J** is an axial vector and **K** a polar vector. In the following we shall be interested in the two *non-unitary* representations (1.23) and (1.23a) of $SL(2C)$, and we shall identify (1.22) with space reflection; in a mathematical sense, but unconnected so far with any physics, this will give us a representation of the Lorentz group including space reflection.

Let us label by $\xi_\sigma(\sigma = \pm\frac{1}{2})$, the components of a vector in the two-dimensional space acted on by the group $SL(2C)$; then we see that space reflection is represented by the transformation

$$\xi_\sigma' = (\zeta)_{\sigma\sigma'}\xi^*_{\sigma'} \qquad \text{or} \qquad \xi' = \zeta\xi^*. \tag{1.26}$$

Since the matrices A and A^* are inequivalent, it follows that space reflection cannot be represented within a single 2-dimensional representation of the Lorentz group. In order to do this one must introduce two different two-dimensional spaces, whose respective coordinates we shall denote by ξ_σ and η_σ. In this new four-dimensional space we denote the coordinates by $(\xi_{1/2}, \xi_{-1/2}, \eta_{1/2}, \eta_{-1/2})$; a Lorentz transformation acts on ξ through A and on η through $\zeta A^*\zeta^{-1}$, so that

$$\begin{pmatrix}\xi' \\ \eta'\end{pmatrix} = \begin{pmatrix}A & 0 \\ 0 & \zeta A^*\zeta^{-1}\end{pmatrix}\begin{pmatrix}\xi \\ \eta\end{pmatrix} \tag{1.27}$$

Space reflection is represented by the four-dimensional matrix

$$\begin{pmatrix} \xi' \\ \eta' \end{pmatrix} = \begin{pmatrix} 0 & \mathbb{1} \\ \mathbb{1} & 0 \end{pmatrix} \begin{pmatrix} \xi \\ \eta \end{pmatrix} \tag{1.28}$$

To sum up, for the Lorentz group plus space reflection one can find a non-unitary representation in a space with four dimensions. The 2×2 matrix representing a Lorentz transformation Λ will be denoted by $\mathscr{D}^{(1/2)}(\Lambda)$, by an extension of the nomenclature for rotations. Accordingly, the representation in the four-dimensional space is given by the matrix

$$\begin{pmatrix} \mathscr{D}^{(1/2)}(\Lambda) & 0 \\ 0 & \bar{\mathscr{D}}^{(1/2)}(\Lambda) \end{pmatrix} \tag{1.29}$$

where

$$\bar{\mathscr{D}}^{(1/2}(\Lambda) = \zeta \mathscr{D}^{(1/2*}(\Lambda)\, \zeta^{-1}. \tag{1.30}$$

2. The Dirac equation

(a) *Modification of* (1.1)

From the foregoing, the rotation matrix which occurs in the representation (1.1) of the Poincaré group can be factored into matrices representing Lorentz transformations, i.e.

$$\mathscr{D}^{(1/2)}(L^{-1}(\Lambda p)\,\Lambda L(p)) = [\mathscr{D}^{(1/2)}(L(\Lambda p))]^{-1}\, \mathscr{D}^{(1/2)}(\Lambda)\, \mathscr{D}^{(1/2)}(L(p)) \tag{2.1}$$

Similarly, one can write

$$\begin{aligned} \mathscr{D}^{(1/2)}(L^{-1}(\Lambda p)\,\Lambda L(p)) &= \bar{\mathscr{D}}^{(1/2)}(L^{-1}(\Lambda p)\,\Lambda L(p)) \\ &= [\bar{\mathscr{D}}^{(1/2)}(L(p))]^{-1}\, \bar{\mathscr{D}}^{(1/2)}(\Lambda) \bar{\mathscr{D}}^{(1/2)}(L(p)) \end{aligned} \tag{2.2}$$

and in general the factors appearing on the right of (2.1) and (2.2) are different even though the left hand sides are the same.

The new state vectors $|\sigma, p, 1\rangle$ and $|\sigma, p, 2\rangle$ defined by

$$|\,\sigma, p, 1\rangle = [\mathscr{D}^{(1/2)}(L(p))]^{-1}_{\sigma'\sigma}\,|\,\sigma', p\rangle \tag{2.3}$$

$$|\,\sigma, p, 2\rangle = [\bar{\mathscr{D}}^{(1/2)}(L(p))]^{-1}_{\sigma'\sigma}\,|\,\sigma', p\rangle \tag{2.4}$$

transform under the Poincaré group according to the new rules

$$\begin{aligned} U\{a, \Lambda\}\,|\,\sigma, p, 1\rangle &= e^{ia\cdot\Lambda p}\, \mathscr{D}^{(1/2)}_{\sigma'\sigma}(\Lambda)\,|\,\sigma', p, 1\rangle \\ U\{a, \Lambda\}\,|\,\sigma, p, 2\rangle &= e^{ia\cdot\Lambda p}\, \bar{\mathscr{D}}^{(1/2)}_{\sigma'\sigma}(\Lambda)\,|\,\sigma', p, 2\rangle. \end{aligned} \tag{2.5}$$

We see that the transformation, acting on these new vectors, is expressible directly in terms of the 2×2 matrices $\mathscr{D}$ and $\bar{\mathscr{D}}$ representing Λ.

In this sense the new formulae are simpler than (2.1), since their dependences on p and on Λ have been separated and are manifestly covariant. However, in several other respects they are more complicated; in particular, the matrices are no longer unitary. Moreover, one is now dealing with four interdependent vectors instead of two; and space reflection, instead of acting directly on the vector $|\sigma, p\rangle$, now interchanges $|\sigma, p, 1\rangle$ and $|\sigma, p, 2\rangle$:

$$\begin{aligned} \Pi \mid \sigma \mathbf{p} 1\rangle &= \epsilon \mid \sigma \Pi \mathbf{p} 2\rangle \\ \Pi \mid \sigma \mathbf{p} 2\rangle &= \epsilon \mid \sigma \Pi \mathbf{p} 1\rangle \end{aligned} \tag{2.6}$$

where ϵ is the intrinsic parity.

At this point we introduce a new notation for the four-component state vectors, writing these as $|p\alpha\rangle$ where $\alpha = (\sigma, i)$. It is even simpler to write $\alpha = 1, 2, 3, 4$, corresponding respectively to $(\sigma, i) = (\frac{1}{2}, 1), (-\frac{1}{2}, 1), (\frac{1}{2}, 2), (-\frac{1}{2}, 2)$. This is essentially Dirac's notation.

(b) *Derivation of the Dirac equation*

The two two-component vectors (2.3) and (2.4) are not independent. Hence they are connected by a linear relationship which is called the Dirac equation and which we now proceed to determine.

To ease the notation we shall denote the transpose of the matrix $\mathscr{D}^{(1/2)}(\Lambda)$ by $D(\Lambda)$, and the transpose of $\bar{\mathscr{D}}^{(1/2)}(\Lambda)$ by $\bar{D}(\Lambda)$. The transposes are useful in rewriting (2.3) and (2.4) as matrix products

$$| p1\rangle = D^{-1}(L(p) \mid p\rangle \tag{2.3a}$$

$$| p2\rangle = \bar{D}^{-1}(L(p) \mid p\rangle \tag{2.3b}$$

Then one has

$$| p1\rangle = D^{-1}(L(p))\, \bar{D}(L(p)) \mid p2\rangle. \tag{2.7}$$

In order to calculate the matrix occurring in equation (2.7), we consider first the special case where $\mathbf{p}$ is parallel to the 3 axis; then the boost $L(p)$ can be written explicitly in the form

$$L(p) = \begin{bmatrix} 1 & 0 & 0 & 0 \\ 0 & 1 & 0 & 0 \\ 0 & 0 & \mathrm{ch}\chi & \mathrm{sh}\chi \\ 0 & 0 & \mathrm{sh}\chi & \mathrm{ch}\chi \end{bmatrix} \quad \text{where : } \begin{aligned} \mathrm{ch}\chi &= \gamma, \\ \mathrm{sh}\chi &= \gamma v \end{aligned} \tag{2.8}$$

Let us now show that

$$D(L(p)) = \begin{pmatrix} e^{\chi/2} & 0 \\ 0 & e^{-\chi/2} \end{pmatrix} \tag{2.9}$$

In terms of the matrices X and X' one has

$$X' = \begin{pmatrix} e^{\chi/2} & 0 \\ 0 & e^{-\chi/2} \end{pmatrix} \begin{pmatrix} x_0 + x_3 & x_1 - ix_2 \\ x_1 + ix_2 & x_0 - x_3 \end{pmatrix} \begin{pmatrix} e^{\chi/2} & 0 \\ 0 & e^{-\chi/2} \end{pmatrix}$$

$$= \begin{pmatrix} (x^0 + x^3)e^{\chi} & x_1 - ix_2 \\ x_1 + ix_2 & (x_0 - x_3)e^{-\chi} \end{pmatrix} = \begin{pmatrix} x'^0 + x'^3 & x_1' - ix_2' \\ x_1' + ix_2' & x_0' - x_3' \end{pmatrix}$$

where we have put

$$x_\mu' = (L(p))_{\mu\nu} x_\nu \text{ , which proves (2.9).}$$

Hence one has

$$\overline{D}(L(p)) = \zeta D^{*(1/2)}(L(p))\, \zeta^{-1} = \begin{pmatrix} e^{-\chi/2} & 0 \\ 0 & e^{+\chi/2} \end{pmatrix}$$

and

$$D^{-1}(L(p))\, \overline{D}(L(p)) = \begin{pmatrix} e^{-\chi} & 0 \\ 0 & e^{\chi} \end{pmatrix} = \text{ch}\chi - \text{sh}\chi\, \sigma_3 = \frac{p_0}{m} - \frac{\mathbf{p} \cdot \boldsymbol{\sigma}}{m} \tag{2.10}$$

This last expression is evidently the general form of the matrix, for any arbitrary value of $\mathbf{p}$. Thus one has the relation

$$|\, p1\rangle = \left(\frac{p_0}{m} - \frac{\mathbf{p} \cdot \boldsymbol{\sigma}}{m}\right) |\, p2\rangle \tag{2.11}$$

Conversely, it is easy to show that

$$|\, p2\rangle = \frac{p_0}{m} + \frac{\mathbf{p} \cdot \boldsymbol{\sigma}}{m}\, |\, p1\rangle. \tag{2.12}$$

Going over to a four-dimensional notation, we define four 4×4 matrices γ_0 and $\boldsymbol{\gamma}$ by

$$\gamma_0 = \begin{pmatrix} 0 & 1 \\ 1 & 0 \end{pmatrix} \qquad \gamma_i = \begin{pmatrix} 0 & -\sigma_i \\ +\sigma_i & 0 \end{pmatrix} \tag{2.13}$$

Then, keeping track of the order of the indices, (2.11) and (2.12) become

$$(\gamma_\mu{}^+ p_\mu - m)_{\alpha'\alpha}\, |\, p\alpha'\rangle = 0. \tag{2.14}$$

We now define the four components of a single-particle state $|u\rangle$ by the scalar products

$$u_\alpha(\mathbf{p}) = \langle p\alpha \,|\, u\rangle \tag{2.15}$$

and obtain the *Dirac equation* for these components:

$$(\gamma_\mu p_\mu - m)_{\alpha'\alpha} u_\alpha(\mathbf{p}) = 0 \tag{2.16}$$

$u_\alpha(p)$ is the Dirac wavefunction in momentum space for a one-electron state.

(c) *Invariance of the Dirac equation*

Since the representation (2.1) is valid in any reference frame, the Dirac equation (2.16) is itself invariant under Lorentz transformations. When the momentum p undergoes a Lorentz transformation Λ, the Dirac spinor transforms according to

$$u_\alpha(\mathbf{p}) \to \begin{pmatrix} \mathscr{D}^{(1/2)}(\Lambda) & 0 \\ 0 & \bar{\mathscr{D}}^{(1/2)}(\Lambda) \end{pmatrix}_{\alpha\beta} u_\beta(\mathbf{p}'). \tag{2.17}$$

By equation (2.6), space reflection takes effect simply through the transformation

$$\Pi u(\mathbf{p}) = \epsilon\gamma_0 u(-\mathbf{p}). \tag{2.18}$$

Comment: The historical method by which the Dirac equation is normally introduced raises some very nice problems, such as the connection between the number of components and the spin, and the apparent existence of states with negative energy. This is why it is preferable to use the method adopted here, due essentially to Bargmann and Wigner, which is more general and which does not trail red herrings. Moreover it gives a much clearer understanding of the properties of the Dirac equation and of the degree of arbitrariness which enters in generalising it to higher spins, though the latter problem will not be pursued here.

There exists another method, due to Foldy and Wouthuysen, for analysing the physical meaning of the Dirac equation, and for identifying spin, momentum, etc. in this formalism. It consists basically in inverting the above method, so that eventually one arrives back at Wigner's state vector as it appears in (1.1).

(d) *The adjoint spinor*

Recall that we started with state vectors subject to invariant normalisation:

$$\langle \mathbf{p}'\sigma' \mid \mathbf{p}\sigma\rangle = \delta_{\sigma\sigma'}\,\delta(\mathbf{p} - \mathbf{p}')E \tag{2.19}$$

where $E = \sqrt{\mathbf{p}^2 + m^2}$.

By (2.3a) one deduces from this that

$$\langle \sigma'\mathbf{p}'1 \mid \sigma p1\rangle = \langle \sigma'\mathbf{p}' \mid D^+(L^{-1}(p))\, D(L^{-1}(p)) \mid \sigma p\rangle. \tag{2.20}$$

But for $\mathbf{p}$ parallel to the 3 axis, one has according to (2.9)

$$D^+(L^{-1}(p))\, D(L^{-1}(p)) = \begin{pmatrix} e^{-x} & 0 \\ 0 & e^{x} \end{pmatrix} = \frac{p_0}{m} - \frac{\mathbf{p}\cdot\boldsymbol{\sigma}}{m}$$

whence

$$\langle \sigma'p'1 \mid \sigma p1\rangle = \langle \sigma'p' \mid \frac{p_0}{m} - \frac{\mathbf{p}\cdot\boldsymbol{\sigma}}{m} \mid \sigma p\rangle \tag{2.21}$$

Similarly, one can show that

$$\langle \sigma'p'2 \mid \sigma p2\rangle = \langle \sigma'p' \mid \frac{p_0}{m} + \frac{\mathbf{p}\cdot\boldsymbol{\sigma}}{m} \mid \sigma p\rangle \tag{2.22}$$

whence

$$\sum_{\alpha=1}^{4} \langle p'\alpha' \mid p\alpha\rangle = \frac{2p_0}{m} \sum_{\sigma=1}^{2} \langle \sigma' p' \mid \sigma p\rangle = \frac{2p_0{}^2}{m}\,\delta(\mathbf{p}-\mathbf{p}'). \qquad (2.23)$$

The expression (2.23) defines a non-invariant inner product in the 4-dimensional spinor space. Actually (2.23) behaves under Lorentz transformations like the fourth component of a four-vector (like p^0). There does exist an invariant inner product, namely $u^*\gamma_0 u$. Indeed,

$$\begin{aligned}\langle p'\sigma'\alpha' \mid \gamma^0_{\alpha'\alpha} \mid p\sigma\alpha\rangle &= \langle p'\sigma' 2 \mid p\sigma 1\rangle + \langle p'\sigma' 1 \mid p\sigma 2\rangle \\ &= \langle p'\sigma' \mid \bar{D}^{+}(L^{-1}(p))\, D(L^{-1}(p)) + D^{+}(L^{-1}(p))\, \bar{D}(L^{-1}(p)) \mid p\sigma\rangle \qquad (2.24)\end{aligned}$$

A calculation analogous to the above shows that every matrix in (2.24) is now equal to 1, whence

$$\langle p'\alpha' \mid \gamma^0_{\alpha'\alpha} \mid p\alpha\rangle = 2\delta(\mathbf{p}'-\mathbf{p})\frac{p_0}{m}. \qquad (2.25)$$

This suggests that we should introduce an adjoint spinor defined by

$$\langle \overline{p\alpha} \mid = \langle p\alpha' \mid \gamma^0_{\alpha'\alpha} \qquad (2.26)$$

and

$$\bar{u}(p) = \langle \bar{u} \mid p\alpha\rangle = u^*(p)\,\gamma^0.$$

Then it is easy to prove from (21) and (22) that

$$\begin{aligned}\bar{u}(p)\,\gamma_0 u(p) &= \frac{2p_0}{m}\,\bar{u}(p)\,u(p) \\ \bar{u}(p)\,\gamma_i u(p) &= \frac{2p_i}{m}\,\bar{u}(p)\,u(p)\end{aligned} \qquad (2.27)$$

which shows that $\bar{u}\gamma_\mu u$ transforms under Lorentz transformations like a 4-vector.

3. Algebra of the γ matrices

(a) *Representations*

The representation (2.13) of the γ matrices can be used to determine their products. Defining

$$\gamma_5 = i\gamma_0\gamma_1\gamma_2\gamma_3 \qquad (3.1)$$

one has

$$\gamma_0 = \begin{pmatrix}0 & 1\\ 1 & 0\end{pmatrix} \quad \gamma_i = \begin{pmatrix}0 & -\sigma_i\\ +\sigma_i & 0\end{pmatrix} \quad \gamma_0\gamma_i = \begin{pmatrix}\sigma_i & 0\\ 0 & -\sigma_i\end{pmatrix} \quad \gamma_i\gamma_j = \begin{pmatrix}-i\sigma_k & 0\\ 0 & -i\sigma_k\end{pmatrix}$$

$$\gamma_5 = \begin{pmatrix}1 & 0\\ 0 & -1\end{pmatrix} \quad \gamma_5\gamma_0 = \begin{pmatrix}0 & -1\\ 1 & 0\end{pmatrix} \quad \gamma_5\gamma_i = \begin{pmatrix}0 & -\sigma_i\\ -\sigma_i & 0\end{pmatrix} \quad I = \begin{pmatrix}1 & 0\\ 0 & 1\end{pmatrix} \qquad (3.2)$$

It is easy to see that all these matrices are independent; since there are 16 of them, every 4×4 matrix can be written as a linear combination of these 16.

(b) *Anticommutation relations*

From (3.2) one readily derives the very important anticommutation relations

$$\boxed{\gamma_\mu\gamma_\nu + \gamma_\nu\gamma_\mu = 2g_{\mu\nu}} \tag{3.3}$$

which could be taken as the starting point for recasting the theory of these matrices in a general form, though we shall not do so here. Similarly, one has

$$\gamma_5\gamma_\mu + \gamma_\mu\gamma_5 = 0. \tag{3.4}$$

(c) *Transformation properties*

In the last paragraph we saw that $\bar{u}u$ behaves like a scalar and $\bar{u}\gamma_\mu u$ like a four-vector. The same method can be used to show, for instance, that $\bar{u}\gamma_5 u$ transforms like a pseudoscalar. Thus, from (2.18) we have

$$\bar{u}(\mathbf{p})\,\Pi^{-1}\gamma_5\Pi u(\mathbf{p}) = \bar{u}(-\mathbf{p})\,\gamma_0\gamma_5\gamma_0 u(-\mathbf{p}) = -\bar{u}(-\mathbf{p})\,\gamma_5 u(-\mathbf{p}). \tag{3.5}$$

where in the last step we have used (3.4) and $\gamma_0{}^2 = 1$. In this way one can draw up a table of transformation properties under Lorentz transformations:

$\bar{u}u$	scalar
$\bar{u}\gamma_\mu u$	four-vector
$\bar{u}\sigma_{\mu\nu}u$	antisymmetric tensor of rank 2
$\bar{u}\gamma_5 u$	pseudoscalar
$\bar{u}\gamma_5\gamma_\mu$	pseudovector

where $\sigma_{\mu\nu} = \frac{i}{2}(\gamma_\mu\gamma_\nu - \gamma_\nu\gamma_\mu)$ (3.6)

(d) *Traces*

It follows directly from (3.2) that

$$\text{Trace}\ \gamma_\mu = 0 \quad \text{Trace}\ \sigma_{\mu\nu} = 0 \quad \text{Trace}\ \gamma_5 = 0 \quad \text{Trace}\ \gamma_5\gamma_\mu = 0 \tag{3.7}$$

$$\text{Trace}\ \gamma_\mu\gamma_\nu = 4g_{\mu\nu} \tag{3.8}$$

(e) *Another representation*

We can define the two-component spinor

$$\chi_0(\mathbf{p}) = \langle \mathbf{p}\sigma \mid u\rangle \tag{3.9}$$

and have, from §2,

$$u(\mathbf{p}) = \begin{bmatrix} \mathscr{D}^{(1/2)}(L^{-1}(\mathbf{p})) & 0 \\ 0 & \bar{\mathscr{D}}^{(1/2)}(L^{-1}(\mathbf{p})) \end{bmatrix} \begin{pmatrix} \chi(\mathbf{p}) \\ \chi(\mathbf{p}) \end{pmatrix}$$

In view of equation (2.10) this gives

$$u(\mathbf{p}) = \frac{1}{\sqrt{2m(E+m)}} \begin{bmatrix} E + m - \boldsymbol{\sigma}\cdot\mathbf{p} & 0 \\ 0 & E + m + \boldsymbol{\sigma}\cdot\mathbf{p} \end{bmatrix} \begin{pmatrix} \chi(\mathbf{p}) \\ \chi(\mathbf{p}) \end{pmatrix}. \tag{3.10}$$

This expression is often quoted as a solution of the Dirac equation. To derive it we have used (2.10) and the formulae

$$\operatorname{ch}\frac{\chi}{2} = \sqrt{\frac{E+m}{2m}} \qquad \operatorname{sh}\frac{\chi}{2} = \sqrt{\frac{E-m}{2m}} \qquad \operatorname{ch}\chi = \frac{E}{m} \qquad \operatorname{sh}\chi = \frac{p}{m}.$$

A nonrelativistic particle can be described by a two-component spinor, as we saw at the end of chapter 4. As one lets $\mathbf{p}$ tend to 0, this property is obvious from (1.1) but not from (3.9).

Indeed, as $\mathbf{p}$ tends to 0, the Dirac equation becomes

$$(\gamma^0 - 1)\, u(\mathbf{p}) = 0,$$

in other words the solution diagonalises γ^0. When one is interested in the non-relativistic limit of the Dirac equation, it is therefore convenient to diagonalise γ^0; this can be done by applying to the representation (3.2) the transformation

$$\begin{pmatrix} 1/\sqrt{2} & 1/\sqrt{2} \\ -1/\sqrt{2} & 1/\sqrt{2} \end{pmatrix}$$

One obtains

$$\gamma^0 = \begin{pmatrix} 1 & 0 \\ 0 & -1 \end{pmatrix} \qquad \gamma_i = \begin{pmatrix} 0 & \sigma_i \\ -\sigma_i & 0 \end{pmatrix} \qquad \gamma_5 = \begin{pmatrix} 0 & I \\ I & 0 \end{pmatrix} \qquad \gamma_5\gamma_0 = \begin{pmatrix} 0 & -I \\ I & 0 \end{pmatrix}$$

$$\gamma_5\gamma_i = \begin{pmatrix} -\sigma_i & 0 \\ 0 & \sigma_i \end{pmatrix} \qquad \gamma_{0k} = i\begin{pmatrix} 0 & \sigma_k \\ \sigma_k & 0 \end{pmatrix} \qquad \sigma_{ij} = \begin{pmatrix} \sigma_k & 0 \\ 0 & \sigma_k \end{pmatrix}. \tag{3.11}$$

In the new representation the solution of the Dirac equation becomes

$$u(\mathbf{p}) = \sqrt{\frac{E+m}{2E}} \begin{bmatrix} 1 & 0 \\ 0 & \dfrac{\boldsymbol{\sigma}\cdot\mathbf{p}}{E+m} \end{bmatrix} \begin{pmatrix} \chi(\mathbf{p}) \\ \chi(\mathbf{p}) \end{pmatrix}. \tag{3.12}$$

4. Antiparticles and charge conjugation

In discussing the representations (1.1) of the Poincaré group we dealt only with Lorentz transformations continuously connected to the identity, and ignored space reflection Π and time reversal T. Each of these two transformations associates with any state another state of equal mass, which, itself, consequently transforms under the representation (1.1). The last-mentioned property is shared by charge conjugation, which, as we saw in chapter 6, associates with any particle an antiparticle of equal mass.

Thus, one can formulate the following problem: given the commutation relations of Π, C, and T with the operations of the Poincaré group, find all possible representations of these operations. In particular, how many different states $|\sigma p\lambda\rangle$, $\lambda = 1, 2, \ldots$? are needed to form a basis for this representation? The solution, given by Bargmann, Michel, Wightman, and Wigner, is that there do indeed exist several representations, and that one must chose between them in the light of experiment. Every such representation corresponds to a very different physical situation: particles and antiparticles identical or distinct; relative parity of particles and antiparticle at rest + or −, etc.

With every state vector $|u\rangle$ one can associate four-component spinor wavefunctions $u_\alpha(p \cdot \lambda)$, each of which obeys the Dirac equation. Thus, if particle and antiparticle are distinct, one can introduce two four-component spinors $u_\alpha(p)$ and $u_\alpha{}^c(p)$, obeying

$$(\gamma_\mu p_\mu - m)\, u(p) = 0 \tag{4.1}$$

$$(\gamma_\mu p_\mu - m)\, u^c(p) = 0. \tag{4.2}$$

For reasons which depend basically on convenience in field theory, it is customary to link $u(p)$ and $u_c(p)$ as follows:

The Dirac equation (4.1) is a condition which picks out a 2-dimensional subspace from the 4-dimensional space of the $u(p)$. Indeed, the operator $\gamma_\mu p_\mu$ has also the eigenvalue $-m$, with degeneracy 2, as can be seen from the fact that its square is given by

$$(\gamma_\mu p_\mu)^2 = (\gamma_\mu p_\mu)(\gamma_\nu p_\nu) = \tfrac{1}{2}(\gamma_\mu\gamma_\nu + \gamma_\nu\gamma_\mu)\, p_\mu p_\nu = g_{\mu\nu} p_\mu p_\nu = p^2 = m^2 \tag{4.3}$$

In other words, the equation

$$(\gamma_\mu p_\mu + m)\, v(p) = 0. \tag{4.4}$$

has two independent solutions $v(p)$. Define now a 4×4 matrix C which obeys

$$C^{-1}\gamma_\mu C = -\gamma_\mu{}^T \tag{4.5}$$

where the superfix T denotes the transposed matrix. In the representation (2.13) one has

$$-\gamma_0{}^T = -\gamma_0 \qquad -\gamma_1{}^T = +\gamma_1 \qquad -\gamma_2{}^T = -\gamma_2 \qquad -\gamma_3{}^T = +\gamma_3$$

whence, *in this representation:*

$$C = \gamma_0 \gamma_2. \tag{4.6}$$

If we now write

$$v_\alpha(p) = C_{\alpha\beta} \gamma^0{}_{\epsilon\beta} u_\epsilon{}^* \tag{4.7}$$

or

$$\widetilde{C(\bar{u}^c)} = C\tilde{\gamma}_0 u^*$$

then it is easy to check that this relation is compatible with equations (4.4) and (4.2). In fact, (4.2) entails

$$(\gamma_\mu{}^* p_\mu - m)\, u_c{}^*(p) = 0$$

whence

$$\tilde{\gamma}_0^{-1} C^{-1} (C \gamma_0{}^T \gamma_\mu{}^* \gamma_0^{T-1} C^{-1} - m)\, C \gamma_0{}^T u_c{}^*(p) = 0.$$

In the representation (2.13) it is easy to see that

$$C \gamma_0{}^T \gamma_\mu{}^* \gamma_0^{T-1} C^{-1} = -\gamma_\mu\,,$$

which shows that (4.2), (4.4), and (4.7) are compatible.

In the light of (4.7) we can thus give a simultaneous physical interpretation of both the solutions (4.1) and (4.4).

If we now adopt the *convention* that space reflection is represented by

$$\Pi v(\mathbf{p}) = \gamma_0 v(-\mathbf{p}) \tag{4.8}$$

then the parity of a particle-antiparticle pair at rest is negative, for one has

$$\Pi u u_c = \Pi u \gamma_0^{*T} C^{-1*} v^* = \gamma_0 u \gamma_0^{*T} C^{-1*} \gamma_0{}^* v^* = -u u_c. \tag{4.9}$$

In actual fact, e^+ and e^- do indeed have opposite parities, as is shown by analysing the decay into photons of positronium (the atom containing an electron and a positron interacting through the Coulomb potential). The same applies to p and $\bar{p}$.

Note that the choice (4.8) is actually obligatory in field theory, if the particles are to be represented by a four-component local field $\psi_\alpha(x)$.

Generalization: the method used above to obtain the Dirac equation from the representations of the Poincaré group can be applied for other spins. For spin 1 and zero mass, it leads to Maxwell's equations.

5. Neutrinos

In this paragraph we show that the description of states by the Dirac equation can be extended to neutrinos, in spite of the fact that Wigner's representation (1.1) which we took as our starting point is evidently not applicable to zero mass particles. Nevertheless, one can think of such a particle as a limiting case with the rest mass tending to zero; this is the point of view which we shall adopt in the following.

To begin with, note that in contrast to the case of protons and neutrons, there is no reason whatever to include space reflection amongst the operations that can act on the states of a neutrino. If we start from Wigner's states $|\sigma\mathbf{p}\rangle$, there is consequently no reason to introduce both of the states $|\mathbf{p}1\rangle$ and $|\mathbf{p}2\rangle$ defined by (2.3) and (2.4). Only one is needed, the choice between them being purely a matter of convention. We shall take $|\sigma\mathbf{p}1\rangle$; with the notation of equation (2.3a), this leads to

$$|\ \mathbf{p}1\rangle = D^{-1}(L(p))\ |\ \mathbf{p}\rangle \tag{5.1}$$

At this stage, the particle mass is still finite.

As the mass tends to zero, the state of the particle must be completely specified by the quantum number σ; here we shall denote this by λ, in order to avoid confusing it with the spin indices that have been used previously. Thus we can write

$$\lambda = \lim_{m\to 0} \frac{W_0}{p_0} \tag{5.2}$$

In view of the explicit expression (5.6), chapter 4, for W^0, and of the representation by the matrices $\frac{1}{2}\boldsymbol{\sigma}$ of the generators $\mathbf{J}$ of the rotation group, one has

$$\lambda = -\frac{1}{2}\frac{\mathbf{p}\cdot\boldsymbol{\sigma}}{p} \tag{5.3}$$

where $p = p^0 = \sqrt{\mathbf{p}^2}$.

In terms of the unit vector $\hat{\mathbf{p}}$ parallel to the momentum, one gets

$$\lambda = -\tfrac{1}{2}\,\hat{\mathbf{p}}\cdot\boldsymbol{\sigma} \tag{5.4}$$

which shows immediately that λ can assume only the values $\pm\frac{1}{2}$. It follows that a neutrino or antineutrino state must satisfy the condition

$$(\boldsymbol{\sigma}\cdot\hat{\mathbf{p}})\ |\ \mathbf{p}\rangle = \pm\ |\ \mathbf{p}\rangle \tag{5.5}$$

One can, as is customary, introduce the wave-function, which for the neutrino state $|u\rangle$ is the two-component spinor

$$\chi\sigma(\mathbf{p}) = \langle\sigma\mathbf{p}\ |\ u\rangle \qquad \sigma = \pm\tfrac{1}{2} \tag{5.6}$$

This spinor satisfies one of the conditions

$$(\sigma\cdot\hat{\mathbf{p}} - 1)\,\chi(\mathbf{p}) = 0 \tag{5.6a}$$

or

$$(\boldsymbol{\sigma}\cdot\hat{\mathbf{p}} + 1)\,\chi(\mathbf{p}) = 0 \tag{5.6b}$$

Often one adopts a convention describing neutrinos in a formalism with four-component spinors. In that case one introduces both of the states

$|\sigma\mathbf{p}1\rangle$ and $|\sigma\mathbf{p}2\rangle$ before taking the zero mass limit. Similarly, one introduces the four-component wavefunction

$$\langle\sigma\mathbf{p}i \mid u\rangle = u_\alpha(\mathbf{p}) \qquad \alpha = 1 \cdots 4 \tag{5.7}$$

In view of the explicit representation (3.2) of the matrix γ_5, it is clear that even in a four-component formalism one can put oneself in a position to consider only $|\sigma\mathbf{p}1\rangle$ and to ignore $|\sigma\mathbf{p}2\rangle$, by imposing the condition

$$\gamma_5 u(\mathbf{p}) = u(\mathbf{p}) \tag{5.8}$$

which eliminates those components of $u_\alpha(p)$ which are proportional to $|\sigma\mathbf{p}2\rangle$. Alternative ways of writing the same condition are

$$(1 - \gamma_5)\, u(\mathbf{p}) = 0$$

or

$$u(\mathbf{p}) = \frac{1 + \gamma_5}{2}\, u(\mathbf{p}) \tag{5.9}$$

as is obvious from (5.8).

The condition (5.6), specifying the polarisation of the neutrino (i.e. its helicity, as discussed in the next paragraph), now takes the form

$$(\gamma_\mu p_\mu)\, u(p) = 0 \tag{5.10}$$

while (5.6b) becomes

$$(\gamma_0 p_0 + \boldsymbol{\gamma} \cdot \mathbf{p})\, u(\mathbf{p}) = 0 \tag{5.11}$$

Most often one choses to work with the representation ($|\sigma\mathbf{p}1\rangle$ or $|\sigma\mathbf{p}2\rangle$) which leads to the simplest form of the polarisation condition, namely

$$\begin{cases} (\gamma \cdot p)\, u(\mathbf{p}) = 0 & (5.12\text{a}) \\ (1 - \gamma_5)\, u(\mathbf{p}) = 0 & (5.12\text{b}) \end{cases}$$

or

$$\begin{cases} (\gamma \cdot p)\, u(p) = 0 & (5.13\text{a}) \\ (1 + \gamma_5)\, u(p) = 0. & (5.13\text{b}) \end{cases}$$

Note that in this case the polarisation condition (5.12a) or (5.13a) is the natural limit of the Dirac equation as the mass tends to zero.

6. Helicity

We wish to define a physical quantity which allows one to describe the states of massive and of massless particles in very similar ways.

So far we have used Wigner's formalism, where the spin states of a particle at rest are defined by the spin component along a z-axis fixed once and for all, while the spin states of a moving particle are found by applying Lorentz boosts $L(\mathbf{p})$ to the states at rest.

Actually, nothing prevents us from chosing the spin quantisation axis for particles at rest to suit our own convenience. Apart from choosing a direction fixed once and for all, as we have done hitherto, the only other natural choice is to take the spin quantisation axis along the momentum of the particle whose states we wish to describe eventually. With $\hat{\mathbf{p}}$ the unit vector in this direction, we specify the states of the particle at rest by the quantum number

$$\Lambda = \mathbf{J} \cdot \hat{\mathbf{p}} \tag{6.1}$$

called the *helicity*. If we denote by $|\lambda\mathbf{0}\rangle$ the state of a particle at rest corresponding to the value λ of the quantum number Λ, then the state of a moving particle corresponding to the same value λ is defined by

$$|\,\lambda\mathbf{p}\rangle = \mathscr{D}[L(\mathbf{p})]\,|\,\lambda\mathbf{0}\rangle \tag{6.2}$$

The following properties of the helicity should be noted:

1. It can take the value $\lambda = -j, -j+1, ..., j$, where j is the spin of the particle.
2. The helicity of a single-particle state is unchanged by a pure Lorentz transformation with velocity parallel to $\mathbf{p}$, provided that the sign of $\mathbf{p}$ is unchanged. (If the velocity of the Lorentz transformation is opposite to $\mathbf{p}$ and exceeds the velocity of the particle, then $\mathbf{p}$ changes sign and the unit vector $\hat{\mathbf{p}}$ occurring in (6.1) corresponds not to the new momentum, but to its opposite. Thus the new helicity, appropriately defined, has the value $-\lambda$, even though Λ is unchanged). The invariance of Λ depends on the commutation relations of the Lorentz group of the type
$$[J_1, K_1] = 0 \tag{6.3}$$
which ensure that a Lorentz transformation leaves unchanged the spin component along its velocity.
3. Since $\mathbf{J}$ is an axial vector and $\hat{\mathbf{p}}$ a vector, Λ is a pseudoscalar and changes sign under space reflection.

By referring to the expression (5.6), chapter 4, for the vector W_μ, one sees that the helicity is simply the ratio $-(W^0/|\mathbf{p}|)$; in the limit as the particle mass tends to zero, this becomes the ratio $-(W^0/p^0)$, which is just the ratio of corresponding components of W_μ and p_μ referred to any frame, and which determines the state of a zero-mass particle. One has, finally,

$$W_\mu\,|\,\lambda\mathbf{p}\rangle = -\lambda p_\mu\,|\,\lambda\mathbf{p}\rangle \tag{6.4}$$

Problems

1. Show that the trace of the product of an odd number of γ matrices vanishes. Show that the trace of the product of γ_5 and an odd number of γ matrices vanishes.

2. Directly from the properties of the matrices $\mathscr{D}^{(1/2)}$ and $\bar{\mathscr{D}}^{(1/2)}$ show that $S^{-1}\gamma_\mu S = \Lambda_{\mu\nu}\gamma_\nu$, where

$$S = \begin{bmatrix} \mathscr{D}^{(1/2)}(\Lambda) & 0 \\ 0 & \bar{\mathscr{D}}^{(1/2)}(\Lambda) \end{bmatrix}$$

3. Show that for any solution $u(p)$ of the Dirac equation, one has

$$\frac{\gamma_\mu p_\mu + m}{2m} u(p) = u(p).$$

4. Use the 3-axis for measuring the spin-component of a particle at rest; let S be the four-vector obtained by acting on the unit vector along the 3-axis with the boost $L(p)$, and let $u(p)$ be the state given by this transformation applied to the spin state $\sigma = \epsilon/2$, ($\epsilon = \pm 1$) for a particle at rest. Show that

$$u(p) = \frac{1}{4m}(1 + \gamma_5\gamma_\mu S_\mu)(\gamma_\mu p_\mu + m)\, u(p) = P(p, S)\, u(p).$$

Show further that the operator $P(p, S)$ is a projection operator ($P^2 = P, P^+ = P$).

Part II

The second part of this book outlines the most important techniques of particle theory. The ideas that are involved will not be discussed in sufficient depth and detail to make them immediately applicable; and even quite fundamental ones, like Feynman's rules, will be described with too little justification to make them fully understood. Our chief aim has been to present a relatively complete panoramic view of the many different aspects of particle physics, and to be deliberately superficial since we lack space for a study in depth. In spite of all its unavoidable drawbacks, we believe that such an attempt answers a genuine need for an overall survey of a subject so various, so difficult, and so complex.

CHAPTER 11

QUANTUM ELECTRODYNAMICS

After recalling the classical Maxwell equations, we give a description of the atom, based on the Dirac equation.

Quantisation of the electromagnetic field introduces the concept of photons. We display the equivalence between the notions of a quantised field and of particles, and are thus led to introduce the quantised electron field.

The quantisation of the electromagnetic field is applied to the emission of radiation by atoms.

Having established the Hamiltonian for the interaction between electrons and photons, we investigate the problems arising in perturbation theory; they lead us to postulate the Feynman rules, which are treated by way of examples.

We sketch some especially striking applications, to the anomalous magnetic moment of the electron, to vacuum polarisation, and to the Lamb shift; and we discuss the problems attending renormalisation.

1. Classical electrodynamics

We start from classical electrodynamics, summed up with admirable economy and elegance by Maxwell's equations. In vacuo, a classical electrodynamic field is specified by the electric field vector $\mathbf{E}(\mathbf{x}, t)$ and by the axial vector field of magnetic induction $\mathbf{B}(\mathbf{x}, t)$; they are determined by the current density $\mathbf{j}$ and the charge density ρ of matter through Maxwell's equations*

$$\begin{aligned} \operatorname{rot}\mathbf{E} &= -\frac{1}{c}\frac{\partial \mathbf{B}}{\partial t} \qquad & \operatorname{rot}\mathbf{B} &= \frac{1}{c}\frac{\partial \mathbf{E}}{\partial t} + 4\pi\frac{\mathbf{j}}{c} \\ \operatorname{div}\mathbf{B} &= 0 \qquad & \operatorname{div}\mathbf{E} &= 4\pi\rho. \end{aligned} \tag{1.1}$$

* Translator's note: in English, the vector operator $\nabla \wedge$ is more commonly denoted by curl, though in this book the notation rot is retained for typographical reasons.

These equations are compatible only if the current and charge densities satisfy the condition

$$\frac{\partial \rho}{\partial t} + \text{div}\, \mathbf{j} = 0 \tag{1.2}$$

which expresses the conservation of charge.

We know that Maxwell's equations are invariant under Lorentz transformations. This invariance can be displayed explicitly if the fields **E** and **B** are merged into an antisymmetrix tensor $F_{\mu\nu}(x)$ in 4-dimensional space by writing

$$F_{ij} = \epsilon_{ijk} B_k \qquad (i, j, k = 1, 2, 3) \tag{1.3}$$

$$F_{0j} = -F_{j0} = E_j\,.$$

Then Maxwell's equations assume the manifestly covariant form

$$\frac{\partial F_{\mu\nu}}{\partial x_\nu} = \frac{4\pi}{c} j_\mu \tag{1. 4a}$$

$$\epsilon_{\mu\nu\rho\sigma} \frac{\partial}{\partial x_\nu} F_{\rho\sigma} = 0. \tag{1.4b}$$

The first of these equations shows how the material sources generate the field. The four-vector j_μ appearing on the right has time-component ρ and space components **j**. The second equation, (1.4b), governs the structure of the field. It makes $F_{\mu\nu}$ into the curl (see footnote on page 211) of a four-vector $A_\mu(x)$, or in other words

$$F_{\mu\nu} = \frac{\partial A_\mu}{\partial x_\nu} - \frac{\partial A_\nu}{\partial x_\mu}\,. \tag{1.5}$$

The time component of this four-vector is simply the electric potential Φ, while its space component is the vector potential **A**. These potentials determine the fields **E** and **B** through the equations

$$\mathbf{E} = -\text{grad}\, \Phi - \frac{1}{c}\frac{\partial \mathbf{A}}{\partial t} \tag{1.6a}$$

$$\mathbf{B} = \text{rot}\, \mathbf{A}. \tag{1.6b}$$

which follow from (1.5).

Note finally the relativistic form of the condition for charge conservation (1.2):

$$\frac{\partial j_\mu}{\partial x_\mu} = 0. \tag{1.7}$$

There are many different ways to establish Maxwell's equations. In practice, Coulomb's law plus relativistic invariance lead naturally to (1.1), as soon as one accepts that an electromagnetic field is fully specified by prescribing two physical quantities like **E** and **B**, and that there exist no magnetic monopoles in nature.

These results challenge us to harmonise them with quantum mechanics. At first sight the prospect of having to explore electrodynamics on a scale appropriate to particles appears rather formidable. Many difficulties must certainly be expected: we shall have to transcend the description of the fields by means of the classical vectors **E** and **B**, in order to exhibit the corpuscular structure of the photon. We shall also need a much refined description of matter, in order to allow for the quantal behaviour of charges (electrons, protons, etc.), and for new phenomena like the creation of particle-antiparticle pairs. We shall have to interpret the equations governing the propagation of the field and its generation by charges down to subnuclear distances (10^{-13} cm), and eventually to distances smaller still.

When all these difficulties are surmounted, and all these effects included, how much will still be left of the miraculous simplicity of Maxwell's equations? A priori, we might well expect the worst kinds of complications. Why should the linearity of Maxwell's equations be maintained for the very intense field close to and inside a proton? Should we not expect that Coulomb's law will be modified at very small distances? In any case, we should be prepared to find that every change in the scale of length and the inclusion of every new effect is reflected by new aggravations, not only in the mathematical formalism, but also in the number of terms to be retained in the equations; every such new term being shaped in detail by new experimental results.

It is staggering that in fact there are no such complications whatever; finally, after appropriate reinterpretation and clarification, Maxwell's equations apply just as well down to the smallest distances and up to the largest attainable energies as they do in our everyday macroscopic world. Better still, we can understand them more thoroughly. Thus, as we plunge towards phenomena further and further removed from our customary intuition, we need merely indicate how Maxwell's equations are to be interpreted, exploited, and refined: in other words, how we are to read them.

2. The relativistic electron

In the nonrelativistic theory of atoms, the Coulomb field of the nucleus is taken as a classical field through which the electrons move while they themselves are described by quantum mechanics. It is known that in this way one can calculate the energy levels say of the hydrogen atom. They are specified by the value ℓ of the orbital angular momentum, and by a principal (or nodal) quantum number n, which we will meet again when considering Regge poles. At this stage the electron spin is irrelevant, at least for hydrogen. Recall that the energy of the level (n, ℓ) is given by

$$E_n = -\frac{m\alpha^2}{2\hbar^2 n^2} \qquad \alpha = \frac{e^2}{\hbar c} \simeq \frac{1}{137} \tag{2.1}$$

where m is the electron mass (or better, the reduced mass), and α is the fine structure constant.

Our next job is to take into account relativistic effects on the motion of the electron, while continuing to describe the nuclear Coulomb field classically. To this end, we look for inspiration to the dynamics of a charged relativistic particle in classical physics. We know that in an electromagnetic field such a particle, with velocity $\mathbf{v}$, is acted on by the Lorentz force

$$\mathbf{F} = e\left(\mathbf{E} + \frac{\mathbf{v} \wedge \mathbf{B}}{c}\right). \tag{2.2}$$

This force can be described within the framework of analytical mechanics which, as we know, is the form of classical mechanics easiest to quantise. A law of motion corresponding precisely to the effects of the Lorentz force ensues if we attribute to the charged particle the Lagrangian

$$L(\mathbf{x}, \mathbf{v}) = -mc^2\left(1 - \frac{v^2}{c^2}\right)^{1/2} + \frac{e}{c}\mathbf{A}(\mathbf{x}) \cdot \mathbf{v} - e\Phi(\mathbf{x}) \tag{2.3}$$

In order to quantise we must find the Hamiltonian corresponding to the Lagrangian (2.3). One begins by finding the momentum $\boldsymbol{\pi}$ canonically conjugate to the position vector:

$$\boldsymbol{\pi} = \frac{\partial L}{\partial \mathbf{v}} = \frac{m\mathbf{v}}{\sqrt{1 - v^2/c^2}} + \frac{e}{c}\mathbf{A}(\mathbf{x}) \tag{2.4}$$

Then the energy is given by

$$W = \mathbf{v} \cdot \frac{\partial L}{\partial \mathbf{v}} - L = \frac{mc^2}{\sqrt{1 - v^2/c^2}} + e\Phi = \left[m^2c^4 + c^2\left(\boldsymbol{\pi} - \frac{e}{c}\mathbf{A}\right)^2\right]^{1/2} + e\Phi \tag{2.5}$$

If we compare (2.5) with the energy of a free particle, namely

$$E = c[m^2c^2 + p^2]^{1/2} \tag{2.6}$$

we see that the energy of a particle in a field can be obtained from the energy of a free particle by the substitution

$$W \to W - e\Phi \qquad \mathbf{p} \to \boldsymbol{\pi} - \frac{e}{c}\mathbf{A} \tag{2.7}$$

In relativistic notation, this is

$$p_\mu \to p_\mu - \frac{e}{c}A_\mu. \tag{2.8}$$

Even in quantum mechanics it seems appropriate to use the same simple transformation to obtain the Hamiltonian of a charged particle from that of a free (i.e. uncharged) one. In this way one obtains an interaction between field and particle which has been dubbed the "minimal coupling"; as we shall see, it suffices to describe the interaction between electrons or muons and photons, and possibly also between photons and more complex particles.

It is not known at present whether all electromagnetic couplings, including those "induced" by the strong interactions, can be considered to be minimal in the sense of (2.8).

In chapter 10 we saw that the quantum states of a free relativistic electron can be described by the Dirac equation

$$(\gamma_\mu p_\mu - m)\, u(p) = 0 \tag{2.9}$$

where the wavefunction $u(p)$ is a 4-component spinor. This form ensures manifest relativistic covariance, including covariance under space reflection. Then the assumption of minimal coupling leads to the following equation for the quantised motion of an electron in an electromagnetic field:

$$\left[\gamma_\mu p_\mu - \frac{e}{c}\gamma_\mu A_\mu(x) - m\right] u(p) = 0. \tag{2.10}$$

Introducing the Fourier transform of $u(p)$ by

$$\psi(x) = \int e^{+ipx} u(p)\, d^4p, \tag{2.11}$$

we obtain the Dirac equation in x-space:

$$\left[i\gamma_\mu \frac{\partial}{\partial x_\mu} - \frac{e}{c}\gamma_\mu A_\mu(x) - m\right] \psi(x) = 0. \tag{2.12}$$

The consequences of the Dirac equation (2.12) agree very well with experiment. The two most striking results are obtained in the special cases of a uniform magnetic field and an electrostatic Coulomb field, respectively. Thus one predicts the magnetic moment of the electron, and an expression for the energy levels of the hydrogen atom including relativistic effects.

If we take for $\mathbf{A}(x)$ a potential which generates a uniform magnetic field $\mathbf{B}$, and if we investigate the resulting motion of the electron in the nonrelativistic limit, then we find that the electron displays a magnetic moment μ_0 whose value is given by

$$\mu_0 = \frac{e\hbar}{mc} \tag{2.13}$$

which is the Bohr magneton for the electron.

If we take $\mathbf{A}(\mathbf{x}) = 0$, and for $\Phi(\mathbf{x})$ the Coulomb field e/r, then the Dirac equation determines the energy levels of the hydrogen atom, taking into account relativistic corrections to the electron motion. One needs merely to look for square integrable solutions of (2.14). The Schroedinger equation predicts a degeneracy between all states with the same principal quantum number n but with different orbital angular momenta; this degeneracy is now lifted, and the energy depends explicitly on the total angular momentum J of the electron, through the formula

$$E_{mJ} = m\left[1 + \frac{e^4}{(n - \epsilon_J)^2}\right]^{-1/2}, \qquad \hbar = c = 1$$

$$\epsilon_J = J + \tfrac{1}{2} - \sqrt{(J + \tfrac{1}{2})^2 - e^4}, \qquad n = 1, 2, \ldots, \infty, \qquad J = \tfrac{1}{2}, \tfrac{3}{2}, \ldots, n - \tfrac{1}{2} \tag{2.14}$$

Note that the total angular momentum $\mathbf{J}$, which satisfies $\mathbf{J}^2 = J(J+1)\hbar^2$, is the sum of the orbital angular momentum $\mathbf{L}$ and of the electron spin. To any given value of J there correspond in general two possible values of ℓ (defined by $\mathbf{L}^2 = \ell(\ell + 1)\hbar^2$), namely $\ell = J \pm \frac{1}{2}$. The corresponding states have opposite parities $(-)^\ell$. The formula (2.14) shows that these states, though different, have the same energy, so that the energy levels of hydrogen are still degenerate with respect to parity. Thus, the ground state,* $1s_{1/2}$, which was unique, remains so; but the next higher states corresponding to $n = 2$, $\ell = 0$ and 1, which were degenerate as solutions of the Schroedinger equation, are partially split according to their J value as solutions of the Dirac equation: the states $2s_{1/2}$ and $2p_{1/2}$ remain degenerate, but are separated in energy from the state $2p_{3/2}$ (Figure 11.2.1).

3. Quantisation of the electromagnetic field. Photons

We shall now consider the problem of photons; in order to unmask the particles latent in the electromagnetic field, we must obviously quantise Maxwell's equations. It is easiest to start by observing that from a dynamical point of view the electromagnetic field, like all mechanical systems with linear equations of motion, is just a superposition of simple harmonic oscillators, and that we know perfectly well how to quantise these. This superposition property becomes obvious if the potential for a free field is expressed as a Fourier integral.

(a) *Fourier expansion of the field*

Start from the Fourier integral

$$A_\mu(x) = \int \tilde{A}_\mu(k)\, e^{ikx}\, d^4k. \tag{3.1}$$

* The notation is $n = 1$, $\ell = 0$, $J = \frac{1}{2}$.

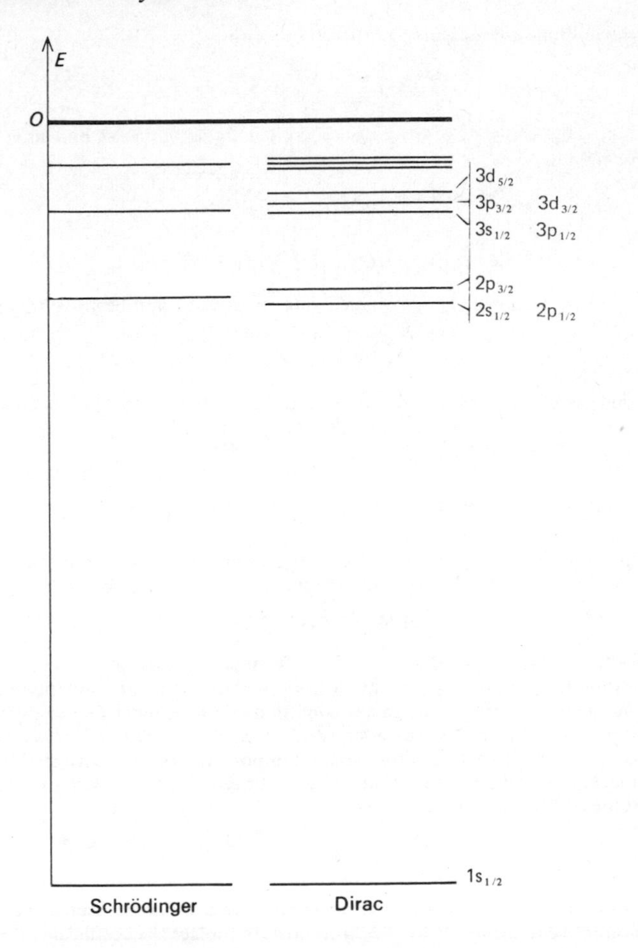

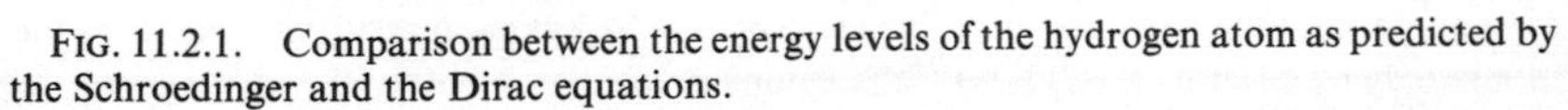

FIG. 11.2.1. Comparison between the energy levels of the hydrogen atom as predicted by the Schroedinger and the Dirac equations.

Next, note that d'Alembert's equation for the free field

$$\Box A_\mu = \left(\frac{\partial^2}{\partial t^2} - \nabla^2\right) A_\mu(x) = 0 \tag{3.2}$$

entails the following condition on the Fourier transform $\tilde{A}_\mu(k)$:

$$k^2 \tilde{A}_\mu(k) = 0. \tag{3.3}$$

It follows at once that $\tilde{A}_\mu(k)$ is proportional to $\delta(k^2)$:

We substitute (3.4) into (3.1) and can then perform the k^0-integration; this yields two contributions to $A_\mu(x)$, deriving from the positive and negative roots, respectively, of the equation for k_0, namely $k^2 = 0$:

$$k^0 = \pm \mid \mathbf{k} \mid. \tag{3.5}$$

The two contributions are separated explicitly by writing

$$A_\mu(x) = \int (\bar{A}_\mu^{(+)}(\mathbf{k})\, e^{ik_0x - i\mathbf{k}\cdot\mathbf{x}} + \bar{A}_\mu^{(-)}(\mathbf{k})\, e^{-ik_0x - i\mathbf{k}\cdot\mathbf{x}}) \frac{d^3k}{2k^0} \tag{3.6}$$

Here, and later, k_0 is always positive: $k_0 = |\mathbf{k}|$. Changing variables and intercharging $\mathbf{k}$ and $-\mathbf{k}$, one obtains

$$A_\mu(x) = \int (A_\mu^{(+)}(\mathbf{k})\, e^{ikx} + A_\mu^{(-)}(\mathbf{k})\, e^{-ikx}) \frac{d^3k}{2k^0} \tag{3.7}$$

where

$$\bar{A}_\mu^{(+)}(\mathbf{k}) = A_\mu^{+}(\mathbf{k}) \qquad A_\mu^{(-)}(\mathbf{k}) = -\bar{A}_\mu^{(-)}(\mathbf{k}).$$

Equation (3.7) has separated $A_\mu(x)$ explicitly into its positive and negative frequency components, since

$$kx = k_0 t - \mathbf{k} \cdot \mathbf{x}. \tag{3.8}$$

At this point, we confine ourselves to describing the radiation field in the gauge where

$$A_0(x) = 0 \qquad \text{div } \mathbf{A}(x) = 0 \tag{3.9}$$

Hence one has

$$A_0^{(\pm)}(\mathbf{k}) = 0 \qquad \mathbf{k} \cdot \mathbf{A}^{(\pm)}(\mathbf{k}) = 0. \tag{3.10}$$

The second condition shows that the vectors $\bar{A}^{(\pm)}(\mathbf{k})$ are orthogonal to $\mathbf{k}$. It is possible to formulate this in a simple way if, instead of using a reference frame fixed once and for all, we introduce, at each point of $\mathbf{k}$ space, orthogonal axes along three orthonormal vectors

$$(\mathbf{k}/|\mathbf{k}|, \boldsymbol{\epsilon}^{(1)}(\mathbf{k}), \boldsymbol{\epsilon}^{(2)}(\mathbf{k})).$$

Thus, the vectors $\boldsymbol{\epsilon}^{(1)}(\mathbf{k})$ and $\boldsymbol{\epsilon}^{(2)}(\mathbf{k})$ form a basis for three-vectors orthogonal to $\mathbf{k}$, and the second condition in (3.10) shows that the vectors $\mathbf{A}^{(\pm)}(\mathbf{k})$ are linear combinations of $\boldsymbol{\epsilon}^{(1)}(\mathbf{k})$ and $\boldsymbol{\epsilon}^{(2)}(\mathbf{k})$. Actually this form of the gauge condition can be extended so as to embrace the first constraint in (3.10): we now consider two four-vectors $\epsilon_\mu^{(1)}(\mathbf{k})$ and $\epsilon_\mu^{(2)}(\mathbf{k})$ whose time-components $\epsilon_0^{(1,2)}(\mathbf{k})$ are zero, and whose space-components coincide with $\boldsymbol{\epsilon}^{(1,2)}(\mathbf{k})$. Then all the conditions of this 'radiation gauge' are obeyed if the $A_\mu^{(\pm)}(\mathbf{k})$ are taken to be linear combinations of the $\epsilon_\mu^{(1,2)}(\mathbf{k})$, so that

$$A_\mu(x) = \sum_{i=1,2} \int [\epsilon_\mu^{(i)}(\mathbf{k})\, \alpha_i^{(+)}(\mathbf{k})\, e^{ikx} + \epsilon_\mu^{(i)}(\mathbf{k})\, \alpha_i^{(-)}(\mathbf{k})\, e^{-ikx}] \frac{d^3k}{2k^0}. \tag{3.11}$$

In view of the fact that the field $A_\mu(x)$ must be real, we see that the coefficients $\alpha_i^{(+)}(\mathbf{k})$ and $\alpha_i^{(-)}(\mathbf{k})$ are complex conjugates. It will be convenient to replace the coefficients $\alpha_i^{(\pm)}(\mathbf{k})$, which have appeared quite spontaneously, by new coefficients $a_i(\mathbf{k})$ and $a_i^*(\mathbf{k})$ differing from the former only by a factor of $\sqrt{2k_0}(2\pi)^{3/2}$. Then one has

$$A_\mu(x) = \sum_{i=1,2} \int \left[\frac{\epsilon_\mu^{(i)}(\mathbf{k})}{\sqrt{2k^0}} a_i(\mathbf{k})\, e^{ikx} + \frac{\epsilon_\mu^{(i)}(\mathbf{k})}{\sqrt{2k^0}} a_i{}^*(\mathbf{k})\, e^{-ikx}\right] \frac{d^3\mathbf{k}}{(2\pi)^{3/2}} \tag{3.12}$$

Recall that $k^0 = |\mathbf{k}|$, and that the two four-vectors $\epsilon_\mu^{(1,2)}(\mathbf{k})$ have space-components only and are orthogonal to $\mathbf{k}$.

So far we have simply Fourier-analysed the field. To understand how the new variables $a_i(\mathbf{k})$ are relevant to the dynamics, we shall express the Hamiltonian in terms of these parameters. To this end we exploit the expressions (1.6) for the fields $\mathbf{E}$ and $\mathbf{B}$ in terms of the vector potential; this allows them to be written as functions of the coefficients $a_i(\mathbf{k})$. The energy of the electromagnetic field, namely

$$\mathrm{H} = \frac{1}{8\pi} \int (\mathbf{E}^2(\mathbf{x}, t) + \mathbf{B}^2(\mathbf{x}, t))\, d^3x \tag{3.13}$$

can thus be written in the form

$$\mathrm{H} = \sum_{i=1,2} \int k^0 a_i{}^*(\mathbf{k})\, a_i(\mathbf{k}) \frac{d^3k}{(2\pi)^3} \tag{3.14}$$

This form makes it immediately obvious that from the point of view of its dynamics the electromagnetic field is a superposition of harmonic oscillators. Indeed, $k^0 a_i^*(\mathbf{k}) a_i(\mathbf{k})$ is just the familiar expression for the energy of an oscillator with frequency k^0, while the integral

$$\sum_i \int \frac{d^3k}{(2\pi)^3} \tag{3.15}$$

is evidently a sum over all kinematic and polarisation states (**k** and (i) respectively) in unit volume.

(b) *Review of the harmonic oscillator*

The electromagnetic field is easy to quantise now that we know it to be a superposition of harmonic oscillators. It will be useful to begin by recalling some properties of the harmonic oscillator in quantum mechanics. Its Hamiltonian can be written as

$$\mathrm{H} = \tfrac{1}{2}(P^2 + \omega^2 Q^2) \tag{3.16}$$

where Q and P are conjugate variables, and consequently satisfy the canonical commutation rule

$$[Q, P] = i. \tag{3.17}$$

One traditional method of quantising the harmonic oscillator introduces the operators

$$a = \frac{P + i\omega Q}{\sqrt{2}} \qquad a^+ = \frac{P - i\omega Q}{\sqrt{2}} \tag{3.18}$$

which satisfy the commutation rule

$$[a, a^+] = 1 \tag{3.19}$$

Then the Hamiltonian factorises in the form

$$\mathrm{H} = (\tfrac{1}{2} + a^+ a)\, \omega. \tag{3.20}$$

One knows that in this formalism the ground state $|0\rangle$ of the harmonic oscillator has energy $\frac{1}{2}\omega$, and satisfies the condition

$$a\,|\,0\rangle = 0. \tag{3.21}$$

The excited states, properly normalised, are then given by

$$|n\rangle = \frac{1}{\sqrt{n!}} (a^+)^n\,|0\rangle \tag{3.22}$$

the energy of the state $|n\rangle$ being $(\frac{1}{2} + n)\omega$. Hence the operator a^+a is a 'number' operator, in the sense that it is a Hermitean operator whose eigenstates are the states $|n\rangle$, and whose eigenvalues are all the nonnegative integers $n = 0, 1, 2, 3, \ldots$.

(c) *Quantisation of the electromagnetic field*

We shall quantise the electromagnetic field by quantising separately each of its constituent oscillators. To each value of the polarisation index i and to every value of the wavenumber $\mathbf{k}$ there corresponds one oscillator. We make sure that these different dynamical systems are all independent by imposing on the corresponding operators a and a^+ the commutation rules

$$[a_i(\mathbf{k}), a_j(\mathbf{k}')] = [a_i(\mathbf{k}), a_j^+(\mathbf{k}')] = 0 \tag{3.23}$$

for $\mathbf{k}' \neq \mathbf{k}$ and $i \neq j$. For $\mathbf{k} = \mathbf{k}'$, the fact that $\mathbf{k}$ varies continuously leads us to replace the right hand side of (3.19) by a δ-function; thus we write

$$\begin{aligned}[a_i(\mathbf{k}), a_j(\mathbf{k}')] &= 0 \\ [a_i(\mathbf{k}), a_j^+(\mathbf{k}')] &= \delta(\mathbf{k} - \mathbf{k}')\,\delta_{ij}\,.\end{aligned} \tag{3.24}$$

Comparing the Hamiltonians of the electromagnetic field and of the harmonic oscillator, we see that the operator $a_i(\mathbf{k})$ is simply the quantised version of the dynamical variable denoted by the same symbol. Its Hermitean adjoint $a_i^+(\mathbf{k})$ is the quantised version of the complex conjugate variable $a_i^*(\mathbf{k})$. They have been written in the Heisenberg picture where these operators are time-independent.

To every oscillator there corresponds a number operator

$$N_i(\mathbf{k}) = a_i^+(\mathbf{k})\, a_i(\mathbf{k})$$

whose eigenvalues will enable us to label the states and their energies. In particular, the Hamiltonian can be put in the form

$$\mathrm{H} = \sum_{i=1,2} \int k^0 a_i^+(\mathbf{k}) a_i(\mathbf{k}) \frac{d^3k}{(2\pi)^3}\,. \tag{3.25}$$

The ground state of this Hamiltonian is the direct product of the oscillator ground states; in other words it is a state, again written as $|0\rangle$, obeying the conditions

$$a_i(\mathbf{k})\,|0\rangle = 0 \tag{3.26}$$

for all i and $\mathbf{k}$. We call it the vacuum.

The state

$$|\,\mathbf{k}_i\rangle = a_i^+(\mathbf{k})\,|\,0\rangle \tag{3.27}$$

has all the physical properties of a state representing one free photon with momentum $\mathbf{k}$ and polarisation $\epsilon^{(i)}(\mathbf{k})$. To see this, note first that by (3.25) its energy is k^0. The operators for the total momentum and angular momentum of the field as well as for the energy can be found in explicit form; it can then be verified that (3.27) is indeed one of their eigenstates and that it belongs to the appropriate eigenvalues.

Similarly, a state containing r photons with momenta $\mathbf{k}_1 \cdots \mathbf{k}_r$ and polarisations $\epsilon^{i_1} \cdots \epsilon^{i_r}$ respectively can be written as

$$| \mathbf{k}_1 i_1 ,..., \mathbf{k}_r , i_r \rangle = a^+_{i_1}(\mathbf{k}_1) \cdots a^+_{i_r}(\mathbf{k}_r) \,|\, 0\rangle \tag{3.28}$$

Thus, states containing 1, 2, 3, ... photons can be constructed by acting on the vacuum with the operators a^+, which it is therefore natural to call creation operators. Conversely, the a are called annihilation operators because in general they remove one photon from any state on which they act.

Since all creation operators commute, we note that the state described by (3.28) is symmetric (invariant) under the interchange of any pair of photons. Thus, the construction we have adopted has led us automatically to photons obeying Bose–Einstein statistics.

4. Photons and Maxwell's equations

It is pleasing to have recovered the photon by quantising Maxwell's equations. Nevertheless it would be just as useful to understand how, conversely, the electromagnetic field and Maxwell's equations can emerge as macroscopic manifestations of photons.

We begin by recalling that Maxwell's equations are simply the Dirac equation for the photon, in the sense of chapter 10; and we sketch this approach in outline, as follows.

Let a photon be defined as a particle of spin 1 and zero mass. We saw in chapter 4 that a state of such a particle is labeled by its momentum $\mathbf{p}$ and its helicity $\lambda = \pm 1$, so that we write it as $|\mathbf{p}, \lambda\rangle$. In this instance, the helicity is invariant under all the operations of the Poincaré group, and the two helicity states are connected only by space reflection:

$$\Pi \,|\, \mathbf{p}\lambda\rangle = |\, -\mathbf{p}, -\lambda\rangle. \tag{4.1}$$

Exactly how this state transforms under a Lorentz transformation Λ depends, a priori, on the momentum, just as for massive particles. In view of the fact that for a zero-mass particle the helicity is invariant under Lorentz transformations, the state $|\mathbf{p}\lambda\rangle$ transforms into $|\mathbf{p}'\lambda\rangle$, where $p' = \Lambda p$. The only other possible effect is multiplication by a coefficient; because the transformation is unitary, such a coefficient must have the form given by

$$U\{\Lambda\} \,|\, \mathbf{p}\lambda\rangle = e^{+i\lambda\theta(\mathbf{p},\Lambda)} \,|\, \Lambda\mathbf{p}, \lambda\rangle \tag{4.2}$$

where the angle θ depends explicitly on $\mathbf{p}$ and on Λ. Just as for massive spin $\frac{1}{2}$ particles, it is possible to eliminate the $\mathbf{p}$-dependence of equation (4.1) by introducing extra state vectors. For spin $\frac{1}{2}$ we had to introduce $2 \times (2s + 1) = 4$ such state vectors; the present case is analogous, and the number needed now is $2 \times (2 + 1) = 6$. Under the Lorentz group these six state vectors transform like the six components of an antisymmetric tensor. But in actual fact there are only two independent states, whence the new vectors thus introduced must satisfy supplementary conditions, which have the form of Maxwell's equations in vacuo.

Next, let us see how to construct a quantised field by starting from particles. One inverts the construction given in the last paragraph, which allowed one to build photon states by quantising the electromagnetic field. Our starting point now is that we are given the state vectors describing arbitrary numbers of photons with momenta $\mathbf{k}_1, \ldots \mathbf{k}_r$ and helicities $\lambda_1, \ldots \lambda_r$; and we can, to begin with, replace the notation specifying helicities by one specifying instead the polarisation vectors $\epsilon_\mu^{(i)}$. To do this we need merely construct, for every photon, two unit three-vectors $\boldsymbol{\epsilon}^{(1)}(\mathbf{k})$ and $\boldsymbol{\epsilon}^{(2)}(\mathbf{k})$ orthogonal to its momentum $\mathbf{k}$; their direction will indicate the transverse polarisation of the light. A state with helicity λ represents a photon with its spin collinear with its momentum; in practice such a photon is associated with an intrinsic rotational motion about its momentum. Such motion can correspond only to light in a state of circular polarisation, implying that the relation between the states of transverse polarisation and helicity is the same as the familiar optical relation between, on the one hand, the electric field components of transversely polarised waves along two directions normal to the wavenumber, and, on the other hand, the electric field of a circularly polarised wave. Thus,

$$| \mathbf{k}\lambda\rangle = \frac{1}{\sqrt{2}} [| \mathbf{k}\epsilon^{(1)}\rangle \pm i \mid \mathbf{k}\epsilon^{(2)}\rangle]. \tag{4.3}$$

Applying this procedure to every photon, we recover the states $|\mathbf{k}_1 i_1, \ldots \mathbf{k}_r i_r\rangle$ of the last section.

In the Hilbert space spanned by the states $|\mathbf{k}_1 i_1, \ldots \mathbf{k}_r i_r\rangle$ we can then formally define creation and annihilation operators $a_i^+(\mathbf{k})$ and $a_i(\mathbf{k})$ through their matrix elements, which are to coincide with those of the operators, introduced in the last section. The operators thus defined automatically satisfy the commutation rules (3.24). From them one could construct the photon number operators $N_i(\mathbf{k}) = a_i^+(\mathbf{k})a_i(\mathbf{k})$.

From these creation and annihilation operators we can build up a new object, namely the potential operator

$$A_\mu(x) = \frac{1}{(2\pi)^{3/2}} \sum_{(i)} \int \frac{\epsilon_\mu^{(i)}(\mathbf{k})}{\sqrt{2k^0}} [a_i(\mathbf{k})\, e^{ikx} + a_i^+(\mathbf{k})\, e^{-ikx}]\, d^3\mathbf{k} \tag{4.4}$$

or in other words a field in terms of which we can construct the electromagnetic fields $\mathbf{E}$, $\mathbf{B}$ as operators satisfying the usual definitions

$$\mathbf{E} = -\mathrm{grad}\, \Phi - \frac{1}{c}\frac{\partial \mathbf{A}}{\partial t} \qquad \mathbf{B} = \mathrm{rot}\, \mathbf{A}. \tag{4.5}$$

Finally, one can link the quantised and the classical fields, or in other words elucidate the connection between the operators (4.5) and the classical fields observed in optics or in radio transmission. One would expect that the wave nature of the field will make itself felt to the full when one deals with physical states containing very many photons. Since the operators (4.5) are Hermitean, the corresponding dynamical variables are observable; clearly they are just the components of the electromagnetic field, and if they are measured in a state $|\alpha\rangle$, then the result is given by

$$\begin{aligned} \mathbf{E}_{\text{physical}} &= \langle\alpha\,|\mathbf{E}_{\text{operator}}|\,\alpha\rangle \\ \mathbf{B}_{\text{physical}} &= \langle\alpha\,|\mathbf{B}_{\text{operator}}|\,\alpha\rangle \end{aligned} \tag{4.6}$$

The state $|\alpha\rangle$ represents a system behaving classically if the quantum fluctuations of the field are negligible, i.e. if the difference

$$\langle\alpha\,|\mathbf{E}^2_{\text{operator}}|\,\alpha\rangle - [\langle\alpha\,|\mathbf{E}_{\text{operator}}|\,\alpha\rangle]^2 \tag{4.7}$$

is small compared to either term separately.

To sum up, there is a two-way connection between photons and electromagnetic fields. Quantising Maxwell's equations leads one to introduce particles with helicity ± 1 and with zero mass. Conversely, if such particles are described in a natural way by the Dirac formalism of chapter 10, then Maxwell's equations appear as consequences of the zero value of the photon mass and of the value 1 of its spin.

5. Fields and particles

The construction of the electromagnetic field from photon states can be generalised to any particles, and in this way one can associate a quantised field with particles of any kind.

(a) *The* π^0

Consider for instance the case of the neutral pion. We can introduce a Hilbert space spanned by the states of noninteracting π^0's, or in other words by the states $|\mathbf{q}_1 \ldots \mathbf{q}_r\rangle$ of such mesons with sharply defined momenta. In the light of experiment we assume that pions obey Bose–Einstein statistics, and therefore consider only states which are symmetric under exchanges of pions.

Just as for photons, we can define annihilation and creation operators through their matrix elements. In practice this amounts to postulating

$$\langle\underbrace{\mathbf{q}_1 \cdots \mathbf{q}_1}_{N\text{ factors}}|\,a^+(q_1)\,|\underbrace{\mathbf{q}_1 \cdots \mathbf{q}_1}_{N-1\text{ factors}}\rangle = \sqrt{N} \tag{5.1}$$

In view of the normalisation of the states, this implies that the annihilation and creation operators once again satisfy the commutation rules (3.24).

From the operators $a(\mathbf{q})$ and $a^+(\mathbf{q})$ we can construct a pion *field* according to the definition

$$\varphi(x) = \frac{1}{(2\pi)^{3/2}} \int \frac{d^3\mathbf{q}}{\sqrt{2q^0}} [a(\mathbf{q})\, e^{iqx} + a^+(\mathbf{q})\, e^{-iqx}] \tag{5.2}$$

where $(q^0)^2 = \mathbf{q}^2 + m^2$, m is the pion mass, and $e^{iqx} = e^{iq^0x^0 - i\mathbf{q}\mathbf{x}}$. This operator is invariant under proper Lorentz transformations. We can see that the pion field $\varphi(x)$ must change sign under space reflection, because the pion is pseudoscalar, and because by virtue of the definition of the annihilation operators we have a non-zero matrix element $\langle 0|\varphi(x)|\mathbf{q}\rangle$ of $\varphi(x)$ between the vacuum and a single-pion state. In other words, if we let $U(\Lambda)$ and Π be the unitary operators which represent, in Hilbert space, the Lorentz transformation Λ and space reflection respectively, then one must have

$$\begin{aligned} U(\Lambda)\, \varphi(x)\, U^{-1}(\Lambda) &= \varphi(\Lambda x) \\ \Pi\varphi(\mathbf{x}, t)\, \Pi^{-1} &= -\varphi(-\mathbf{x}, t). \end{aligned} \tag{5.3}$$

Such a field can be interpreted classically. One can envisage pion states where the field fluctuations are very weak:

$$\langle\alpha \mid \varphi^2 \mid \alpha\rangle - [\langle\alpha \mid \varphi \mid \alpha\rangle]^2 \ll [\langle\alpha \mid \varphi \mid \alpha\rangle]^2. \tag{5.4}$$

States like this necessarily contain very large numbers of pions and cannot be produced in the laboratory. But nothing prevents us for instance from contemplating the possibility that an expanding universe was once so small and so full that the mesons in it were packed tightly enough to form a fluid, (a very viscous one because of their strong interactions), and that the pion field then behaved classically.

(b) *Electrons and protons*

Apart from a few changes, practically the same construction can be implemented for particles with spin $\frac{1}{2}$. The only essential difference is that they obey Fermi–Dirac statistics, and that in their Hilbert space we cannot define creation operators $a^+(\mathbf{q})$ which commute, because from the vacuum we could then construct the *symmetric* state

$$\mid \mathbf{q}_1\mathbf{q}_2\rangle = a^+(\mathbf{q}_1)\, a^+(\mathbf{q}_2) \mid 0\rangle. \tag{5.5}$$

In order that this state be antisymmetric, we must have

$$\mid \mathbf{q}_1\mathbf{q}_2\rangle + \mid \mathbf{q}_2\mathbf{q}_1\rangle = 0$$

whence

$$[a^+(\mathbf{q}_1)\, a^+(\mathbf{q}_2) + a^+(\mathbf{q}_2)\, a^+(\mathbf{q}_1)] \mid 0\rangle = 0 \tag{5.6}$$

This suggests that the operators should now be subjected to *anticommutation* rules instead of the commutation rules (3.19). Moreover, the state of a spinning particle must be labeled by a spin index μ, in the same way as a photon state had to be labeled by the index j giving its polarisation vector. For the creation and annihilation operators $a_\mu^+(\mathbf{p})$ and $a_\mu(\mathbf{p})$ of an electron, one is thus tempted to write the anticommutation rules

$$\begin{aligned} \{a_\mu(\mathbf{p}), a_\nu(\mathbf{p}')\} &= 0 \\ \{a_\mu(\mathbf{p}), a_\nu^+(\mathbf{p}')\} &= \delta_{\mu\nu}\, \delta(\mathbf{p} - \mathbf{p}') \end{aligned} \tag{5.7}$$

where the anticommutator bracket is defined by

$$\{a, b\} = ab + ba. \tag{5.8}$$

On the other hand, suppose we start with *antisymmetric* states $|\mathbf{p}_1\mu_1, \ldots, \mathbf{p}_r\mu_r\rangle$ and define annihilation operators through their matrix elements

$$\langle \mathbf{p}_1\mu_1 \cdots \mathbf{p}_r\mu_r \mid a_\mu(\mathbf{q}) \mid \mathbf{p}_1\mu_1 \cdots \mathbf{p}_r\mu_r\mathbf{q}\mu\rangle = 1 \tag{5.9}$$

Then it is easy to check that the operator $a_\mu(\mathbf{q})$ thus defined, and its adjoint, satisfy the equation (5.7).

How does one build a field with these annihilation and creation operators? Actually, one has some freedom of choice; but we would not learn much physics even if we were to explore and compare all the possibilities in detail, and to chose between them. Hence we confine ourselves to quoting the result.

One begins by repeating for positrons the same construction of creation and annihilation operators as we have followed for electrons. These operators are now denoted by $b_\mu^+(\mathbf{q})$ and $b_\mu(\mathbf{q})$; they obey anticommutation rules similar to (5.7), in order to secure Fermi–Dirac statistics. Moreover we assume that

$$\{a_\mu(\mathbf{p}), b_\alpha(\mathbf{p}')\} = \{a_\mu(\mathbf{p}), b_\mu^+(\mathbf{p}')\} = 0. \tag{5.10}$$

The field can now be constructed along the lines apparent from (4.4). To this end, we chose a geometrical object whose behaviour under Lorentz transformations serves to specify the transformation properties of the single-particle states with given momentum. For the photon, this object was its polarisation vector; here we take the Dirac wavefunction $u^{(\mu)}(\mathbf{p})$ of an electron with momentum $\mathbf{p}$ in a spin state (μ). This geometric object is multiplied by the product $a_\mu^+(\mathbf{p})e^{-ipx}$ or $a_\mu(\mathbf{p})e^{ipx}$ of a creation or an annihilation operator and the corresponding space-time wavefunction. We sum over polarisation states (Σ_μ) and over the kinematic states; and we chose normalisation factors in such a way that the resulting field has simple transformation properties under the Lorentz group. These factors depend on the normalisation previously chosen for the wavefunctions; thus,

$$\psi_a(x) = \sum_\mu \int \frac{d^3p}{(2\pi)^{3/2}} \frac{u_\alpha^{(\mu)}(\mathbf{p})}{\sqrt{p^0}} a_\mu(\mathbf{p})\, e^{i\mathbf{p}\cdot x} + \cdots \tag{5.11}$$

If the spinor u is subject to invariant normalisation (e.g., $\bar{u}u = m$), then $\psi(x)$ does transform simply under Lorentz transformations:

$$U(\Lambda)\, \psi_\alpha(x)\, U^{-1}(\Lambda) = S_{\alpha\beta}(\Lambda)\, \psi_\beta(\Lambda^{-1}x) \tag{5.12}$$

where the 4×4 matrix $S(\Lambda)$ is the one we met when investigating the Lorentz transformation of Dirac wavefunctions.

It will prove useful to complement (5.11) by a negative-frequency contribution constructed with positron creation operators, in order that the field should have simple causality properties; this will be discussed in the next section. Accordingly, we adopt the following expression for the field:

$$\psi(x) = \int \frac{d^3p}{(2\pi)^{3/2}} [e^{ipx}u^{(\mu)}(\mathbf{p})\, a_\mu(\mathbf{p}) + e^{-ipx}v^{(\mu)}(\mathbf{p})\, b_\mu{}^+(\mathbf{p})] \frac{1}{\sqrt{E}} \tag{5.13}$$

This implies

$$\bar{\psi}(x) = \int \frac{d^3p}{(2\pi)^{3/2}} [e^{ipx}\bar{v}^{(\mu)}(\mathbf{p})\, b_\mu(\mathbf{p}) + e^{-ipx}\bar{u}_\mu(\mathbf{p})\, a_\mu{}^+(\mathbf{p})] \sqrt{\frac{1}{E}}\,. \tag{5.14}$$

In these expressions, the wavefunction $v^{(\mu)}(\mathbf{p})$ of the positron has the form $\mathscr{C}u^{(\mu)}(\mathbf{p})$, where $\mathscr{C}$ is the charge-conjugation matrix defined in chapter 10. With this choice, the fields $\psi(x)$ and $\bar{\psi}(x)$ transform in a simple way under the action of the charge conjugation operator. From the definition of charge conjugation, we have

$$C \mid \text{electron } \mathbf{p}\mu\rangle = \mid \text{positron } \mathbf{p}\mu\rangle. \tag{5.15}$$

This implies the following transformation law for the creation operators:

$$Ca_\mu{}^+(\mathbf{p})\, C^{-1} = b_\mu{}^+(\mathbf{p}) \tag{5.16}$$

whence

$$\begin{aligned} C\psi(x)\, C^{-1} &= \mathscr{C}\bar{\psi}(x) \\ C\bar{\psi}(x)\, C^{-1} &= \mathscr{C}\psi(x). \end{aligned} \tag{5.17}$$

Any other choice of field $\psi(x)$ would lead to a more complicated behaviour under C. As we shall see, it would also lead to the fields being nonlocal.

6. Measurability of the fields

The expression (5.2) for the π^0 field has another important property which we shall now discuss: it is causal.

From the known commutation rules for the creation and annihilation operators, one can easily establish the commutation rules for the operators $\varphi(x)$; for instance

$$[\varphi(x), \varphi(y)] = D(x - y) \tag{6.1}$$

where $D(x - y)$ is a distribution expressible in terms of Bessel functions. One fundamental property of this commutator is that for spacelike $(x - y)$ one finds (Figure 11.6.1)

$$D(x - y) = 0 \qquad \text{for} \qquad (x - y)^2 < 0. \tag{6.2}$$

This is evidently a necessary condition if the field $\varphi(x)$ is to have a classical limit. It shows that we can measure the values of the field at two space-like separated points x and y without any quantum mechanic interference between the results of the two measurements. If it were otherwise, then

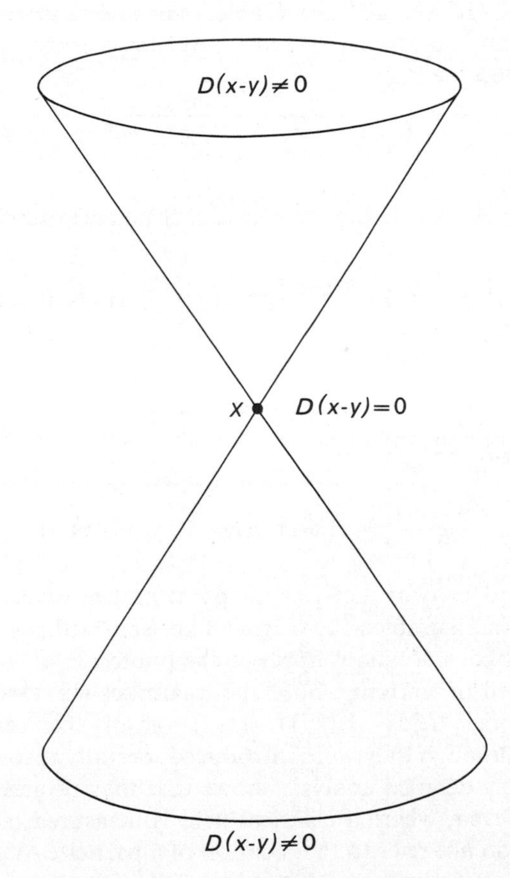

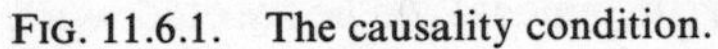
FIG. 11.6.1. The causality condition.

such interference would show the existence of a signal propagating faster than light. It was in order to satisfy this condition that we chose the expression appearing in brackets in (5.2), and not, for example,

$$\varphi(x) = \frac{1}{(2\pi)^{3/2}} \int \frac{d^3q}{\sqrt{2q^0}} [a(\mathbf{q})\, e^{iqx} - a^+(\mathbf{q})\, e^{-iqx}].$$

Similarly, choosing the expression (5.13) for the spinor field entails the anticommutation rules for the fields:

$$\left.\begin{array}{l} \{\psi(x)\, \psi(y)\} = 0 \\ \{\psi(x)\, \bar{\psi}(y)\} = 0 \end{array}\right. \quad \text{for} \quad (x-y)^2 < 0. \tag{6.3}$$

It is clear that these relations do not entail the simultaneous measurability of the fields; but this should not be demanded in any case, because a spinor field changes sign under a rotation through 2π and cannot therefore be measurable. Indeed, all measurable quantities associated with the electron field are quadratic in the field. One example is the energy operator whose eigenvalues we know:

$$W \mid p_1\mu_1\,, p_2\mu_2 \cdots\rangle = (\sqrt{p_1{}^2 + m_1{}^2} + \sqrt{p_2{}^2 + m_2{}^2} + \cdots) \mid p_1\mu_1, p_2\mu_2\rangle \tag{6.4}$$

It can be re-expressed as follows in terms of the creation and annihilation operators:

$$W = \int \sum_{(\mu)} \sqrt{\mathbf{p}^2 + m^2} [a_\mu{}^+(\mathbf{p})\, a_\mu(\mathbf{p}) + b_\mu{}^+(\mathbf{p})\, b_\mu(\mathbf{p})] \frac{d^3p}{(2\pi)^3} \tag{6.5}$$

and in terms of the fields by

$$W = \int \bar{\psi}(x) \gamma_0 (i\boldsymbol{\gamma} . \nabla + m)\, \psi(x)\, d^3x \tag{6.6}$$

If we now identify the quantity

$$\frac{dW(\mathbf{x}, t)}{d^3x} = \bar{\psi}(\mathbf{x}, t)\, \gamma_0 (i\boldsymbol{\gamma} \cdot \nabla + m)\, \psi(\mathbf{x}, t) \tag{6.7}$$

with the energy density at time t at the point $\mathbf{x}$, then we do find that it is simultaneously measurable at two space-like separated points.

All the above considerations apply to the photon.

Finally a word of warning about the quantities $(\mathbf{x}, t)$ which occur as arguments of the fields $A_\mu(\mathbf{x}, t)$, $\varphi(\mathbf{x}, t)$ or of the energy density $dW(\mathbf{x}, t)/d^3x$. Although they were introduced formally through a Fourier transformation, a detailed analysis shows that they do indeed represent the space-time event where a classical field is measured in the classical limit. But they do not refer to the location of a particle. At a fixed time t, the variable $\mathbf{x}$ *cannot* in general be identified with the operator which quantum mechanics associates with the measurement of the position of a particle.

7. Emission and absorption of photons by atoms

To gain a better appreciation of the quantised electromagnetic field as discussed in the last few sections, we shall now show how it can be used to discuss the emission and absorption of photons by atoms (or nuclei). We shall see that a complete description of these effects is made possible by the creation and annihilation operator formalism.

In order to investigate a quantum transition it helps to know the Hamiltonian. Suppose that it can be written in the form

$$\mathrm{H} = \mathrm{H}_0 + \mathrm{H}_1 \tag{7.1}$$

where the small term H_1 is responsible for the transition under study, which takes place between an initial state $|i\rangle$ and a final state $|f\rangle$, both being eigenstates of H_0. Then the transition amplitude is given by first order perturbation theory as

$$T_{i\to f} = \langle f \,|\, \mathrm{H}_1 \,|\, i\rangle. \tag{7.2}$$

We are interested in the case where $|i\rangle$ consists of an excited atom in a state 1, and $|f\rangle$ consists of the same atom in a state 2, plus a photon with momentum **k** and polarisation ϵ. Our first task is to specify the Hamiltonian H.

The Hamiltonian for a relativistic electron in a classical field can easily be found from the Dirac equation (2.12), if the latter is written in the form

$$Eu(\mathbf{p}) = \gamma_0[\boldsymbol{\gamma}\cdot\mathbf{p} + e\gamma_\mu A_\mu^{\mathrm{cl}}(x) + m]\, u(\mathbf{p}) \tag{7.3}$$

which gives

$$\mathrm{H}_{\mathbf{atom}} = \gamma_0[\boldsymbol{\gamma}\cdot\mathbf{p} + m + e\gamma_\mu A_\mu^{\mathrm{cl}}(x)] \tag{7.4}$$

For an electron in an atom, $A_\mu^{\mathrm{cl}}(x)$ is the Coulomb field of the nucleus.

In particular, we see that the Hamiltonian for a free electron is

$$\mathrm{H}_{\textbf{free electrons}} = \gamma_0[\boldsymbol{\gamma}\cdot\mathbf{p} + m] \tag{7.5}$$

and that (7.5) is obtained by adding to this the interaction Hamiltonian with the classical field

$$\gamma_0 e\gamma_\mu A_\mu^{\mathrm{cl}}(x). \tag{7.6}$$

In our case there are also photons described by a quantised field. Ignoring their interaction with the electrons, their Hamiltonian is given by (3.20) as

$$\mathrm{H}_{\textbf{free photons}} = \sum_i \int k^0 a_i^+(\mathbf{k})\, a_i(\mathbf{k}) \frac{d^3k}{(2\pi)^3}\,. \tag{7.7}$$

Thus $|i\rangle$ and $|f\rangle$ are eigenstates of the unperturbed Hamiltonian

$$\mathrm{H}_0 = \mathrm{H}_{\mathrm{atom}} + \mathrm{H}_{\mathrm{free\ photons}}. \tag{7.8}$$

For the interaction Hamiltonian between the electrons and the quantised field we write, by analogy to (7.6),

$$\mathrm{H}_1 = \gamma_0 e \gamma_\mu A_\mu(x). \tag{7.9}$$

With this choice, the canonical formalism derives equations of motion showing that the electromagnetic field $A_\mu(x)$ continues to obey Maxwell's equations (1.1). The current-density four-vector occurring in them is given by

$$j_\mu(\mathbf{x}, t) = \bar{\psi}(\mathbf{x}, t)\, \gamma_\mu \psi(\mathbf{x}, t) \tag{7.10}$$

where $\psi(\mathbf{x}, t)$ is the electron wavefunction

$$\psi(\mathbf{x}, t) = \int e^{ip\cdot x} u(p) \frac{d^4p}{(2\pi)^{3/2}}\,. \tag{7.11}$$

The transition amplitude can now be calculated at once. The initial state $|i\rangle$ is the direct product of the electron state 1, with wavefunction $\psi_1(x)$, and the state with no photons, $|0\text{ ph}\rangle$. The final state is the direct product of the electron state 2, with wavefunction $\psi_2(x)$, and the one-photon state $|\boldsymbol{\epsilon}\mathbf{k}\rangle$. Therefore the matrix element of the interaction Hamiltonian (7.9) factorises as follows

$$T_{i\to f} = \int \psi_2{}^*(x)\, \gamma_0 \gamma_\mu \psi_1(x) \langle \boldsymbol{\epsilon}\mathbf{k} \mid A_\mu(\mathbf{x}) \mid 0\text{ ph}\rangle\, d^3x \tag{7.12}$$

Expanding the electromagnetic field as in (4.4), we see that its matrix element in (7.12) can draw a contribution only from the creation operator $a_i{}^+(\mathbf{k})$ whose index i refers precisely to polarisation along $\boldsymbol{\epsilon}$. In view of (3.22), one has

$$\langle \boldsymbol{\epsilon}\mathbf{k} \mid A_\mu(x) \mid 0\text{ ph}\rangle = \frac{1}{(2\pi)^{3/2}} \frac{1}{\sqrt{2k_0}}\, \epsilon_\mu e^{-ikx} \tag{7.13}$$

whence

$$T_{i\to f} = \int \bar{\psi}_2(\mathbf{x}, t)\, \gamma_\mu(\mathbf{x}, t)\, \psi_1(\mathbf{x}, t) \frac{1}{(2\pi)^{3/2}}\, \epsilon_\mu \frac{e^{-ikx}}{\sqrt{2k_0}}\, d^3x. \tag{7.14}$$

Finally, we can eliminate the variable t by using

$$\psi_1(\mathbf{x}, t) = e^{-iE_1 t} \psi_1(\mathbf{x})$$

$$\psi_2(\mathbf{x}, t) = e^{-iE_2 t} \psi_2(\mathbf{x})$$

and by noting that we are interested only in transitions which conserve energy, so that $E_1 = E_2 + k_0$. We get

$$T_{i\to f} = \int \bar{\psi}_2(\mathbf{x})\, \gamma_\mu \psi_1(\mathbf{x}) \frac{1}{(2\pi)^{3/2}}\, \epsilon_\mu \frac{e^{i\mathbf{k}\cdot\mathbf{x}}}{\sqrt{2k_0}}\, d^3x. \tag{7.15}$$

For practical applications of this formula to calculate radiative transitions, the reader is referred to the texts on quantum electrodynamics listed in the references. Here we shall confine ourselves to showing how it leads to the selection rules for such transitions.

Since the vector $\boldsymbol{\epsilon}$ can be taken as purely spacelike, $(\epsilon_0 = 0)$, the matrix element in (7.15) is simply

$$\int \bar{\psi}_2(\mathbf{x})\, \boldsymbol{\gamma} e^{i\mathbf{k}\cdot\mathbf{x}} \psi_1(\mathbf{x})\, d^3x. \tag{7.16}$$

But $\boldsymbol{\gamma}$ is a vector operator. Hence, by the Wigner–Eckart theorem, the matrix elements can be nonzero only between states whose angular momenta differ by 0 or 1. Moreover, since $\boldsymbol{\gamma}$ has negative parity, contributions can come only from opposite-parity components of ψ_2 and of $\psi_1 e^{i\mathbf{k}\cdot\mathbf{x}}$.

There is an immediate generalisation to atoms with any numbers of electrons, which we omit since we are not concerned with atomic physics as such. Note only that in general there is excellent agreement with experiment.

8. The Coulomb field

It may seem rather unsatisfactory to treat electromagnetic radiation from an atom according to quantum mechanics, while at the same time we give a classical description of the nuclear Coulomb field, which is also after all an electromagnetic field. It would be desirable to include the latter, too, in the framework of quantum mechanics.

To see what is involved, we shall study the Fourier transform of the Coulomb potential, since such transforms were our starting point for quantising the radiation field. For the Fourier transform of the potential one finds easily

$$\int e^{ikx} \frac{1}{|x|}\, d^4x = \frac{\delta(k^0)}{\mathbf{k}^2} \tag{8.1}$$

and for that of the electric field

$$\frac{i\mathbf{k}\, \delta(k_0)}{k^2}. \tag{8.2}$$

We see that the field is directed along the 'wavenumber' vector $\mathbf{k}$; this is often summed up by saying that the Coulomb field is longitudinally polarised.

To put ourselves in a position to quantise the Coulomb field, we should therefore have to introduce, ab initio, polarisations $\boldsymbol{\epsilon}^{(i)}$ parallel to $\mathbf{k}$ into the Fourier decomposition of the field. This can be and has been done. But, unfortunately, the resulting formalism is either much more cumbersome than the simple arguments we have employed so far; or, if it is made elegant, it has to tread very delicately. We shall return to this question when we have to but do not pursue it further for the moment.

Here, we shall not attempt any further generalisations of the formalism of quantum electrodynamics. Instead, we shall consider in some detail a striking experimental effect, namely the Compton effect; we do this so as to strengthen our intuitive rather than our formal motivation.

9. The Compton effect

To find out in what form quantum electrodynamics can be developed further, we shall now consider the Compton effect, and see what problems it raises.

The Compton effect is simply the scattering of a photon by an electron. The electron will be considered as free, in spite of the fact that actual experiments are evidently performed on electrons bound in atoms. Nevertheless, if the photon energy is much greater than the binding energy of the electron, then the effects of binding are negligible except possibly for scattering in the forward direction, which is difficult to detect in any case. Accordingly, we are interested in the following process: a photon 1 of momentum $\mathbf{k}_1$ and polarisation $\boldsymbol{\epsilon}_1$ impinges an electron of momentum $\mathbf{p}_1$ in a spin state (μ_1). To be definite, let the electron spin component along the z axis be μ_1 in its rest frame. In the final state, there is a photon of momentum $\mathbf{k}_2$ and polarisation $\boldsymbol{\epsilon}_2$, plus an electron of momentum $\mathbf{p}_2$ in a spin state (μ_2). The problem is to calculate the probability amplitude for this process.

We saw in section 7 how quantised photons interact with electrons. Since the electrons here can be considered as free, they are described by the Dirac equation for a free particle. Hence the wavefunction of the initial electron is the spinor $u^{(\mu_1)}(\mathbf{p}_1)$, and that of the final electron is $u^{(\mu_2)}(\mathbf{p}_2)$. It seems natural to use the formalism which was successful in studying photon emission from atoms, and accordingly to describe the electron plus photon system by the Hamiltonian

$$\mathrm{H} = \mathrm{H}_0 + \mathrm{H}_1 \tag{9.1}$$

where H_0 contains the free electron Hamiltonian (7.5) and the free photon Hamiltonian (7.7), while H_1 is the interaction Hamiltonian (7.9). The transition amplitude is given by perturbation theory as

$$T_{i\to f} = \langle f \mid \mathrm{H}_1 \mid i \rangle + \sum_n \frac{\langle f \mid \mathrm{H}_1 \mid n \rangle \langle n \mid \mathrm{H}_1 \mid i \rangle}{E_n - E - i\epsilon} + \cdots \tag{9.2}$$

where $|i\rangle$ and $|f\rangle$ denote the initial and final states respectively, E is their energy (the same for both), and $|n\rangle$ runs over a complete set of eigenstates of the Hamiltonian H_0, with E_n the corresponding eigenvalues.

The first term in (9.2) vanishes. To see this, note that the interaction Hamiltonian H_1 is linear in the quantised electromagnetic field $A_\mu(x)$, and consequently linear also in the photon creation and annihilation operators. Since these operators necessarily increase or diminish by one the number of photons in the state on which they act, H_1 has matrix elements only between states differing by one photon. But $|i\rangle$ and $|f\rangle$ each contain one photon, whence the first term of (9.2) does indeed vanish.

From what we have just seen, the only states $|n\rangle$ which can contribute to the second term of (9.2) are those containing either two photons or none. We examine the two cases in turn.

The contribution to the transition amplitude of a state $|n\rangle$ with one electron of momentum **p** in spin state (μ) can be written as

$$\frac{e^2}{(2\pi)^6\sqrt{4k_0^{\,1}k_0^{\,2}}}\sum_{\mu=\pm1/2}\frac{[\bar{u}^{(\mu_2)}(\mathbf{p}_2)\,\gamma_\mu\epsilon_\mu^{(2)}u^{(\mu)}(\mathbf{p})][\bar{u}^{(\mu)}(\mathbf{p})\,\gamma_\rho\epsilon_\nu^{(1)}u^{(\mu_1)}(\mathbf{p})]}{E(\mathbf{p})-E} \tag{9.3}$$

where

$$\mathbf{p}=\mathbf{p}_1+\mathbf{k}_1=\mathbf{p}_2+\mathbf{k}_2\,. \tag{9.4}$$

This expression follows immediately from

$$|\,i\rangle = |\,\mathbf{p}_1\mu_1\,,\,\mathbf{k}_1\epsilon_1\rangle$$
$$|\,f\rangle = |\,\mathbf{p}_2\mu_2\,,\,\mathbf{k}_2\epsilon_2\rangle$$
$$|\,n\rangle = |\,\mathbf{p}\mu,\,0\text{ photon}\rangle$$
$$\langle\epsilon\mathbf{k}\,|\,A_\mu(\mathbf{x})\,|\,0\text{ photon}\rangle = \frac{1}{(2\pi)^{3/2}}\frac{1}{\sqrt{2k_0}}\,\epsilon_\mu\,e^{+i\mathbf{kx}}$$

(If H_1 is written in the Schroedinger picture, it is time-independent, whence we have dropped the factor $e^{-ik_0x_0}$ from the last line.)

Now we have, for instance,

$$\langle\mathbf{p}\mu,\,0\text{ photon}\,|\,\mathrm{H}_1\,|\,\mathbf{p}_1\mu_1\,,\,\mathbf{k}_1\epsilon_1\rangle = \frac{1}{(2\pi)^{3/2}}\frac{e}{\sqrt{2k_0^{\,1}}}\,\epsilon_\nu^{(1)}\int\bar{\psi}(x)\,\gamma_\nu\psi_1(\mathbf{x})\,d^3x\,e^{+i\mathbf{k}_1\mathbf{x}}. \tag{9.5}$$

and for the wavefunctions $\psi_1(x)$ and $\psi_2(x)$ we can substitute their explicit Fourier transforms. Thus,

$$\psi(\mathbf{x}) = \frac{1}{(2\pi)^{3/2}}\,u^{(\mu_1)}(\mathbf{p}_1)\,e^{i\mathbf{p}_1\cdot\mathbf{x}}. \tag{9.6}$$

From this we see that the integral occurring in (9.5) reduces to the δ-function $\delta(\mathbf{p}-\mathbf{p}_1-\mathbf{k})$, which expresses momentum conservation. Hence we can rewrite (9.5) as

$$\frac{1}{(2\pi)^3}\frac{e}{\sqrt{2k_0^{\,1}}}\,\bar{u}^{(\mu)}(\mathbf{p})\,\gamma_\nu\epsilon_\nu^{(1)}\,u^{(\mu_1)}(\mathbf{p}) \tag{9.7}$$

from which (9.3) follows immediately.

If $|n\rangle$ contains two photons, then they must be just the initial photon $\mathbf{k}_1\epsilon_1$ and the final photon $\mathbf{k}_2\epsilon_2$, because H_1 can create or annihilate only one photon at a time but cannot do both these things simultaneously.

Therefore the state $|n\rangle$ must necessarily have the form

$$| n\rangle = | \mathbf{p}'\mu', \mathbf{k}_1\epsilon_1, \mathbf{k}_2\epsilon_2\rangle. \tag{9.8}$$

The matrix elements are evaluated much as before. For instance, one has

$$\langle \mathbf{p}'\mu', \mathbf{k}_1\epsilon_1, \mathbf{k}_2\epsilon_2 | \mathrm{H}_1 | \mathbf{p}_1\mu_1, \mathbf{k}_1\epsilon_1\rangle$$
$$= \frac{1}{(2\pi)^{3/2}} \frac{e}{\sqrt{2k_2{}^0}} \epsilon_\nu^{(2)} \int \bar{\psi}'(\mathbf{x})\, \gamma_\nu \psi_1(\mathbf{x})\, e^{-i\mathbf{k}_2\cdot\mathbf{x}} \tag{9.9}$$

Rewritten in terms of wavefunctions in momentum space, this becomes

$$\frac{1}{(2\pi)^3} \frac{e}{\sqrt{2k_2{}^0}} \bar{u}^{(\mu')}(\mathbf{p}')\, \gamma_\nu \epsilon_\nu^{(2)} u^{(\mu_1)}(\mathbf{p}_1) \tag{9.10}$$

subject to the condition

$$\mathbf{p}' = \mathbf{p}_1 - \mathbf{k}_2\,. \tag{9.11}$$

which follows from the integration over $\mathbf{x}$.

Finally, the contribution from two-photon states becomes

$$\frac{1}{(2\pi)^6} \frac{e^2}{\sqrt{4k_1{}^0 k_2{}^0}} \sum_{\mu'=1/2} \frac{(\bar{u}^{(\mu_2)}(\mathbf{p}_2)\, \gamma_\rho \epsilon_\rho^{(1)} u^{(\mu')}(\mathbf{p}'))(\bar{u}^{(\mu')}(\mathbf{p}')\, \gamma_\nu \epsilon_\nu^{(2)} u^{(\mu_1)}(\mathbf{p}_1))}{E(\mathbf{p}') - E} \tag{9.12}$$

The contributions (9.3) and (9.12) can be interpreted intuitively. Thus, (9.3) describes a process where the initial photon is absorbed by an electron in state $\mathbf{p}_1$, which thereby changes to the state $\mathbf{p}$. The constraint (9.4) simply imposes momentum conservation on this step. Afterwards the electron in state $\mathbf{p}$ emits the final photon $\mathbf{k}_2$, changing thereby to the state $\mathbf{p}_2$. These successive steps can be represented graphically by the diagram in figure (11.9.1), where wavy lines denote photons and straight lines denote electrons, each propagating freely in a straight line. One could think of this diagram as drawn in space-time.

Note that such an interpretation is possible only thanks to the uncertainty principle. In actual fact, the photon absorption process, i.e. the reaction

$$\text{electron 1} + \text{photon 1} \rightarrow \text{electron } (\mathbf{p})$$

is incompatible with energy conservation. Indeed, one has

$$\Delta E = E(\mathbf{p}) - E(\mathbf{p}_1) - k_1{}^0 \neq 0.$$

This means that in going from the initial to the intermediate state, energy fails to be conserved by an amount ΔE, and that consequently the intermediate state can survive only for a length of time $\Delta t \sim [\Delta E]^{-1}$; after which it must have re-emitted the second photon. The intermediate electron must play its part during a short period and cannot be detected physically because it fails to conserve energy; it is called a *virtual* electron.

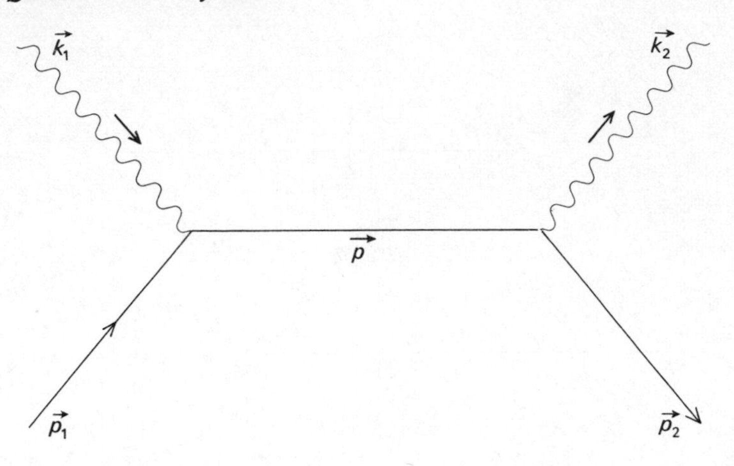

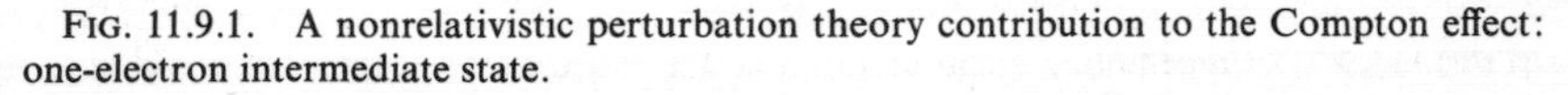

FIG. 11.9.1. A nonrelativistic perturbation theory contribution to the Compton effect: one-electron intermediate state.

Similarly, (9.12) represents a sequence of events which starts by the initial electron emitting the final photon 2, changing thereby to the intermediate state of momentum $\mathbf{p}'$, given by (9.11) so as to conserve momentum. Afterwards, the intermediate electron absorbs the initial photon 1, charging thereby to the final state $\mathbf{p}_2$. This can be represented graphically as in figure (11.9.2).

The form of H_1 shows that each absorption or creation of a photon introduces a definite factor into the transition amplitude.

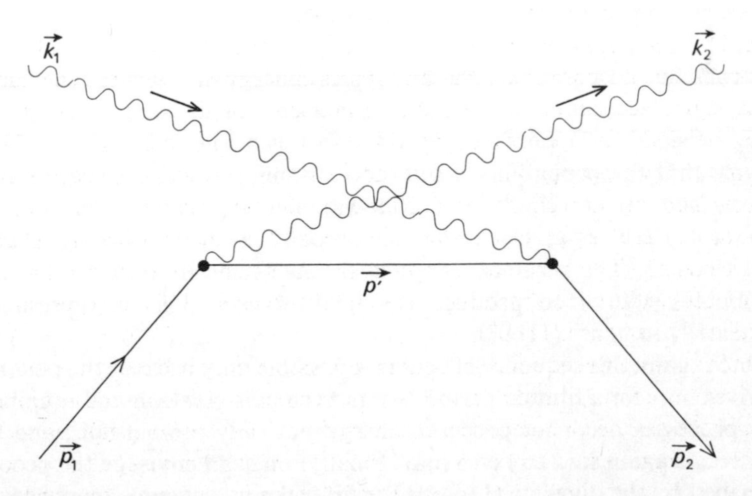

FIG. 11.9.2. Contribution to the Compton effect from the intermediate state with one electron and two photons.

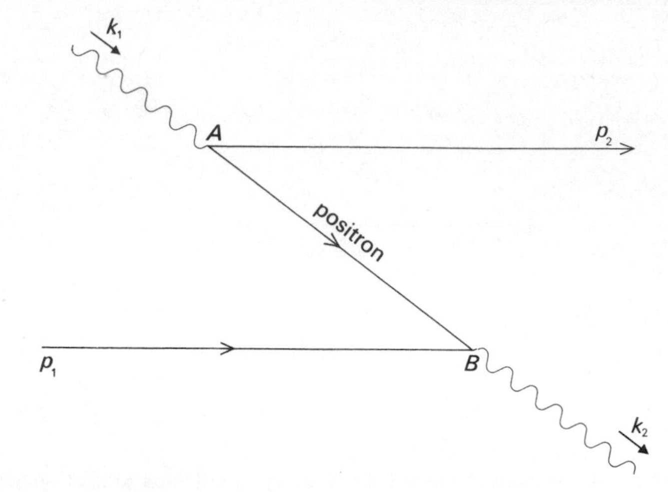

FIG. 11.9.3. Contribution to the Compton effect from intermediate states with two electrons and one positron.

If one now compares experiment with the result for the Compton effect as calculated in second order perturbation theory, one finds that there is reasonable agreement as long as the photon energy in the centre of mass frame is much smaller than the rest energy mc^2 of the electron. As soon as these quantities become comparable there is pronounced disagreement, which can be blamed neither on third order effects (which vanish), nor on fourth order effects (which are very small because they carry an extra factor of $e^2 = \alpha = \frac{1}{137}$). Hence something important must have been neglected in the above calculation.

Up to this point we have admitted intermediate states containing electrons but no positrons. But the typical energy (mc^2) where the effects so far neglected become important is precisely of the same order as the energy needed for a photon to produce an electron-positron pair. If we assume that the probability amplitude for this production process is of order e, then we can think of the following effect of second order in e:

Photon 1 creates an electron-positron pair, the electron being just the final electron. The positron propagates, meets the initial electron, and annihilates with it to produce the final photon. This is represented graphically in figure (11.9.3).

Once again, this sequence of events is possible only because the positron survives only for a limited period, so that the pair-creation and -annihilation processes need not conserve energy but only momentum; and the positron is again said to be virtual. Finally, one can envisage the process described by the diagram (11.9.4); the first step is a curious one, where a pair and a photon are created simultaneously from the vacuum. This is

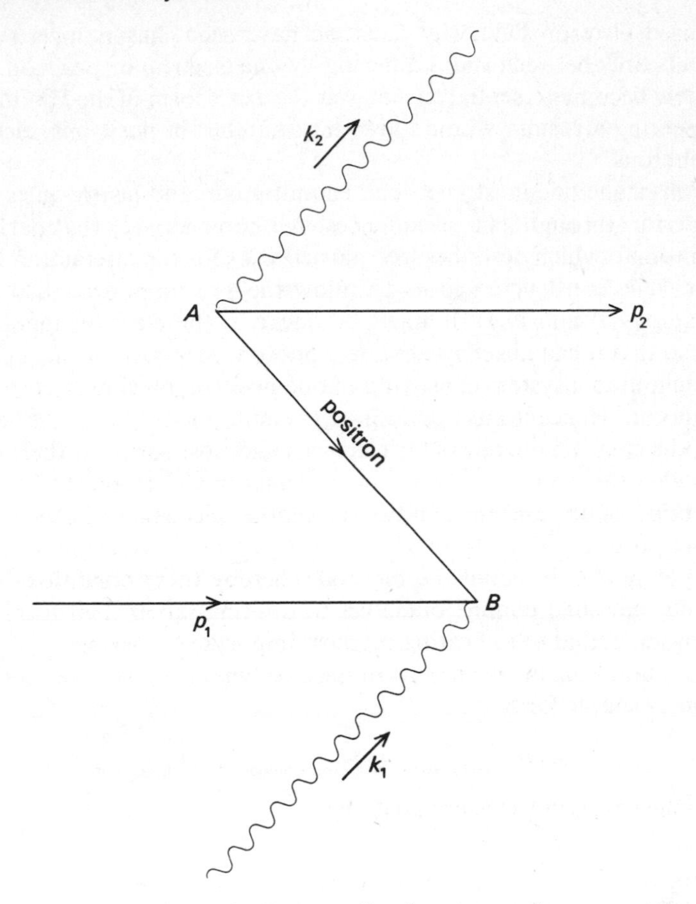

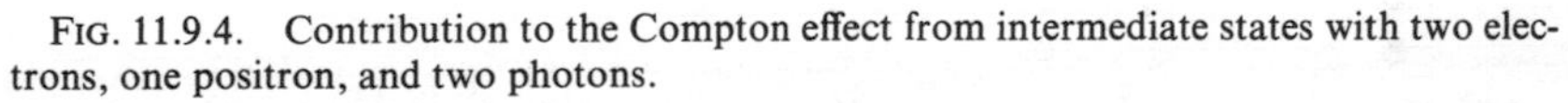
FIG. 11.9.4. Contribution to the Compton effect from intermediate states with two electrons, one positron, and two photons.

not forbidden by any conservation law apart from that for energy; hence it can occur if the positron remains virtual, as it does here.

10. The Hamiltonian for quantised fields

There is no doubt whatever that the two last-mentioned processes must be taken into account, since there is experimental evidence for reactions where photons do give rise to electron-positron pairs. The problem is to find out how they should be described theoretically.

Seeing that we shall have to deal with events where electrons can be created or destroyed, we shall presumably benefit from introducing the quantised electron field which, as we have seen, has non-zero matrix elements only between states differing by one electron or positron.

It now becomes essential to discover the exact form of the Hamiltonian representing a system with an arbitrary number of positrons, electrons, and photons.

At this stage one must guess the Hamiltonian, and justify one's choice a posteriori through its consequences. We know already that part of the Hamiltonian which describes free particles. As for the interaction Hamiltonian, it is constructed so as to allow the processes described by the diagrams (9.3) and (9.4). It must be linear in the electromagnetic field $A_\mu(x)$ so that it can absorb or create a photon. Moreover it must connect the vacuum to a system consisting of one positron plus one electron plus one photon. Hence it must contain $A_\mu(x)$ multiplied by a^+b^+, a^+ and b^+ being the creation operators for electrons and positrons. Further, it must allow also the processes shown in the diagrams (9.1) and (9.2), i.e. the absorption of one and the creation of another electron (**p**); hence it must contain products of the form a^+a so as to reproduce (9.7) and (9.11). We shall not pursue in detail the method whereby these conditions can be satisfied, but shall confine ourselves to quoting the explicit form of the Hamiltonian, and to indicating its most important properties.

The Hamiltonian of quantum electrodynamics has the following extremely simple form:

$$\mathrm{H} = \mathrm{H}_{\text{free electrons}} + \mathrm{H}_{\text{free photons}} + \mathrm{H}_{\text{interaction}}$$

where the first two terms are given by

$$\mathrm{H}_{\text{free photons}} = \sum_{(i)} \int \frac{d^3\mathbf{k}}{(2\pi)^3}\, k^0 a^{+(i)}(\mathbf{k})\, a^{(i)}(\mathbf{k}) \tag{10.1}$$

$$\mathrm{H}_{\text{free electrons}} = \sum_{u=\pm 1/2} \int \frac{d^3\mathbf{p}}{(2\pi)^3}\, E(\mathbf{p})[a^{+(u)}(\mathbf{p})\, a^{(u)}(\mathbf{p}) + b^{+(u)}(\mathbf{p})\, b^{(u)}(\mathbf{p})] \tag{10.2}$$

In other words, the numbers of photons, electrons, or positrons with momenta **k** or **p** are multiplied by their energy k^0 or $E(\mathbf{p}) = (\mathbf{p}^2 + m^2)^{1/2}$.

The interaction Hamiltonian is easily written in terms of the quantised fields, and has the form

$$\mathrm{H}_{\text{interaction}} = \int e d^3\mathbf{x}[\bar{\psi}(\mathbf{x}, t)\, \gamma_\mu \psi(\mathbf{x}, t)\, A_\mu(\mathbf{x}, t)]. \tag{10.3}$$

The choice (10.3) has many desirable features, which include the following.

(a) The fields $A_\mu(\mathbf{x}, t)$ must vary with time according to the fundamental equation

$$i\,\frac{\partial A_\mu(\mathbf{x}, t)}{\partial t} = [\mathrm{H}, A_\mu(\mathbf{x}, t)] \tag{10.4}$$

By exploiting the commutation relations between the fields, one can show that (10.4) leads to quantised fields which obey Maxwell's equations. In this sense, the choice made for the photon part of the Hamiltonian (10.1) amounts to the assumption that Maxwell's equations in the presence of sources remain valid even at energies where pair creation becomes possible.

The equations of motion for the electron field $\psi(\mathbf{x}, t)$ can be found by the same method. They have the Dirac form

$$i\gamma_\mu \frac{\partial \psi}{\partial x_\mu} + m\psi - eA_\mu(x)\,\gamma_\mu \psi(x) = 0. \tag{10.5}$$

This is the reason why the introduction of quantised fields is often referred to as 'second quantisation'. Equation (10.5) applies equally to the wavefunction of an electron in a classical field, and to the quantised fields $\psi(x)$ and $A_\mu(x)$. Hence one could imagine that equation (10.5) is obtained by 'quantising' the wavefunction of the electron, as if the wavefunction itself were a field. This would then amount to quantising the electron a second time. This point of view and nomenclature are not only useless but specious; we have mentioned them only because they occur in the literature.

(b) If one confines oneself to photon energies much smaller than mc^2 and considers single-electron states, then the Hamiltonian (10.3) reduces to the Dirac interaction, which has been successful in accounting for photon emission and absorption by atoms.

In view of all this simplicity, it is especially remarkable that (10.3) by itself suffices to describe completely the electrodynamics of photons and electrons. Every one of its consequences deduced so far has been confirmed within experimental error, often to very great precision, and even when tested down to very small distances (almost down to 10^{-15} cm). Hence we shall have no occasion to attempt any further refinements of electrodynamics as regards its foundations.

Indeed, the challenge at the present time is precisely to find a way to falsify its consequences.

11. The Feynman rules*

How far have we got? We have a well-defined Hilbert-space, spanned by electron, positron, and photon states, at least if we defer consideration of other particles, all of which are much more massive. We have assumed a Hamiltonian operating in this vector space. Hence the time evolution of an arbitrary state and the transitions it undergoes constitute a well-defined problem, which at the very least can be solved in perturbation theory.

* See the examples applying these rules: figures 11.11.3 to 11.12.1.

Accordingly, it is now easy to include in the calculation of the Compton effect the contributions of intermediate states with an electron and a positron or with an electron, a positron, and two photons, like those described by figures (11.9.1) to (11.9.4). The interaction Hamiltonian (10.3) does indeed have matrix elements connecting such states with the initial and final state in Compton scattering.

The terms occurring in the usual perturbation calculation, like (9.3) and (9.12), have the drawback that they are not individually covariant. But the reaction amplitude itself is evidently a relativistic scalar, so that the sum of all the various terms in perturbation theory must be invariant as well. But this invariance is spoiled by our procedure in perturbation theory. This is not surprising, because perturbation theory consists in an iterative solution of the Schroedinger equation for the states, in which time plays a privileged role.

The question arises whether the rules of perturbation theory can be changed so that the individual terms become covariant. This would have the advantage of greater elegance and ultimately of greater simplicity. In fact, we shall see later on that electrodynamics involves difficulties whose solutions are very subtle, (infinite selfmasses, vacuum polarisation, etc.), so much so that they are absolutely impossible to handle in the framework of nonrelativistic perturbation theory. Thus a covariant formulation of perturbation theory is not only aesthetically desirable, but is necessary in practice.

This covariant reformulation of perturbation theory is a well-defined problem which has been solved by Feynman, Schwinger, and Tomonaga, using different methods. The simplest solution consists of a set of rules due to Feynman; they are given in table 11.1.

Table 11.1: *Feynman rules*

The transition probability per second from an initial state with energy E into a final state with the same energy in the continuum is given by

$$P = 2\pi N^{-1} \mid \mathcal{M} \mid^2 \rho(E)$$

where $\rho(E)$ is the number of final states per unit energy (the density of final states), $\mathcal{M}$ is a transition matrix element, and N a normalisation constant.

Values of N:

For a normalised bound state, $N = 1$.

For a state consisting of free particles, N is a product of factors N_i, one for every particle in the initial and the final states. N_i depends on how the particle wavefunctions are normalised. The simplest rule is $N_i = 2E_i$, where E_i is the particle energy.

This corresponds to photon polarisations normalised to unit space-like vectors: $\epsilon^2 = -1$, and electrons normalised by $\bar{u}u = 1$.

Calculation of $\mathcal{M}$ (up to numerical factors)

1. For a virtual electron line $\cdot\xrightarrow[p]{}\cdot$, a factor $(\gamma p - m)^{-1}$.

2. For a virtual photon line $\mu \cdot\!\sim\!\cdot\, \mu$ (k), a factor $1/k^2$ and factors $\gamma_\mu \cdots \gamma_\mu$.
3. For the emission or absorption of a real photon with polarisation vector ϵ_μ, a factor ϵ_μ.
4. Factors $\bar{u}(p)$, $u(p)$, $\bar{v}(p)$, $v(p)$ for an electron leaving or entering, or for a positron entering and leaving, respectively: ↖ ↗ ↖ ↗
5. A classical potential $A_\mu(x)$ imparting momentum q contributes a factor $\gamma \cdot a(q)$, where

$$a_\mu(q) = \int A_\mu(x)\, e^{iq\cdot x}\, d^4x \qquad \cdot\times\times\overset{q}{\times}\times\times\cdot$$

6. The trace in Dirac spinor space is taken over any closed fermion loop.
7. For an electron changing its identity (electron → positron), and for a closed fermion loop, multiply by -1 (Pauli principle).
8. Energy-momentum is conserved at every vertex.
9. Multiply by $(2\pi)^{-4}\, d^4p = (2\pi)^{-4}\, dp_0\, dp_x\, dp_y\, dp_z$ and integrate over all values of four-momenta which are not determined by momentum conservation.

Numerical factors:

1. A factor $e\sqrt{4\pi}$ for every electron-photon vertex.
2. A factor $-i$ for every virtual photon.
3. A factor i for every virtual electron.
4. A factor $-i$ for every vertex involving a photon or an external potential.

We do not attempt to derive these rules in any way, but shall make some comments to try to elucidate their physical significance:

(a) On account of the annihilation and creation operators in it, the interaction Hamiltonian can cause the annihilation or creation of one and only one photon. The factor $\bar{\psi}(x)\gamma_\mu\psi(x)$ contains electronic terms of the form a^+a, b^+b, a^+b, ab^+; simultaneously with photon emission or absorption it can produce any charge-conserving effect which involves two fermions picked from either or both of the initial and final states, i.e.

transmission of an electron
transmission of a positron
creation of a pair
annihilation of a pair.

If these processes are shown diagrammatically as was done for the Compton effect, we see that everything can be constructed from the basic elements of figure 11.11.1.

(b) The terms emerging from non-relativistic perturbation theory, like those occurring for the Compton effect in section 9, are not Lorentz-invariant. By contrast, the overall scattering amplitude must be invariant; the only reason why its individual constituents are not derives from the fact that non-relativistic perturbation theory assigns a privileged role to the time variable, or equivalently to the energy.

If one adds together the non-relativistic perturbation contributions of diagrams (9.1) and (9.3), or of (9.2) and (9.4), one sees that the sums are covariant. This suggests that the processes (9.1) and (9.3) for instance should be considered not separately, but as a single process; and that one

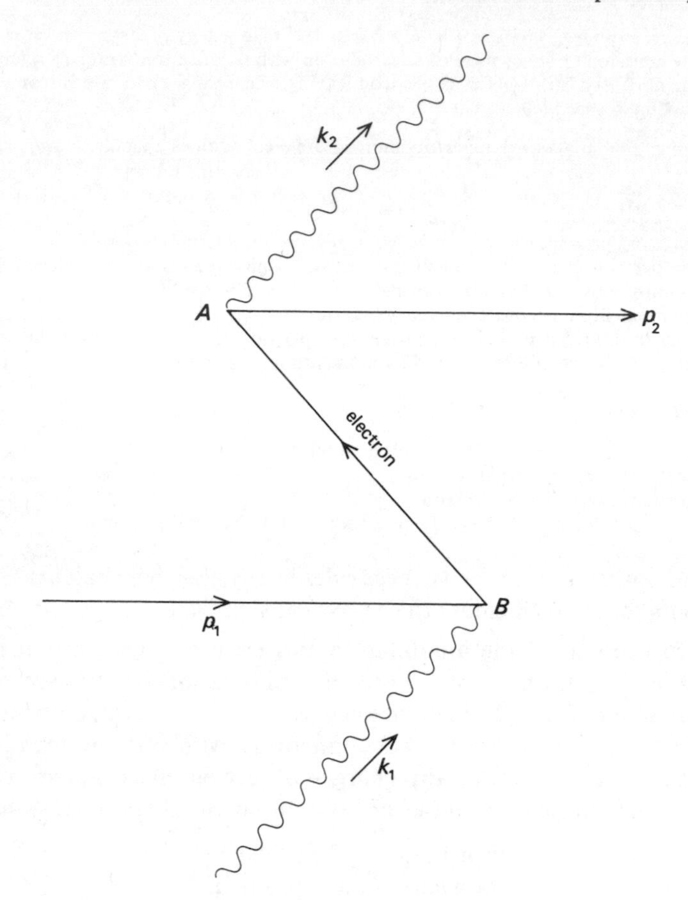

FIG. 11.11.1. Positron exchange considered as exchange of an electron moving backward in time.

should find a rule for expressing the contribution of this single process to the amplitude.

Note that the two diagrams (9.1) and (9.3) become practically identical if the positron propagating in (9.3) is replaced by an electron propagating in the opposite direction. Figuratively, this amounts to considering a positron which propagates normally as identical with an electron propagating backward in time with opposite momentum. (In the next chapter, the same viewpoint will emerge in the PCT theorem.) Be this as it may, there results the prescription to draw all possible diagrams involving the elementary events of figure (11.11.1), and to consider them as identical to those of figure (11.9.4) with the positron arrows reversed on the latter.

In this way one obtains continuous fermion lines which can represent either an electron or a positron. A line of type $\leftarrow$ on the right (the entry) side of the diagram represents an entering electron, while a line $\rightarrow$ on the right represents an exiting positron. The interior lines can be interpreted in different ways, as is illustrated by the diagrams (9.1) and (9.3).

Taking the diagrams representing all possible processes, we see by the light of (11.2a) and (11.2b) that their interiors contain continuous fermion lines which must either transverse the entire diagram or form *closed loops* as in figure (11.11.7).

(c) We saw in section 9 that the terms of the non-relativistic perturbation series conserve momentum at every vertex. It is clear that the relativistic generalisation will conserve both momentum and energy at every vertex, and consequently overall.

(d) The results in section 9 already suggest some connection between the diagrams representing the various processes, and the algebraic form of their contributions to the amplitude. Thus, every interaction introduces a factor e (the electron charge). An entering electron leads to a factor $u(\mathbf{p})$ representing its wavefunction in the initial state, and an exiting electron leads to a factor $\bar{u}(\mathbf{p})$. Similarly, one obtains a factor $\bar{v}(\mathbf{p})$ for an entering positron (at the right hand end of a fermion line, i.e. at the leading end in the sense of the arrow), and a factor $v(\mathbf{p})$ for an exiting positron.

(e) It is a nicer problem to find out what replaces terms of the kind

$$\sum_{(\mu)} \frac{\bar{u}_{(\mu)}(\mathbf{p})\, u_{(\mu)}(\mathbf{p})}{E(\mathbf{p}) - E} \tag{11.1}$$

encountered in the non-relativistic calculation, and which represented a propagating electron. It is through the structure of these terms that the relativistic calculation, by its simplicity, really transcends the non-relativistic procedure. Nevertheless, it is possible to guess the result.

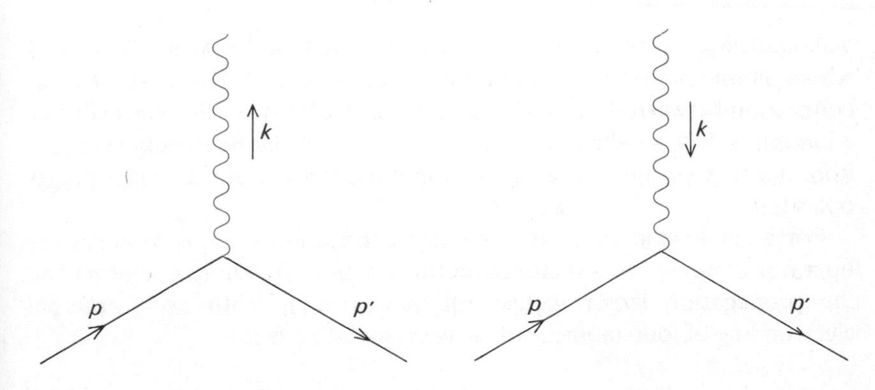

FIG. 11.11.2. Coupling of a photon to an electron line.

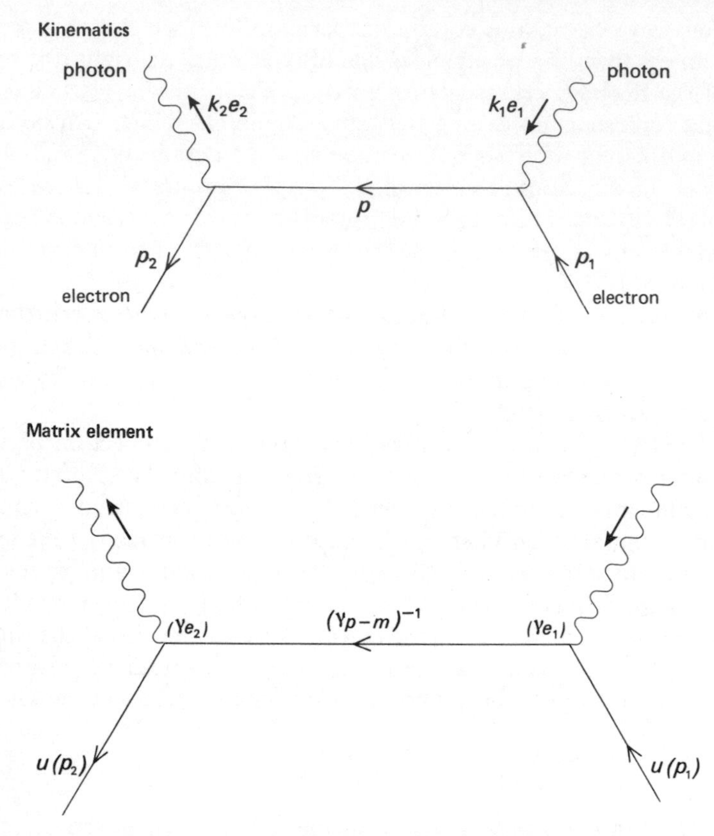

FIG. 11.11.3. A diagram contributing to the Compton effect.

Referring back to non-relativistic collision theory, we recall that a factor

$$\frac{1}{\mathbf{k}^2 - E - i\epsilon} \tag{11.2}$$

was associated with the propagation of a particle between two points where its interaction with the potential takes effect. There is an obvious connection between (11.2) and the operator occurring in the Schroedinger equation. (11.2) is called the Green's function of the Schroedinger equation; apart from the term $i\epsilon$, it is simply the inverse of the Schroedinger operator.

For a relativistic electron, a similar connection obtains between the operator defining the wavefunction (in this case the Dirac operator) and the propagation factor (called the *propagator*). With every internal electron line of four-momentum p, we associate a factor

$$\frac{1}{\gamma p - m - i\epsilon}. \tag{11.3}$$

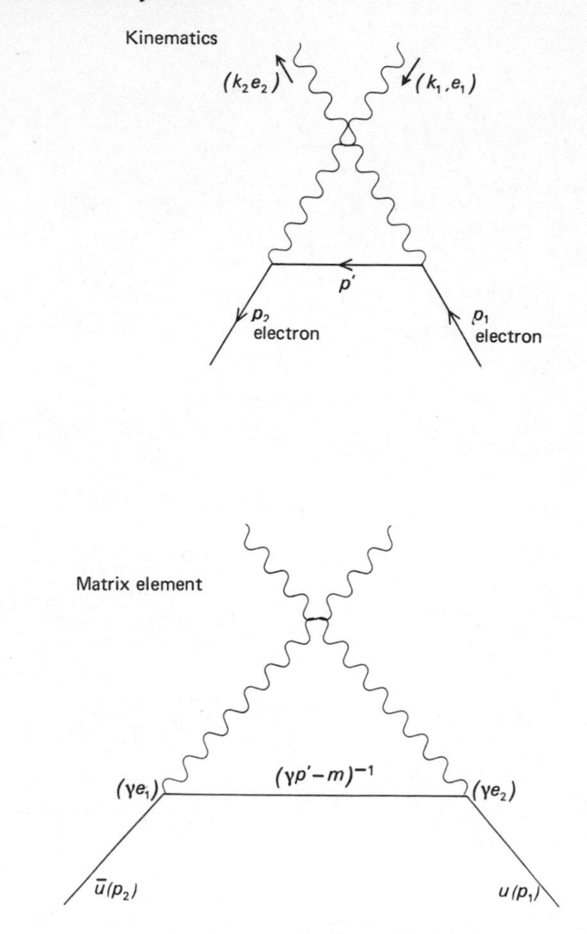

FIG. 11.11.4. A diagram contributing to the Compton effect.

Similarly, with every internal photon line of four-momentum k we associate a factor

$$\frac{1}{k^2 - i\epsilon} \tag{11.4}$$

representing the inverse of the d'Alembertian.

(f) To every photon emitted or absorbed there corresponds a factor $\epsilon_\mu(\mathbf{k})$ multiplied by γ_μ, as we saw in section 9. If a photon is emitted and reabsorbed, it is associated with two factors $\gamma_\mu \cdots \gamma_\mu$ (the same γ_μ in both cases, since the photon is emitted into and reabsorbed from one and the same polarisation state).

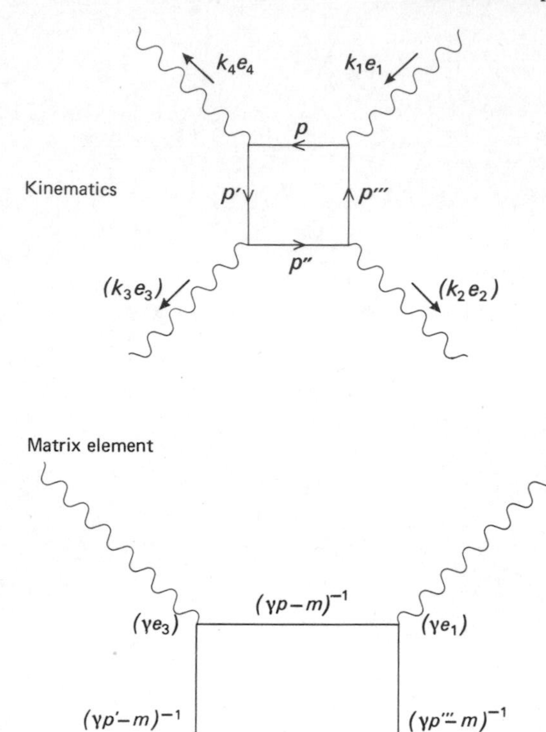

FIG. 11.11.5. Scattering of light by light.

(g) Occasionally it is useful to introduce an unquantised electromagnetic field, like the Coulomb field of a nucleus, or a magnetic field. Such a field can impart four-momentum to particles.

(h) There are cases where the four-momentum q of a virtual particle is not fully determined by energy-momentum conservation at every vertex and by the four-momenta of the initial and final particles. In non-relativistic perturbation theory this was reflected in a summation over all kinematic states of this particle, i.e. by the integration

$$\int \frac{d^3q}{(2\pi)^3}\,. \tag{11.5}$$

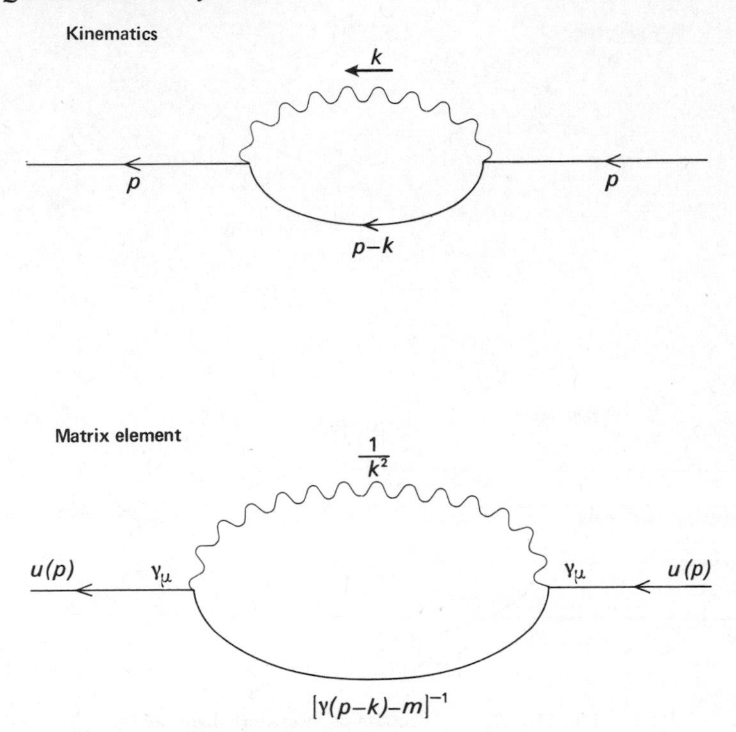

FIG. 11.11.6. An electron self-energy diagram.

In the relativistic case we have analogously

$$\int \frac{d^4q}{(2\pi)^4} \tag{11.6}$$

where the integration over each of the four components of q extends from $-\infty$ to $+\infty$.

The Feynman rules as given in table 11.1 follow naturally on these remarks. The simplest way to exploit them is to draw all the diagrams which can be constructed from the elements occurring in figure 11.11.2, and which can contribute to the effect under study; the initial particles are drawn entering from the right and the final particles leave on the left. The momentum of every particle is determined in terms of those momenta that are independent.

Both ends of internal photon lines are labelled by indices $\mu \cdots \mu$ to help with the book-keeping.

Every electron line is considered separately. Every line having a beginning and an end is followed from right to left, and the various fermion factors (Dirac spinors, propagators (11.3), γ_μ matrices at every vertex)

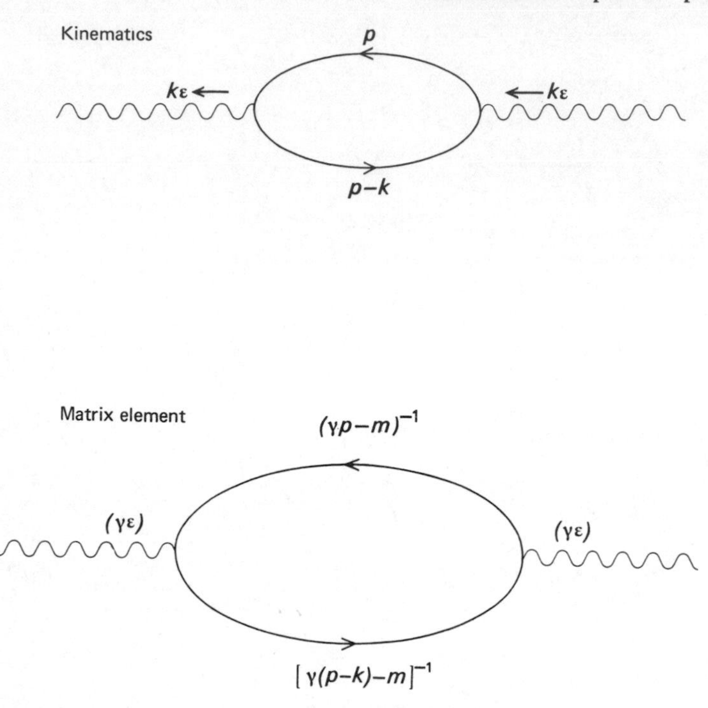

FIG. 11.11.7. A "vacuum polarisation" diagram.

are written down also from right to left, and in the *same order* as they occur on the diagram.

Closed fermion loops are followed similarly from any arbitrary starting point, and one takes the trace in Dirac spinor space of the 4×4 matrix which results.

Next, one considers each photon, and writes down the polarisation vectors ϵ_μ and the propagators $1/k^2$, in any order.

The expression thus obtained is 'premultiplied' by a functional operator

$$\int \frac{d^4q}{(2\pi)^4}\,.$$

for every internal particle momentum q which is undetermined. Finally one multiplies by the numerical factors.

12. Photon exchange and the Coulomb field

It is one of the basic consequences of field theory that the Coulomb attraction or repulsion between two charged particles can be attributed to the exchange of photons emitted by the one and absorbed by the other.

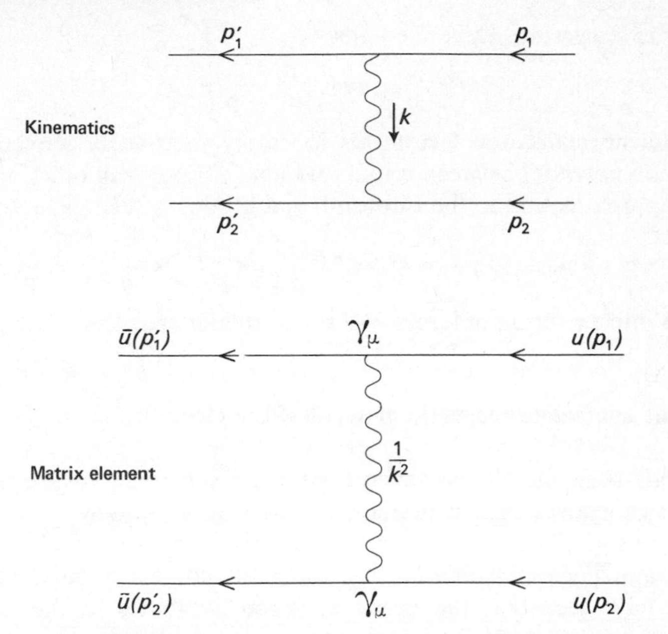

FIG. 11.12.1. The Coulomb potential in field theory: one-photon exchange.

Obviously this opens up a wide field for generalisation; in particle physics forces result from exchange, a slogan which may apply even beyond physics.

Let us look at this phenomenon more closely, and as simply as possible. Consider for instance the scattering of two electrons. The simplest diagram is shown in figure (11.12.1); a single photon is exchanged between the electrons. For the corresponding transition amplitude the Feynman rules give

$$\mathscr{M} = 4\pi e^2 i(\bar{u}(p_2')\,\gamma_\mu u(p_2))(\bar{u}(p_1')\,\gamma_\mu u(p_1))\,\frac{1}{k^2}\,. \tag{12.1}$$

In the non-relativistic approximation to the Dirac spinors only the γ_0 matrices need to be kept, these being the only diagonal ones in formulae (3.11), Chapter 10; we find the scattering amplitude

$$\mathscr{M} = 4\pi e^2\,\frac{i}{k^2}\,. \tag{12.2}$$

By four-momentum conservation, $k = p_1' - p_1$, which in non-relativistic approximation leads to

$$k^2 = -\mathbf{k}^2 = -(\mathbf{p}' - \mathbf{p})^2,$$

where $\mathbf{p}$ and $\mathbf{p}'$ are the centre-of-mass momenta of the electrons before and after scattering. Thus one finds

$$\mathcal{M} = -4\pi e^2 \frac{i}{(\mathbf{p}' - \mathbf{p})^2} \tag{12.3}$$

Up to a normalisation factor this is simply the matrix element of the Coulomb potential between initial and final states, or in other words the Born approximation to the Coulomb amplitude; to see this, note

$$\langle \mathbf{p}' \mid V_{\text{coulomb}} \mid \mathbf{p} \rangle = e^2 \int e^{i\mathbf{p}'\cdot\mathbf{x}} \frac{1}{|\mathbf{x}|} e^{-\mathbf{p}\cdot\mathbf{x}} \, d^3\mathbf{x} = \frac{e^2}{(\mathbf{p}' - \mathbf{p})^2}\,.$$

This interpretation of forces will recur in later chapters.

13. The anomalous magnetic moment of the electron

It has been mentioned several times already that quantum electrodynamics enables one to predict, in practice, accurately, and in detail, the results of all experiments involving only electrons and photons. As a first example we shall consider the magnetic moment of the electron.

We have seen that the Dirac equation attributes to the electron a magnetic moment of 1 Bohr magneton, corresponding to a gyromagnetic ratio of 2. The same applies to the muon. These gyromagnetic ratios have been measured to high precision, and their values are

$$\begin{aligned} \tfrac{1}{2} g_{\text{electron}} &= 1{,}001159657 \pm 0{,}000000035 \\ \tfrac{1}{2} g_{\text{muon}} &= 1{,}00116616 \pm 0{,}00000031 \end{aligned} \tag{13.1}$$

We give a rapid sketch of how these results are obtained.

The electron magnetic moment can be measured by magnetic resonance (Kush and Foley). The muon magnetic moment is measured as follows (Fig. 11.13.1). Low energy muons are fed into a magnetic field B where they would have a tendency to move in circles. The field B is slightly inhomogeneous, and the μ trajectory is actually a spiral. If one studies the spin precession in a magnetic field by either classical or quantum mechanics, one finds that for a gyromagnetic ratio of exactly 2 the muon spin rigidly follows the momentum. By contrast, a small but finite value of $g - 2$ will show up through a variation of the angle between the polarisation and the momentum of the muon beam as it travels through the magnetic field. By measuring this angle one can determine $g - 2$. Now it so happens that in pion decay muons are produced polarised parallel to their momentum, and that this effect is detectable. The final polarisation direction is found by counting electrons from the decay $\mu \to e + \nu + \bar{\nu}$. Because parity is not conserved in weak interactions, the electron distribution is anisotropic, and the anisotropy allows one to determine the polarisation direction of the μ.

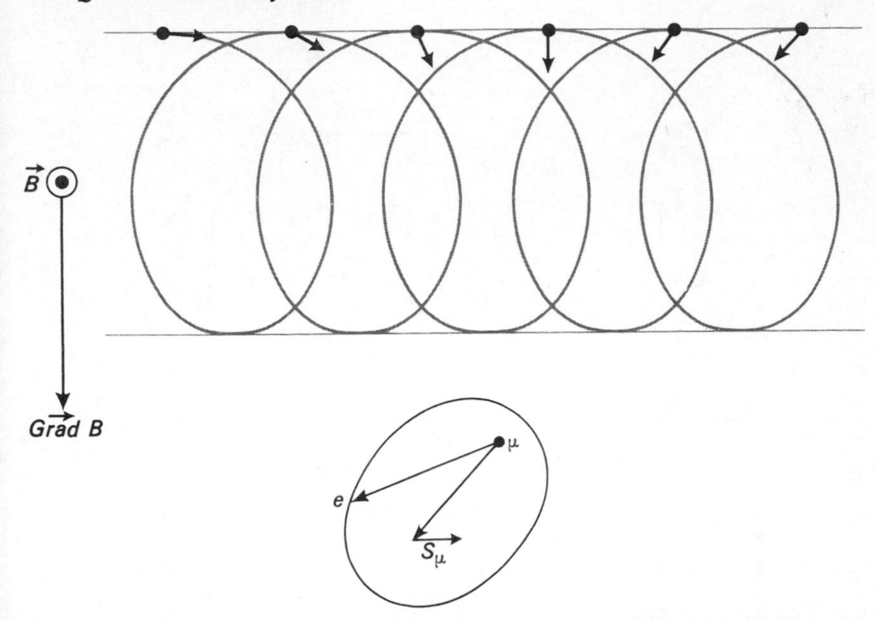

FIG. 11.13.1. Measurement of the muon gyromagnetic ratio. Precession of the muon spin in an inhomogeneous magnetic field, and angular distribution of the decay electrons.

The difference between the experimental value (13.1) and the value 2 predicted by the Dirac equation can easily be explained in terms of Feynman diagrams. In a weak magnetic field the electron is subject to a force which can be split into the Lorentz force $e\mathbf{v} \times \mathbf{B}$ and a magnetic torque $\boldsymbol{\mu} \times \mathbf{B}$. Under the influence of this force it accelerates, and its spin precesses together with the magnetic moment. The resulting motion can be thought of as a sequence of events in each of which the electron is scattered by the classical electromagnetic field; they are represented by diagrams like those of figure (11.13.2), where the broken lines represent a momentum q gained by the electron from the external field. In particular, diagram (a) corresponds to the Born approximation; in the limit as the initial and final electron momenta $\mathbf{p}$ and $\mathbf{p}'$ tend to zero, this becomes directly proportional to the magnetic moment, since in that case it represents merely the precession of the spin under the influence of an external magnetic field. As a result, one can calculate the magnetic moment from the Feynman rules which describe the scattering process.

When virtual photons are taken into account, (in what are called radiative corrections), the diagram (a) is supplemented by the new diagram of figure (11.13.3).

This diagram represents the interaction of the electron with the classical external magnetic field, after the electron has emitted a virtual photon.

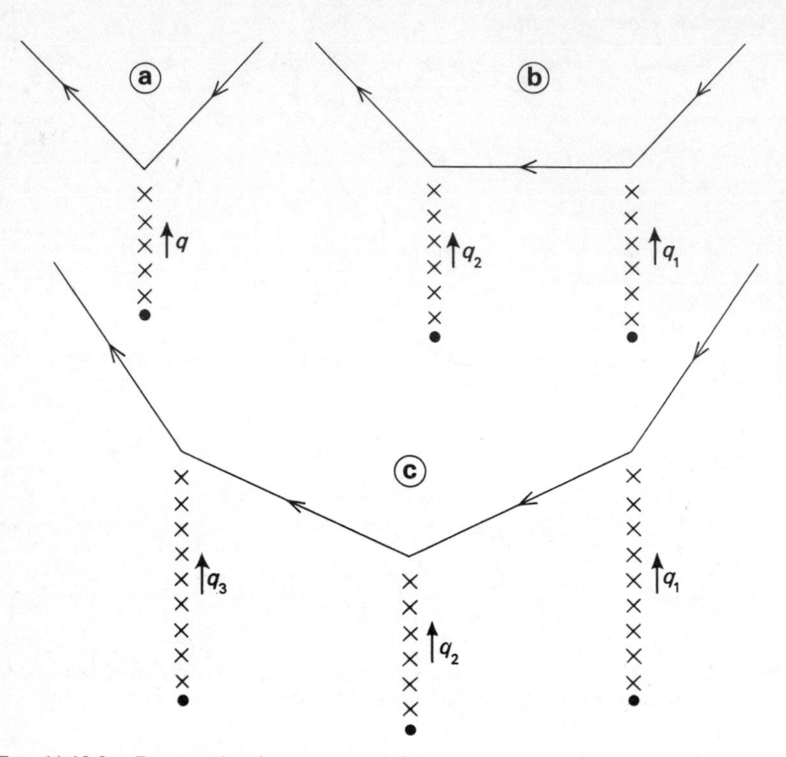

FIG. 11.13.2. Propagation in an external field considered as a sequence of scattering events.

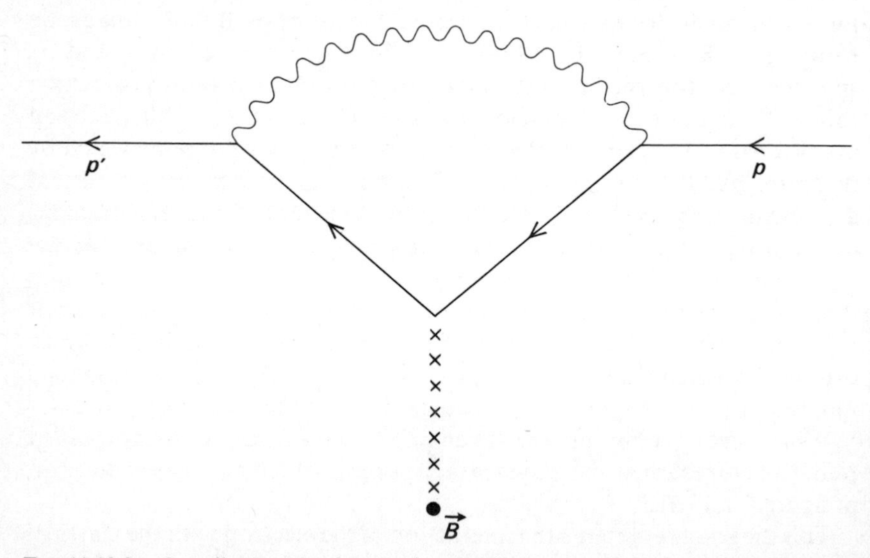

FIG. 11.13.3. Contribution of order α to the anomalous magnetic moment of the electron.

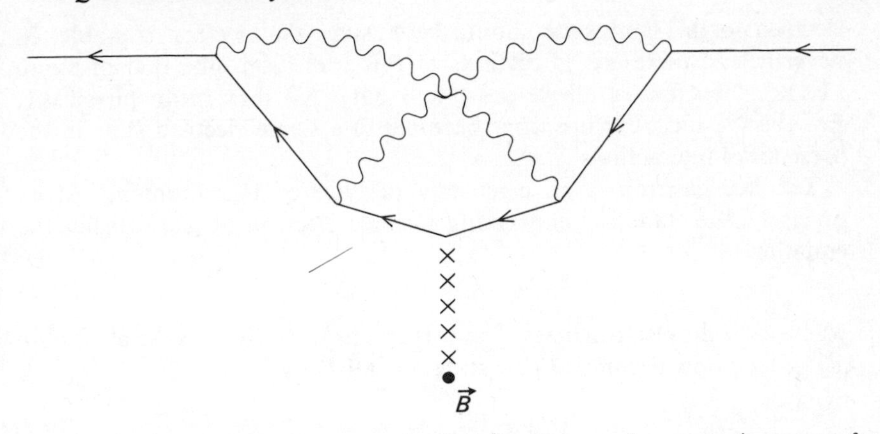

FIG. 11.13.4. A contribution of second order (α^2) to the anomalous magnetic moment of the electron.

This process evidently yields a new contribution to the electron magnetic moment. Indeed, in emitting the virtual photon the electron has recoiled, and has thus acquired an orbital angular momentum relative to the rest frame of the initial electron. But such an orbital angular momentum carries a magnetic moment which will make itself felt under the influence of the external field. Since this is a second order diagram, its contribution has an extra factor α relative to that of diagram (a), and leads to a correction to the gyromagnetic ratio of order α. The calculation eventually yields

$$\Delta g = \frac{\alpha}{\pi}. \tag{13.2}$$

One can go further and take into account also effects of order α^2, like the diagram in figure (11.13.4); then the final calculated values are

$$\begin{aligned} \tfrac{1}{2}g^{\text{th}}_{\text{electron}} &= 1 + \frac{\alpha}{2\pi} - 0{,}328\,\frac{\alpha^2}{\pi^2} \\ \tfrac{1}{2}g^{\text{th}}_{\text{muon}} &= 1 + \frac{\alpha}{2\pi} + 0{,}77\,\frac{\alpha^2}{\pi^2} \end{aligned} \tag{13.3}$$

in excellent agreement with experiment.

14. Renormalisation

We must now investigate the nicer points of quantum electrodynamics. Consider for instance the diagram (11.11.6), which corresponds to one of the simplest possible transitions: a single initial electron emits and reabsorbs a photon, giving rise to a single final electron. A priori, one would expect that the transition probability from a state with one free electron to the same state should be 1, since the electron is stable. If, nevertheless, there are corrections to this transition, like that of figure 11.11.6, then first of all we must find out what they mean physically. For this we must explore what becomes of a single-electron state in the presence of interactions.

The free electron is an eigenstate of the free Hamiltonian.* More precisely, the state $|e_0\rangle$ containing a single electron at rest satisfies the equation

$$\mathrm{H}_0 \mid e_0\rangle = m_0 \mid e_0\rangle \tag{14.1}$$

where m_0 is the electron mass. The perturbation H_1 disturbs the eigenstate $|e_0\rangle$ which now becomes a new state $|e\rangle$ satisfying

$$(\mathrm{H}_0 + \mathrm{H}_1) \mid e\rangle = m \mid e\rangle. \tag{14.2}$$

In general the eigenvalues of the full Hamiltonian $H = H_0 + H_1$ differ from those of the free Hamiltonian, so that m probably differs from m_0. Therefore we must expect the interaction to change the electron mass.

An electron which we actually detect is necessarily in an eigenstate of H and not of H_0, since H_0 is nothing but a convenient mathematical entity while H represents the true Hamiltonian of all electrodynamic systems, including the system consisting of one electron. The physical electron as observed say in a counter is described not by $|e_0\rangle$ but by $|e\rangle$, and its measured mass is not m_0 but m. To sum up, if m differs from m_0, then the electron mass entered in the table of natural constants is m, while m_0 is merely a mathematically convenient quantity.

In principle we can calculate m from m_0 by perturbation theory. On the other hand, through the electron propagator $(\gamma p - m_0)^{-1}$ the Feynman rules involve a parameter, called the 'electron mass,' but which is precisely m_0.† Thus they give us the desired transition amplitudes in terms of the parameter m_0, which causes some embarassment. To avoid this, it is clear that we need merely invert the relation between m and m_0, so that m_0 is expressed in terms of m, and then rewrite the perturbation theory results as functions of the physical mass m. The change of variables (from m_0 to m) whereby this is accomplished is called mass renormalisation. In practice, perturbation theory yields m in the form of a series

$$m = m_0 + \alpha m^{(1)} + \alpha^2 m^{(2)} + \cdots \tag{14.3}$$

which is easily inverted. We see that mass renormalisation is a consequence attending the use of perturbation theory. We can conceive perfectly well that certain transitions may be calculable without recourse to perturbation

* Translator's note: H_0 is often called the 'bare' (rather than the free) Hamiltonian.

† In the Feynman rules given above we wrote m instead of m_0, since in many problems the difference between them has no practical consequences.

theory, for instance by using the methods of chapter 12; and renormalisation ceases to be necessary as soon as one refrains from introducing the free Hamiltonian H_0. Therefore renormalisation is not a necessity intrinsic to quantum electrodynamics, but only an inconvenience of perturbation theory, which remains nevertheless the simplest method to carry out practical calculations.

The electron mass is not the only quantity that must be modified in the course of perturbation calculations. In practice it is just as necessary to take account of the changes in the normalisation of states. Indeed, if one expands the physical electron state $|e\rangle$, then one must write it as

$$|e\rangle = A_1 |e_0\rangle + |\delta e_0\rangle \tag{14.4}$$

where the perturbation $|\delta e_0\rangle$ is orthogonal to $|e_0\rangle$, and where the coefficient A_1 is needed for simultaneously normalising both $|e\rangle$ and $|e_0\rangle$ to 1. Similarly, a normalisation coefficient A_2 must be introduced into the physical photon state. When applying the Feynman rules one must keep track of these factors explicitly, in order to allow for the fact that the participating particles are physical electrons and photons rather than eigenstates of H_0. Here too one expands A_1 and A_2 into series

$$A_1 = 1 + \alpha A_1^{(1)} + \alpha^2 A_1^{(2)} + \cdots \tag{14.5}$$

and introduces a factor A_1^{-1} or A_2^{-1} for every particle entering the interaction. Since this must be done for every particle, it is easy to show that the same overall effect is achieved by taking the following steps.

(i) Multiply the amplitude by $A_1^{-1/2}$ for every external electron, and by $A_2^{-1/2}$ for every external photon.

(ii) Multiply by $A_1^{-1}A_2^{-1/2}$ for every elementary photon-electron vertex since every such vertex involves two electrons and one photon.

The first step amounts simply to renormalising the states of external particles, and is accordingly called wavefunction renormalisation. The second step amounts to replacing by $e_0 A_1^{-1} A_2^{-1/2} = e$ the charge e_0 of the electron as it appears in the interaction Hamiltonian H_1. Once again it is e and not e_0 which is measured experimentally. This second step is called charge renormalisation. All these renormalisations are legacies of perturbation theory.*

15. Beware of infinities

Let us return to diagram (11.11.6). From its contribution to the transition amplitude one can extract the leading correction to the electron mass,

$$\Delta^{(1)}m = \alpha m^{(1)} \tag{15.1}$$

* Note that these are not all the renormalisation effects. The interested reader is referred to the bibliography for more details.

though we shall not try to show this in detail. But on any view one should be disturbed by the Feynman integral associated with this diagram, as given in figure (11.11.6). For one notes that as the integration variables, namely the components of the vector k, tend to infinity, the integrand itself behaves like k^{-4}; accordingly, the integral takes the form

$$\int d^4k k^{-4} \tag{15.2}$$

and diverges. On looking more closely, one notices that the quantities m/m_0, A_1, and A_2, which enter the renormalisation procedure, are all given by sums of divergent integrals. This does not tell us whether these quantities themselves are finite or infinite, since it is possible for a sum of infinite terms to be finite; but in practice we cannot sum the power series in α for m/m_0, A_1, and A_2, and must therefore manipulate them in a form involving infinities.

It is remarkable that this can be done without too much difficulty. The transition probabilities which we wish to calculate are physically measurable quantities. As such they do not involve explicitly the quantities m/m_0, A_1, and A_2, which derive purely from the techniques of perturbation theory. After renormalisation the amplitudes are functions only of the finite and measurable quantities e and m. Hence one need go through the renormalisation procedure only formally, treating the mathematical parameters m/m_0, A_1, and A_2, as if they were finite; they will then disappear automatically from the final expressions. In practice this demands much skill in formal manipulation; indeed the necessary precautions are such that they can be taken successfully only in the framework of relativistic perturbation theory. The great contribution of Feynman, Schwinger, and Tomonaga was to endow quantum electrodynamics with a formulation transparent enough for renormalisation to be carried out as a practical proposition, and for the physically interesting quantities to be extracted with exactitude.

Even so it remains true that the manipulation of infinities can strike one as disagreeable. Actually the infinities appear only in the renormalisation parameters, and nowhere else; hence they cast doubt only on perturbation theory. Basically their presence is due to the fact that perturbation theory, however formulated, relies too much on the details of space-time structure, if only through the arguments ($\mathbf{x}$, t) of the fields.

We shall have occasion to revert to this point when we study axiomatic methods in field theory and in S-matrix theory.

16. Vacuum polarisation

One particular effect in quantum electrodynamics, namely vacuum polarisation, deserves to be discussed separately. As an example, imagine

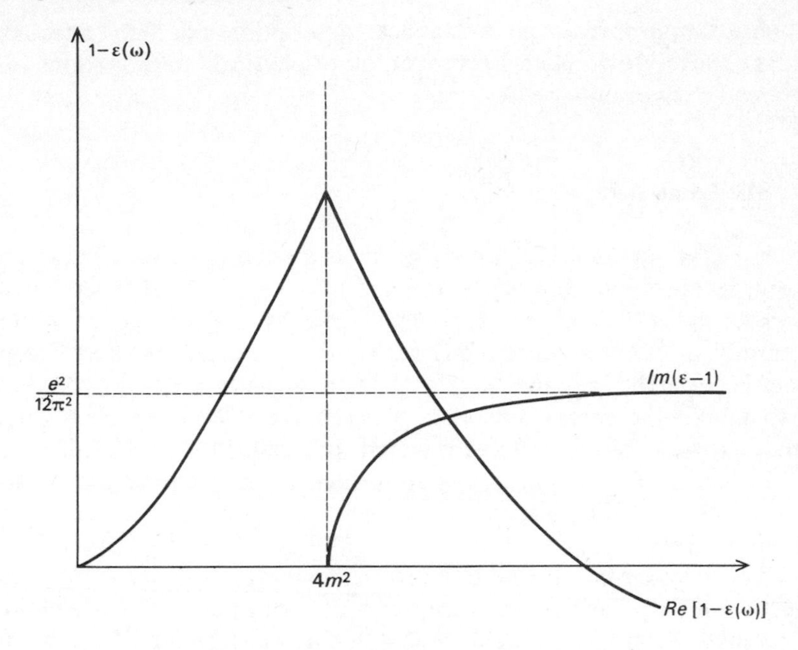

FIG. 11.16.1. Vacuum polarisation as a function of the frequency of the electromagnetic field.

monochromatic classical radiation of a frequency which is not negligible compared with the electron mass, unlike the frequencies of classical radiation realisable in practice. We know that the electromagnetic field operator obeys Maxwell's equations, and that from it we can construct the classical field as its expectation value. What can we say about the properties of this classical field?

In quantum mechanics, a photon forming part of some state of the field can create virtual electron-positron pairs, as is shown explicitly in diagram (11.11.7). Such pairs can survive only for a very short time τ, of the order of $(\Delta E)^{-1}$, where ΔE is the difference between the energy of the pair and the photon frequency ω. Nevertheless τ can become appreciable when ω rises above $2m$. Hence in the presence of such photons the vacuum behaves for part of the time like a pair, which necessarily possesses a dipole moment. From a strictly classical point of view this polarisation implies that at such frequencies the vacuum behaves like a polarisable medium which must be characterised by a dielectric constant $\epsilon(\omega)$ (Fig. 11.16.1).

The vacuum polarisation effect exists, and we shall see in the next section that is has measurable consequences. In spite of this it would be wrong to take this figurative language too literally and to think of the vacuum as an ether full of polarisable pairs. The effect is simply a repercussion of quantum properties on a classical description not fully adequate to take them into account. Moreover, other physical interpretations can be found to fit equally well.

17. The Lamb shift

The Dirac equation has already given us a good understanding of the energy levels of the hydrogen atom, and it is opportune at this point to consider what corrections are introduced by specifically quantum-electrodynamic effect. Since the Dirac equation predicts an exact degeneracy between the $2s_{1/2}$ and $2p_{1/2}$ levels, a particularly clear-cut procedure is to exhibit the energy difference between these levels which, though small, exists, and is accurately measured. It is called the Lamb shift.

The theory of the Lamb shift is dominated by three effects: the change in self-mass, vacuum polarisation, and the anomalous magnetic moment of the electron.

When an electron is bound to a proton, forming a hydrogen atom, one must modify the various terms αm_1, $\alpha^2 m_2$, ... which enter the calculation of its mass according to equation (14.3); this must be done in order to allow for the fact that the electron now experiences the Coulomb field of the proton. We shall not even try to show how the calculation is implemented, but shall establish the main features of the effect by a qualitative argument. To this end we note that the dominant term αm_1 in the mass formula accounts for the emission and reabsorption of a virtual photon by the electron. Alternatively one can say that there is a certain probability for the electron to be found as an electron-photon bound state. For dimensional reasons the spatial extension of this system must be of order $1/m$ ($\hbar/mc$ in conventional units). This can be understood

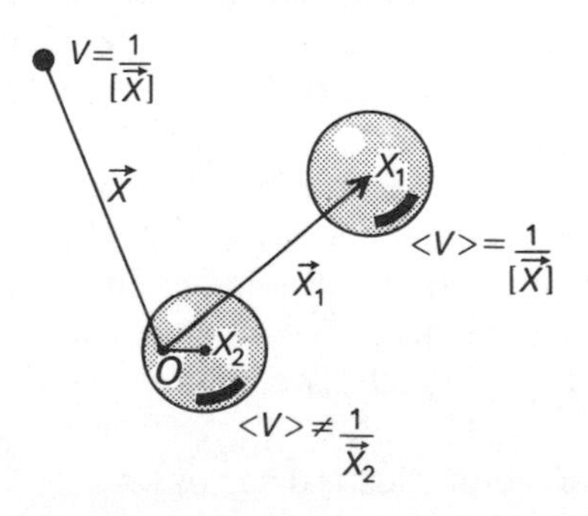

FIG. 11.17.1. The smearing-out of the electron position, due to radiative effects, alters the Coulomb potential seen by the electron.

by realising that in the bound state centre of mass frame the photon energy cannot exceed the mass m of the bound state, and that by the time-energy uncertainty principle the virtual photon cannot survive for longer than roughly $\tau = \hbar/mc^2$. Since its speed is less than c, it cannot move farther than $\tau c = \hbar/mc$, which therefore represents the order of magnitude of the maximum extension of the bound state. Under these conditions the charge of the electron cannot be localised to an accuracy better than $\hbar/mc$, and the overall effect is that the charge is smeared out.

Thus, instead of experiencing the proton's electrostatic potential $V(\mathbf{x})$ at a sharply defined point $\mathbf{x}$, the electron experiences the mean value of the potential $V(\mathbf{x})$ over a volume $v = (\hbar/mc)^3$ of space centered on $\mathbf{x}$; the effective potential acting on such a poorly-localised electron is approximately

$$\frac{1}{v}\int_v V(\mathbf{x} + \boldsymbol{\xi})\, d^3\xi. \tag{17.1}$$

From the explicit form of the Coulomb potential

$$V(\mathbf{x}) = -ze^2/|\,\mathbf{x}\,|,$$

it is easy to check that the average value (17.1) reduces to $V(\mathbf{x})$ itself as long as $|\mathbf{x}|$ exceeds $\hbar/mc$ (Fig. 11.17.1). Hence the resultant effective change $\delta V(\mathbf{x})$ in the potential is nonzero only in the immediate vicinity of the proton; (the Compton wavelength $\hbar/mc$ of the electron is of the order of 10^{-11} cm, while the radius of the hydrogen atom is of the order of 10^{-8} cm). This will shift only those energy levels whose wavefunctions are nonzero at the origin, i.e. only the s states; and in the case we are interested in, only the $2s_{1/2}$ but not the $2p_{1/2}$ state.

The levels are affected also by vacuum polarisation, which effectively replaces the Fourier transform of the Coulomb potential, namely

$$-\frac{Ze^2}{k^2} \tag{17.2}$$

by an expression involving the dielectric constant of the vacuum, namely by

$$-\frac{Ze^2\epsilon(k^2)}{k^2}. \tag{17.3}$$

But $\epsilon(k^2)$ differs appreciably from unity only for values of k comparable with the electron mass; hence, when the potential (17.3) is re-expressed in $\mathbf{x}$-space by a Fourier transformation, it differs from the Coulomb potential only at distances from the proton of the order of $1/m$, i.e. of the Compton wavelength. Therefore only the s states are affected appreciably.

By contrast, in first approximation the anomalous magnetic moment of the electron affects only the $2p_{1/2}$ state. If the electron has an orbital angular momentum **L**, it will generate a magnetic dipole which interacts with the anomalous magnetic moment. This effect exists only if **L** is non-zero, i.e. only in $2p_{1/2}$ state.

A detailed calculation of the Lamb shift predicts a value of 1057.91 ± 0.16 Mc/s (megacycles) in units of frequency; the change in self-mass contributes 1011 Mc/s, vacuum polarisation -27 Mc/s, and the anomalous magnetic moment 68 Mc/s, the balance being due to other smaller effects. The experimental value is 1057.77 ± 0.06 Mc/s. The agreement is manifestly splendid.

18. Outlook

Quantum electrodynamics is most strikingly confirmed by the Lamb shift, by the anomalous magnetic moments of electron and muon, and also by several other quantities measured with comparable precision. It is even more striking that up till now there are no experimental contradictions to the theory. Some such experiments involve spatial distances below 10^{-14} cm, which is one tenth the proton radius.

Not surprisingly, quantum electrodynamics has always served as a source of ideas and of tests for more ambitious theories intended to be applied, hopefully, to strong interactions. So far, neither have its experimental limits been reached, nor does it show any signs of drying up as a source of inspiration. In its present state, and in spite of some infelicities like infinite renormalisations, which are however easily toned down, it constitutes the most complete of all physical theories, one of the most beautiful, and undoubtedly the one with the widest range of applications—even though we have been able to give only a very rough picture of it here. It deserves to be better known beyond the restricted group of specialists, since there is no doubt that it is rich in epistemological and philosophic content which as yet has hardly been touched on.

Even without looking beyond physics, one cannot but admire how phenomena so varied and so unlike are governed by the same few equations due to Maxwell. It is well worth reflecting at length on the amazing power of mathematical language (i.e. simply of the language appropriate to science) which expresses so concisely the characteristics common to light and to the finest details of atomic spectra. There is no need to try to penetrate the central mystery, which is precisely the possibility of condensing so much into so few symbols, in order to recognise in electrodynamics the best of reasons for continuing the search for simplicity in domains where simplicity is still veiled or elusive.

At the present time the most searching experiments involving only electrons and muons are precisely those which look for the limits of validity of electrodynamics. A particularly interesting problem lies in the far-reaching similarity between muon and electron: how can two particles have similar properties (absence of strong interactions, minimal electromagnetic coupling) and yet differ so much in mass? Accordingly, many experiments probe particle structure to the smallest distances (mainly at high energies) in order to penetrate the mystery of the muon, and to discover in what respect, if any, it differs from the electron.

The most delicate problems from a theoretical viewpoint are those where electromagnetic interactions appear not in isolation but combined with strong interactions: electron scattering from protons, calculation of the hadron magnetic moments, electromagnetic mass differences like those between proton and neutron, charged and neutral pions, the three Σ's, etc. At the moment the situation is a rather confused mixture of qualitative understanding and technical difficulties which it is better not to explore here in too much detail.

To conclude, a theory as basic as electrodynamics must serve as a standard for all new theories. This is why there is an increasing tendency to apply S-matrix methods (see chapter 12) to electrodynamics. One of the main advantages of this method is that it does not directly apply perturbation theory in its classical form, and does not involve the detailed structure of space-time. Consequently it runs neither into infinities nor even into the problem of renormalisation. Also, it provides a different interpretation of the classic effects (change of selfmass in the Lamb shift, vacuum polarisation) which often make them more transparent and which show how widely interpretations can differ.

But when all is said and done, the hard core of the theory is still electrodynamics itself. Could it be redesigned so that its most economic expression ceases to be purely formal, as it is now? What determines the value of the fine structure constant? But these are problems for our successors, who, in any case, are likely to raise them in quite a different way.

CHAPTER 12

STRONG INTERACTIONS

After introducing the Yukawa potential via field theory, we investigate the analyticity properties of scattering amplitudes as functions of energy and momentum transfer. This leads us to dispersion relations, to the meson theory of nuclear forces, and to the bootstrap idea.

In the second part of the chapter, we present the fundamentals of axiomatic quantum field theory, and two of its crucial consequences: the connection between spin and statistics, and the ΠCT theorem.*

In the third part, we give a brief sketch of analytic S-matrix methods.

First part: Introduction to dispersion relations

Both historically, and in practice, one encounters the strong interactions most often through the problem of nuclear forces. It is remarkable that our everyday experience is dominated by the weakest, namely by the gravitational interaction. As regards the electromagnetic interactions, they are responsible for most of the essential properties of materials, especially for chemical binding and for the properties of solids; nevertheless their presence is masked on the macroscopic scale by the fact that matter is exactly neutral. But not until the fourth decade of the twentieth century was it discovered that the atomic nucleus itself consists of nucleons bound together by interactions even stronger than electromagnetic forces. Here the forces are masked not by overall neutrality but by their extremely rapid fall-off with increasing distance. Indeed, we shall see later that there is good reason for believing them to decrease with distance exponentially, like e^{-r/r_0}, where r_0 is of the order of one fermi (10^{-13} cm). Thus we can understand why the strong interactions can lead only to the faintest of forces between two atomic nuclei kept apart at distances of the order of 10^{-8} cm by the repulsion between their respective electron clouds.

By contrast, the strong interactions do play a dominant role in stars where the thermal motion has disrupted atomic binding. It is they, through

* Translator's note: For typographical reasons we are adhering in this book to the symbol Π for space reflection, although in the English literature the symbol P is more commonly used, so that the theorem in question is normally called the PCT theorem.

the nuclear reactions which they govern, that allow energy release in the fusion reaction. Note however that it is the weak interactions, which play no important terrestrial role, that are responsible for the initial reaction in hydrogen burning in the interiors of young recently condensed stars on the main sequence:

$$p + p \rightarrow d + e^+ + \nu$$

This reaction is then followed by an electromagnetic process

$$d + p \rightarrow He^3 + \gamma$$

and by a nuclear reaction

$$He^3 + He^3 \rightarrow He^4 + p + p$$

The overall energy release can be written as

$$4p \rightarrow He^4 + 26{,}5 \text{ MeV}.$$

Further, helium catalyses its own production from hydrogen, and burns to form heavier nuclei through combinations of strong, electromagnetic, and weak interactions.

This illustration should suffice to show that the relative importance of the various classes of interactions depends very much on which part of the universe one is considering. It might be important to keep in mind that even our classification of interactions through their strength need not be relevant in all conceivable circumstances, and might simply reflect rather subjective conditions imposed by our present experimental capabilities.

The problem of interactions at very high energies has moved increasingly to the front in recent years, and provides the main reason for the tendency to build more and more powerful accelerators. As for the still more difficult problem of how the different classes of interactions may be related, as yet it has not even been broached.

Returning to the strong interactions, our understanding of them is generally speaking extremely precarious; this makes it reasonable to look for suggestions in the most diverse directions. Until fairly recently such suggestions were drawn from field theory, itself an offspring of quantum electrodynamics. In the last ten years the idea of the S matrix has overtaken field theory in this domain, having used the latter as a springboard initially. Indeed it can be argued that the idea of the S matrix is more basic even than the Hamiltonian formalism, and that as such it provides a good way to approach problems where too many a priori hypotheses are better avoided. On the other hand, one must attribute extremely restrictive and controversial properties to the S matrix if it is to form the basis of a dynamical theory. The most important of these is the axiom of analyticity, by which scattering amplitudes are analytic functions of all their scalar variables. In physics this property made its first appearance in the form of dispersion relations.

In order to orientate the reader in this labyrinth, we begin by recalling the most important properties of nuclear forces; as we shall see, a field theory modeled on quantum electrodynamics suggests that these forces are due to the exchange of mesons between the nucleons. Having realised that field theory is incapable of pursuing this suggestion as far as a practical calculation, we revert to the analytic properties of the scattering amplitude which we have met already in chapter 8; and we investigate the dispersion relations which express the analiticity in energy. Next we consider simultaneous analyticity in energy and in momentum transfer, which underlies Mandelstam theory. We shall see that there emerges a concrete formulation of the meson theory of nuclear forces, built on the fact that the quantum numbers of a particle or resonance imply certain properties for the forces; it will be generalised into the very appealing bootstrap idea, that the hadrons themselves provide the forces responsible for their own existence.

To put contemporary theoretical work more clearly into context, we then proceed in a second part to give the basic ideas of axiomatic field theory, and, in a third part, those of *S*-matrix theory.

1. Nuclear forces

Our knowledge of nuclear forces derives from the properties of the deuteron, from nucleon-nucleon scattering and polarisation experiments, and from the physics of more complex nuclei.

The deuteron is a proton-neutron bound state with isotopic spin 0, spin 1, and parity +. The proton and the neutron are chiefly in an *S* state, with a *D*-state admixture of about 7%; this can be deduced from a study of the magnetic moment and of the electric quadrupole moment. The deuteron is the only two-nucleon bound state.

From a study of nucleon-nucleon scattering one can obtain the phase shifts. Note that on account of the spins of the two nucleons, the scattering amplitude for given angle θ and given energy E is still a 4×4 matrix in spin space. If initial and final spin states are labeled by the values $\lambda_1\lambda_2$, $\lambda_1'\lambda_2'$ of the nucleon helicities, the amplitude tades the form

$$f_{\lambda_1'\lambda_2';\lambda_1\lambda_2}(E, \theta) \tag{1.1}$$

Parity conservation, well established by nuclear spectroscopy, imposes the symmetry

$$f_{-\lambda_1'-\lambda_2',-\lambda_1-\lambda_2}(E, \theta) = f_{\lambda_1'\lambda_2';\lambda_1\lambda_2}(E, \theta) \tag{1.2}$$

Invariance under time reversal imposes

$$f_{\lambda_1\lambda_2;\lambda_1'\lambda_2'}(E, \theta) = f_{\lambda_1'\lambda_2';\lambda_1\lambda_2}(E, \theta) \tag{1.3}$$

Finally therefore one has six independent matrix elements, as shown below

$\lambda_1'\lambda_2' =$ \ $\lambda_1\lambda_2 =$	$\frac{1}{2}\ \frac{1}{2}$	$\frac{1}{2} - \frac{1}{2}$	$-\frac{1}{2}\ \frac{1}{2}$	$-\frac{1}{2} - \frac{1}{2}$
$\frac{1}{2}\ \frac{1}{2}$	f_1	f_2	f_3	f_4
$\frac{1}{2} - \frac{1}{2}$	f_2	f_5	f_6	f_3
$-\frac{1}{2}\ \frac{1}{2}$	f_3	f_6	f_5	f_2
$-\frac{1}{2} - \frac{1}{2}$	f_4	f_3	f_2	f_1

The Pauli principle further reduces the number of independent matrix elements to five.

Problem

Establish this property for the proton-proton system, by using the following expression for the exchange operator P_{12} acting on helicity eigenstates:

$$P_{12} \mid \mathbf{p}\lambda_1\lambda_2\rangle = (-1)^{2s-\lambda_1-\lambda_2}\, e^{i\pi J_y} \mid \mathbf{p}\lambda_2\lambda_1\rangle$$

where $s = \frac{1}{2}$ is the spin of the (identical) particles. Generalise this result to eigenstates of isotopic spin.

Measurements of differential and total cross-sections by themselves are insufficient to determine these five independent matrix elements or, equivalently, the five matrix elements $a^{JP}_{\lambda_1'\lambda_2';\lambda_1\lambda_2}$ for given parity P and total angular momentum J. In addition one must also measure the polarisations of the particles after the collision; this can be done by investigating how they are rescattered by heavier nuclei like C^{12}, since one knows empirically how nucleon $-C^{12}$ scattering depends on nucleon polarisation. Polarised targets, which have become available in the last few years, also allow one to determine the spin-dependence of the amplitude. An analysis of all these experiments yields the reduced amplitudes $a^{JP}_{\lambda_1'\lambda_2';\lambda_1\lambda_2}(E)$ which constitute the information to be discussed theoretically.

Some important qualitative clues are provided by nuclear physics, especially by the spectroscopy of heavier nuclei. In particular, by studying radioactive decays one confirms that the strong interactions conserve parity to very high accuracy. The masses of heavy nuclei show that nuclear forces are effective only between neighbouring nucleons, in contrast to the electrostatic repulsion between protons, which acts over long distances.

In view of the values of nuclear radii, one concludes that nuclear forces decrease considerably over a distance of a few fermis. Although historically nuclear physics has played a major part in the exploration of nuclear forces, the problem has now been inverted, and consists rather in explaining the properties of nuclei in terms of our already detailed knowledge of the forces. Even though in principle these questions do belong to the theory of strong interactions, we shall ignore them here because they deserve to be treated in their own right, and call for special techniques. But before turning away from the problems both of nuclei and of other systems with many hadrons, we underline that we do so not because they are unimportant, but because at the moment we do not have a general and coherent theory for them.

We mentioned above that the basic information consists of the scattering amplitudes and of the properties of the deuteron. For a long time one was in the habit, a legacy from atomic physics, of synthesising this information by calculating potentials which, when inserted into the Schroedinger equation, would reproduce the scattering amplitudes. Such potentials may be of interest in nuclear physics though they can be applicable only at relatively low energies of the two-nucleon system ($E < 100$ MeV). But in particle physics it seems that potentials contribute nothing new, and that they call for a language whose range is too restricted, since it cannot be applied to other hadrons. We shall occasionally mention the concept of potential, since it makes contact with a well established body of knowledge, but shall not dwell on it too insistently. In any case one needs potentials which depend on the nucleon spins, in order to account for the dependence of the $a^{JP}_{\lambda_1'\lambda_2';\lambda_1\lambda_2}(E)$ on the helicities. The general appearance of a typical potential is sketched in figure (12.1.1).

From the figure we note some important properties of nuclear forces. They are very strong, as shown by the depth of the potential. They decrease very fast with distance, and are attractive (negative) at distances of about one fermi. At short distances they are strongly repulsive, and in practice they prevent two nucleons of moderate energy from approaching each other closer than 0.2 fermis. Hence this interior region is called the nucleon "hard core".

Another important property, expected from isospin conservation, is *charge independence*; this means that apart from Coulomb effects, nuclear forces depend only on the total resultant isotopic spin of the two nucleons. It was through this property that isotopic spin was first established experimentally.

The last fundamental property is the necessity for exchange potentials in addition to direct ones. An exchange potential may be defined as the product of an ordinary (direct) potential and of the operator P_{NN} which interchanges two nucleons. Neglecting spin, this exchange operator simply

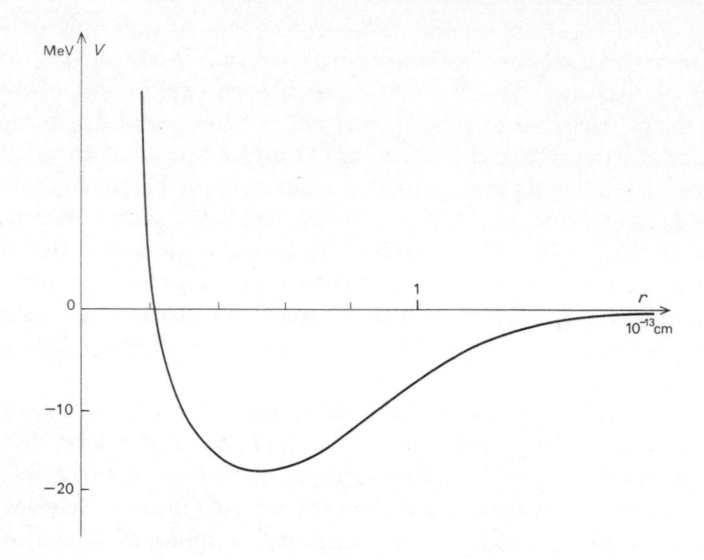

FIG. 12.1.1. A nucleon-nucleon potential.

replaces the relative coordinate vector **x** of the two nucleons by $-\mathbf{x}$; then the exchange potential V_{exch} can be defined by its action on the wave-function $\Psi(\mathbf{x})$:

$$V_{\text{exch}}\Psi(\mathbf{x}) = V^e(\mathbf{x})\,\Psi(-\mathbf{x})$$

Here, V_{exch} is an operator and $V^e(\mathbf{x})$ a function. Acting on the even and odd parts of $\Psi(\mathbf{x})$, respectively, a direct potential $V(\mathbf{x})$ and an exchange potential V_{exch} may be replaced by the potentials

$$V^+(\mathbf{x}) = V(\mathbf{x}) + V^e(\mathbf{x}) \qquad \text{on the even part of } \Psi(\mathbf{x}) \tag{1.4}$$

$$V^-(\mathbf{x}) = V(\mathbf{x}) - V^e(-\mathbf{x}) \qquad \text{on the odd part of } \Psi(\mathbf{x}). \tag{1.5}$$

The first problem of strong interaction theory is to interpret these basic properties at least qualitatively.

2. The Yukawa potential

We saw that the Coulomb force in electrodynamics can be interpreted as due to the exchange of a photon between two charged particles. To first order in α the Coulomb scattering amplitude could be approximated by α/Δ^2, where Δ is the momentum transfer in the charged particle collision. This expression α/Δ^2 is simply the Fourier transform of the potential α/r, or in other words the Born approximation to scattering by the Coulomb potential.

Next, we show that if a particle of mass μ is exchanged between two particles, say between nucleons, then to leading order in perturbation theory the scattering amplitude assumes the form $G/(\Delta^2 + \mu^2)$. We want to find a potential $V(r)$ which yields this expression in first Born approximation, i.e. such that

$$\langle \mathbf{p}' \mid V \mid \mathbf{p} \rangle = \int e^{-i\mathbf{p}'\cdot\mathbf{x}} V(\mathbf{x})\, e^{i\mathbf{p}\cdot\mathbf{x}}\, d^3x = \frac{G}{\mu^2 + \boldsymbol{\Delta}^2} \tag{2.1}$$

where $\mathbf{p}$ and $\mathbf{p}'$ are the centre of mass frame relative momenta of the two nucleons in the initial and final states respectively, and $\boldsymbol{\Delta} = \mathbf{p}' - \mathbf{p}$. Taking the Fourier transform one finds immediately

$$V(x) = 4\pi G \frac{e^{-\mu|\mathbf{x}|}}{|\mathbf{x}|} \tag{2.2}$$

where G is a constant. Such a potential is named after Yukawa.

Accordingly, if we accept this analogy with electrodynamics as a pointer to the origin of nuclear forces, we see that the exchange of a particle between two nucleons corresponds to a nucleon-nucleon potential of the Yukawa type. The lighter the exchanged particle, i.e. the smaller μ, the slower the decrease of the potential with distance. Thus the longest-range component of the nucleon-nucleon potential must be governed by the exchange of the lightest particles.

The relation between the range of the forces and the mass of the exchanged particle can be visualised in a particularly simple way by appeal to the time-energy uncertainty relation. When a proton emits a particle of mass μ, the energy of the new state thus formed differs from the initial energy by at least μ. Hence the new state cannot survive for longer than $\Delta t = \mu^{-1}$ (we are putting $\hbar = 1$). In this time the particle cannot travel farther than a distance $r = \mu^{-1}$, because its speed is less than $c = 1$. Thus it cannot be reabsorbed by a different nucleon unless the latter is within a range of roughly μ^{-1} from the original one. This is what underlies the factor $e^{-\mu r}$ of the Yukawa potential.

Since we are studying strong interactions it is clear that only the exchange of particles having such interactions, i.e. only of hadrons, is relevant. The lightest hadron being the pion, one expects from this point of view that nuclear forces should have a range roughly equal to the inverse pion mass ($1.4 \cdot 10^{-13}$ cm), which is indeed the right order of magnitude. Moreover the exponential decrease of the Yukawa potential agrees well with the extreme weakness of nuclear forces at long distance. Thus the basic notion seems sound and we shall elaborate it somewhat further.

3. The Yukawa potential and field theory

Still pursuing the analogy with electrodynamics, let us try to write down a pion-nucleon interaction Hamiltonian involving the nucleon field $\Psi(x)$ and the pion field $\Phi(x)$. The field $\Psi(x)$ is defined in terms of its creation and annihilation operators just like the electron field, i.e. by equation (5.13), chapter 11, and the pion field is defined by (5.2), chapter 11. The simplest interaction Hamiltonian is of the form

$$\mathrm{H}_1 = g \int \overline{\Psi}(\mathbf{x})\, \gamma_5 \Psi(\mathbf{x})\, \Phi(\mathbf{x})\, d^3x \tag{3.1}$$

where the matrix γ_5 is needed to make H_1 a scalar, seeing that $\Phi(x)$ is a pseudoscalar field because of the pion parity; g is a coupling constant analogous to electric charge. To take isotopic spin into account explicitly we must introduce the nucleon isotopic indices α and β ($\alpha = \pm\frac{1}{2}, \beta = \pm\frac{1}{2}$), the pion indices $i (i = 1, 2, 3)$, and the isospin matrices $\tau^{(i)}$ analogous to the Pauli matrices. Then (3.1) is replaced by

$$\mathrm{H}_1 = g \int \Psi_\beta(\mathbf{x}) \gamma_5 \tau^{(i)}_{\beta\alpha} \Phi_i(\mathbf{x})\, d^3x \tag{3.2}$$

Consider the matrix element of this Hamiltonian between a single-nucleon state with momentum $\mathbf{p}$, represented by the Dirac spinor $u(\mathbf{p})$, and a state consisting of one nucleon with momentum $\mathbf{p}'$ and one meson of momentum $\mathbf{k}$. By virtue of the explicit expressions for the fields this matrix element is given by

$$\langle \mathbf{p}', \mathbf{k} \mid \mathrm{H}_1 \mid \mathbf{p} \rangle = \frac{g}{(2\pi)^{9/2}}\, \bar{u}(p')\, \tau^{(i)} \gamma_5 u(\mathbf{p}) \sqrt{\frac{1}{EE'}} \sqrt{\frac{1}{2\omega}} \tag{3.3}$$

where $E = (m^2 + \mathbf{p}^2)^{1/2}$, $E' = (m^2 + \mathbf{p}'^2)^{1/2}$, and $\omega = (\mu^2 + \mathbf{p}^2)^{1/2}$; m is the nucleon and μ the pion mass. Momentum conservation imposes $\mathbf{p}' + \mathbf{k} = \mathbf{p}$.

By second order perturbation theory, we obtain the following expression for the transition amplitude between an initial state $|i\rangle$ containing two nucleons with momenta $\mathbf{p}_1$ and $\mathbf{p}_2$, and a final state $|f\rangle$ where the momenta are $\mathbf{p}_1'$ and $\mathbf{p}_2'$:

$$\sum_n \frac{\langle f \mid \mathrm{H}_1 \mid n \rangle \langle n \mid \mathrm{H}_1 \mid i \rangle}{E_n - E - i\epsilon}\,. \tag{3.4}$$

The intermediate state $|n\rangle$ could be, (Fig. 12.3.1),

1. nucleon 1′, nucleon 2, and a pion with i spin $\mathbf{i}$ and momentum $\mathbf{k} = -\boldsymbol{\Delta} = \mathbf{p}_1 - \mathbf{p}_1'$;
2. nucleon 2′, nucleon 1, and a pion with momentum $-\mathbf{k} = \boldsymbol{\Delta}$.

in the first case, $$E_{n_1} = E_1' + E_2 + \omega \tag{3.5}$$

in the second case, $$E_{n_2} = E_2' + E_1 + \omega \tag{3.6}$$

and the initial energy is $$E = E_1 + E_2 = E_1' + E_2'. \tag{3.7}$$

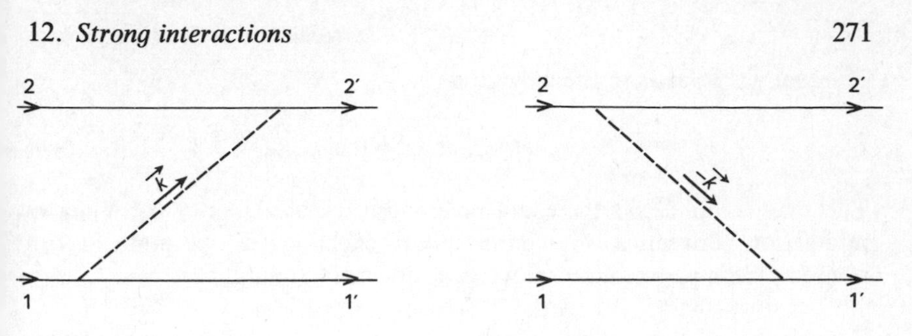

FIG. 12.3.1. The two intermediate states in nonrelativistic perturbation theory.

Thus to second order we have the following explicit expression for the amplitude:

$$\frac{g^2}{(2\pi)^9}[\bar{u}(p_2')\gamma_5\tau_i u(p_2)][\bar{u}(p_1')\gamma_5\tau_i u(p_1)]$$

$$\times\left[\frac{1}{E_{n_1}-E}+\frac{1}{E_{n_2}-E}\right]\sqrt{\frac{1}{4E_1E_2E_1'E_2'\omega^2}}. \tag{3.8}$$

For nonrelativistic nucleons one can put $E_1 = E_2 = E_1' = E_2' = m$;

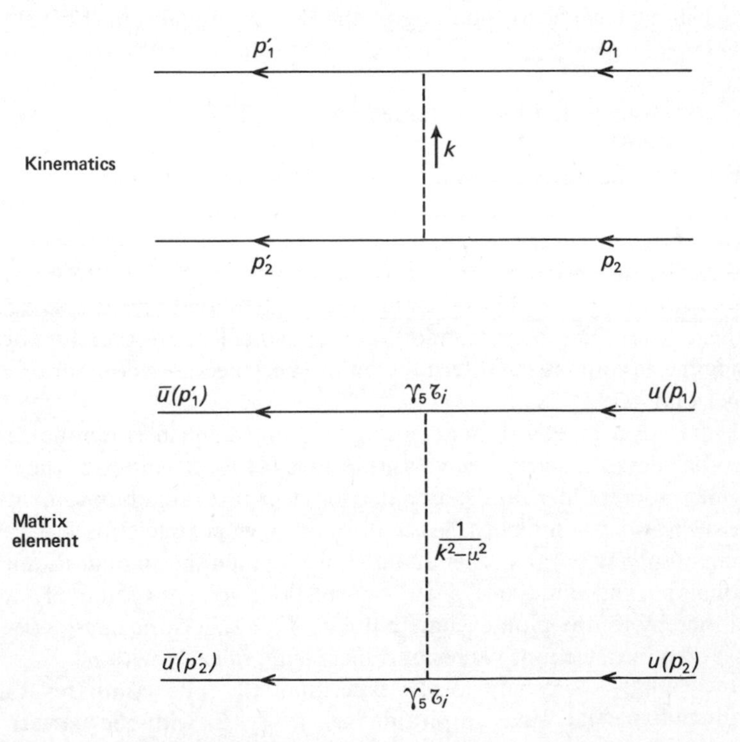

FIG. 12.3.2. Pion exchange in nuclear forces.

eventually this leads to the amplitude

$$\frac{g^2}{(2\pi)^9}\,[\bar{u}(p_2')\,\gamma_5\tau_i u(p_2)][\bar{u}(p_1')\,\gamma_5\tau_i u(p_1)]\,\frac{m^{-2}}{\mu^2+\Delta^2}\,. \tag{3.9}$$

This expression shows the exact meaning to be attached to the Yukawa interaction between two nucleons due to exchange of one pion. Still in nonrelativistic approximation we can replace $\bar{u}(p')\gamma_5 u(p_1)$ by

$$\chi^*(\mathbf{p}_1')\,\frac{(\boldsymbol{\sigma}\cdot\boldsymbol{\Delta})}{2m}\,\chi(\mathbf{p}_1) \qquad \boldsymbol{\Delta}=\mathbf{p}_1'-\mathbf{p}_1 \tag{3.10}$$

where the $\chi(\mathbf{p})$ are two-component spinors specifying the spin states of the nucleons; we obtain the amplitude

$$-\frac{g^2}{(2\pi)^9}\,[\chi^*(p_2')(\boldsymbol{\sigma}\cdot\boldsymbol{\Delta})\,\tau_i\chi(p_2)][\chi^*(p_1')(\boldsymbol{\sigma}\cdot\boldsymbol{\Delta})\,\tau^i\chi(\mathbf{p}_1)]\,\frac{(2\mathrm{m}^2)^{-2}}{\mu^2+\Delta^2} \tag{3.11}$$

From this we see that the interaction depends on the nucleon spins because of the pion parity, and that it depends on the isotopic spin of the nucleons because of the isotopic spin of the pion.

The same results can be obtained more quickly from relativistic perturbation theory. There, both terms of the nonrelativistic calculation merge into a single term corresponding to the Feynman diagram (12.3.2).

4. Verification of the Yukawa potential

How can one verify experimentally that the nucleon-nucleon interaction is really given by (3.11), at least in part? One must note to begin with that particles other than the pion can also be exchanged, for instance two pions that may or may not be merged into a ρ, three pions that may or may not be merged into an ω, etc. The crucial point is that all these systems have much higher masses than the pion, and that their contributions therefore become negligible as soon as the internucleon distance becomes comparable to or larger than μ^{-1}.

There exists a method for detecting the long-range interactions directly; one considers the partial wave amplitudes for large values of the orbital angular momentum ℓ, and hence also for large total angular momentum J. For such ℓ we can think of a classical situation where nucleons with relative momentum p pass each other at a distance b; then the angular momentum ℓ equals pb, and values of $\ell \gg p\mu^{-1}$ correspond solely to values of $b \gg \mu^{-1}$, and thereby to one-pion exchange (figure 12.4.1.). As the energy rises one must take into account values of ℓ increasing linearly with p.

Accordingly, to verify (3.11) experimentally one compares the experimental partial wave amplitudes $a^J_{\lambda_1'\lambda_2';\lambda_1\lambda_2}(E)$ with the partial wave

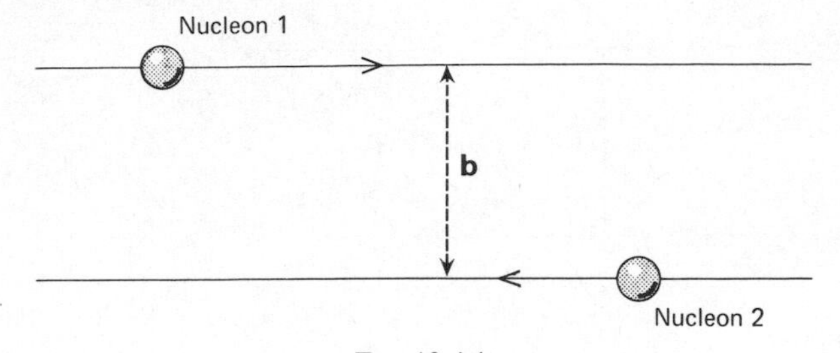

FIG. 12.4.1.

expansion of (3.11) for $J \gg p\mu^{-1}$. There is excellent agreement up to $J \approx 0.7p\mu^{-1}$ which allows one to determine the coupling constant g as

$$\frac{g^2}{4\pi} = 14{,}8. \tag{4.1}$$

5. Field theory and strong interactions

In view of the success of the Yukawa potential it would be reasonable to pursue the analogy between electrodynamics and the strong interactions, and to apply perturbation theory to the latter. Unfortunately, the difficulties of this program are shown up immediately by the large value of g^2, which should be compared to the fine structure constant $\alpha = e^2/4\pi \approx \frac{1}{137}$. Perturbation theory in electrodynamics amounts to expanding scattering amplitudes in powers of α, and this expansion converges precisely because α is so small.*

With strong interactions, all the perturbation terms, for instance all the diagrams of figure (12.5.1), yield contributions of the same order of magnitude to the amplitude.

There is evidently no prospect of evaluating such an infinity of contributions. Moreover the problem in complicated by the need to allow for possible strong interactions between the pions themselves. In particular, pion-pion interactions lead to resonances like the ρ and the ω, and we ought to be able to deal with these first. At this point there arises another delicate problem: while in electrodynamics we know how to write down the interaction Hamiltonian, one knows of no principle like minimal coupling which would determine the hadron interactions. Thus, the prospect for perturbation theory becomes truly terrifying: to evaluate an infinite number of terms with interactions that are not known at the start.

* From a more rigorous viewpoint it seems likely that the perturbation series in electrodynamics does not converge, but that it is an asymptotic series; or in other words a divergent series such that by summing a few low-order terms one obtains a good numerical approximation to the amplitude.

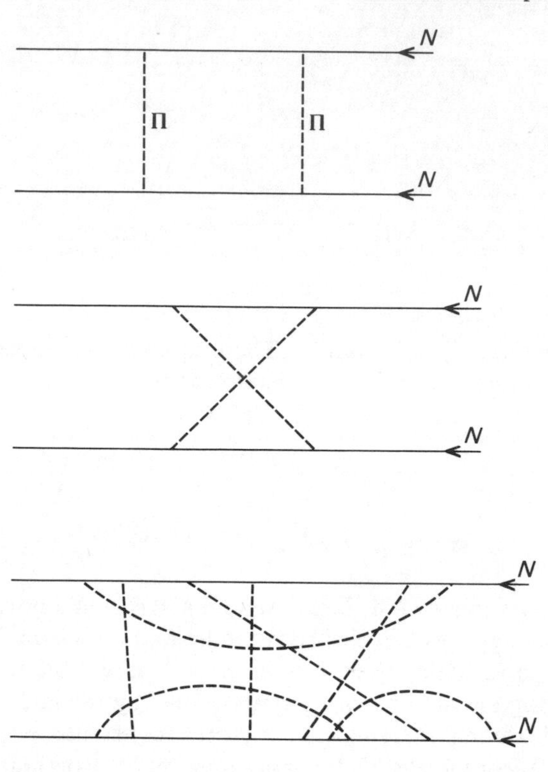

FIG. 12.5.1. Some diagrams in perturbation theory.

We cannot exclude the possibility that the perturbation series may be summed by the classical summation methods for divergent series, but so far this has not been demonstrated convincingly. But even then the problem of the Hamiltonian would remain unresolved. In particular, there is nothing to prevent direct coupling between two nucleons and two or three pions, etc.

In these circumstances the success of (3.9) might seem rather miraculous. Actually it depends on the fact that all other Hamiltonians and all other diagrams lead to interactions whose range is less than μ^{-1}, and that this range is the only feature confirmed by experiment. No such favourable features arise in any other hadronic interactions, for instance in pion-nucleon or pion-pion scattering. It could indeed be the case that the Hamiltonian (3.2) is the only correct one, and that it underlies the true theory of strong interactions; but we are unable to carry out calculations to verify whether this is so. We must quite simply find another method for approaching the problem.

This method, or at least one such possible method, is based on potential theory and on field theory, but abstracts from them ideas very different from those they usually deal with, i.e. very different from the Schroedinger equation and from perturbation theory. By contrast with these, we shall concentrate on the analytic properties of scattering amplitudes, which will eventually lead us to a new way of formulating dynamics.

6. Analytic properties of a nonrelativistic scattering amplitude as function of the energy

We now embark on a study of the analytic properties of scattering amplitudes, sketched already in chapter 8. We shall deal explicitly with nonrelativistic amplitudes derived from a Schroedinger equation, because there the crucial physical ideas emerge more clearly than in field theory, and because this framework is certainly more familiar to the reader.

We begin by defining the kinematics. We consider the scattering of two nonrelativistic particles in their centre of mass frame; the initial momentum is denoted by $\mathbf{k}$ and the final momentum by $\mathbf{k}'$. The momentum transfer is defined by $\mathbf{\Delta} = \mathbf{k}' - \mathbf{k}$, and satisfies

$$\Delta^2 = 2k^2(1 - \cos\theta) \tag{6.1}$$

whence

$$f(k^2, \cos\theta) = \tfrac{1}{2}[f^+(k^2, \cos\theta) + f^+(k^2, -\cos\theta) + f^-(k^2, \cos\theta) - f^-(k^2, -\cos\theta)]$$

We shall suit our convenience by considering the scattering amplitude either as a function $f(k^2, \cos\theta)$ of k^2 and $\cos\theta$, or as a function $F(k^2, \Delta^2)$ of k^2 and Δ^2.

It will be useful to illustrate the discussion by representing the (k^2, Δ^2) plane graphically. In this plane, the physical values of k^2 are positive, while those of $\cos\theta$ lie between -1 and $+1$ so that $0 \leqslant \Delta^2 \leqslant 4k^2$. The corresponding region is shown shaded in figure (12.6.1); note that for later convenience the Δ^2 axis points downwards.

In the following we shall always assume that spin is unimportant, and that the particles can be treated as spinless as regards those features in which we are interested. Finally, we simplify our formulae by choosing units such that the reduced mass of the two particles has the value $\frac{1}{2}$.

Our first task is to investigate the properties of the amplitude $F(k^2, \Delta^2)$ as an analytic function of k^2 for a fixed value of Δ^2. At first sight one might be tempted to do this by exploiting the analytic properties of the partial waves discussed in chapter 8, and to start from the partial wave expansion

$$f(k^2, \cos\theta) = \sum_{\ell} (2\ell + 1)\, a_\ell(k^2)\, P_\ell(\cos\theta). \tag{6.3}$$

But in actual fact this method is useless because of the poor convergence properties of the series (6.3). There do exist very elegant and powerful

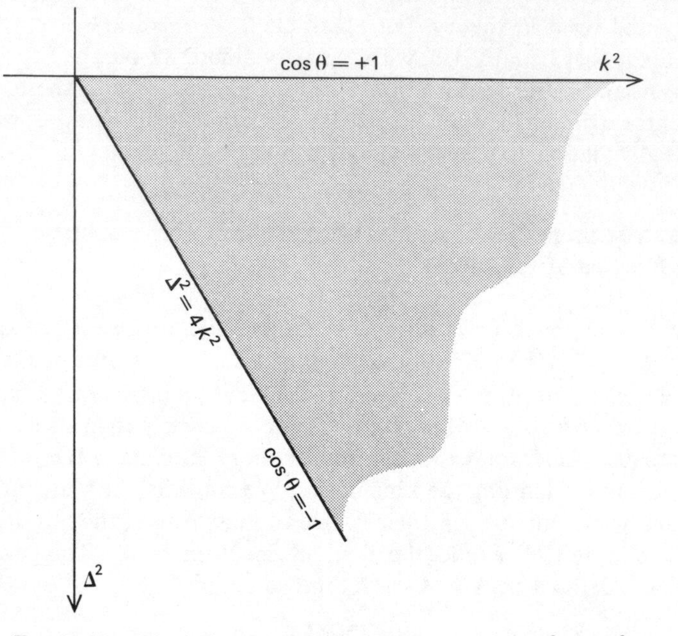

FIG. 12.6.1. Physical region in terms of the coordinates k^2 and Δ^2.

methods for studying this kind of problem, though unfortunately we cannot elaborate on them here, either because they are too lengthy or because they rely on mathematics that is too sophisticated. Hence we confine ourselves to a heuristic method with no thought of rigour, having only the advantage of simplicity and of plausible results.

We start from the Born series for the scattering amplitude as given by (7.4.4), and investigate separately the analytic properties of its successive terms. We shall assume that the analytic properties of the scattering amplitude are the same as those of the sum of the terms in the Born series:

$$
\begin{aligned}
-4\pi f(\mathbf{k}, \mathbf{k}') = \langle \mathbf{k}' \mid V \mid \mathbf{k} \rangle &+ \frac{1}{(2\pi)^3} \int \langle \mathbf{k}' \mid V \mid \mathbf{p} \rangle \frac{d^3p}{(k+i\epsilon)^2 - \mathbf{p}^2} \langle \mathbf{p} \mid V \mid \mathbf{k} \rangle \\
&+ \frac{1}{(2\pi)^6} \int \langle \mathbf{k}' \mid V \mid \mathbf{p}_1 \rangle \frac{d^3p_1}{(k+i\epsilon)^2 - \mathbf{p}_1{}^2} \langle \mathbf{p}_1 \mid V \mid \mathbf{p}_2 \rangle \\
&\times \frac{d^3p_2}{(k+i\epsilon)^2 - \mathbf{p}_2{}^2} \langle \mathbf{p}_2 \mid p \mid \mathbf{k} \rangle + \cdots \qquad (6.4)
\end{aligned}
$$

without worrying whether the series converges.

To begin with we consider the simple special case where V is a Yukawa potential (2.2) To evaluate the Born series explicitly we need first the explicit form of the matrix element $\langle \mathbf{p} | \, V \, | \mathbf{p}' \rangle$, which we have already seen is given by

$$\langle \mathbf{p} \mid V \mid \mathbf{p}' \rangle = \frac{4\pi G}{(\mathbf{p} - \mathbf{p}')^2 + \mu^2} \cdot \tag{6.5}$$

Substituting this into the Born series we get

$$F(k^2, \Delta^2) = -\frac{1}{4\pi} \sum_{n=1}^{\infty} F_n(k^2, \Delta^2) \tag{6.6}$$

where

$$F_1(k^2, \Delta^2) = \frac{4\pi G}{\Delta^2 + \mu^2} \tag{6.7}$$

$$F_2(k^2, \Delta^2) = \frac{2G^2}{\pi} \int \frac{d^3\mathbf{p}}{[(\mathbf{k}' - \mathbf{p})^2 + \mu^2][k^2 - \mathbf{p}^2 + i\epsilon][(\mathbf{p} - \mathbf{k})^2 + \mu^2]} \tag{6.8}$$

$$F_3(k^2, \Delta^2) = \frac{G^3}{\pi^3} \iint \frac{d^3\mathbf{p}_1\, d^3\mathbf{p}_2}{\left[\begin{array}{l}[(\mathbf{k}' - \mathbf{p}_1)^2 + \mu^2][k^2 - \mathbf{p}_1{}^2 + i\epsilon][(\mathbf{p}_1 - \mathbf{p}_2)^2 + \mu^2] \\ \qquad \times [k^2 - \mathbf{p}_2{}^2 + i\epsilon][(\mathbf{p}_2 - \mathbf{k}_2)^2 + \mu^2]\end{array}\right]} \tag{6.9}$$

and so on.

The first term, being independent of k^2, is clearly analytic in k^2. The other terms are analytic functions of k^2 and of Δ^2 at all points where the denominators do not vanish; this conditions shows immediately that there is a singularity on the real positive k^2 axis, since there are denominators $(k^2 - \mathbf{p}^2)$ which vanish as $\mathbf{p}^2$ varies between zero and infinity. This is just the right hand cut which we have encountered already in the partial wave amplitude.

It is equally clear that the denominators of the type $(\mathbf{k} - \mathbf{p})^2 + \mu^2$ deriving from the potential cannot vanish when the imaginary parts of $\mathbf{k}$ and $\mathbf{k}'$ are sufficiently small. To see this, note that in order to vanish they would require

$$(\mathbf{p} - \mathbf{k}_1)^2 - \mathbf{k}_2{}^2 + \mu^2 = 0 \qquad \mathbf{p} \cdot \mathbf{k}_2 = 0 \tag{6.10}$$

where $\mathbf{k} = \mathbf{k}_1 + i\mathbf{k}_2$; but the first equality can be satisfied only when $\mathbf{k}_2{}^2$ exceeds μ^2. Unfortunately this is not a strong restriction on $\mathbf{k}_2{}^2$. In trying to discover the analytic properties over a larger domain we should be led to mathematical methods that we have no intention of developing here: for instance, the integral over $\mathbf{p}_1$ along a real contour may be deformed to lie along a complex contour, or some of the integrations may be performed explicity. Instead, we shall be satisfied with merely quoting the very simple final result established by such methods: *for fixed nonnegative values of Δ^2, every term of the Born series is analytic in k^2 except on the right hand cut.*

It can be shown that the same is true in the general case, up to one simple modification. We saw in chapter 8 that if the potential is attractive enough to have a bound state with angular momentum ℓ and binding energy B, (i.e. with energy $-B$), then the corresponding partial wave $a_\ell(k^2)$ has a pole at $k^2 = -B$. The partial wave expansion (6.3) shows that this pole must appear also in the full amplitude $F(k^2, \Delta^2)$. If $a_\ell(k^2)$ behaves like $N/(k^2 + B)$ in the vicinity of the pole, then for a single bound state one has

$$f(k^2, \cos\theta) = (2\ell + 1)\frac{N}{k^2 + B} P_\ell\left(1 + \frac{\Delta^2}{2B}\right) + \varphi(k^2, \Delta^2) \qquad (6.11)$$

where the only singularity of $\varphi(k^2, \Delta^2)$ is the right hand cut. To sum up, for fixed Δ^2 the amplitude has only the right hand cut plus poles corresponding to bound states.

7. Analyticity in momentum transfer

It is a more delicate matter to investigate the analytic properties of $F(k^2, \Delta^2)$ as a function of Δ^2 for fixed k^2; they are not easy to discover from the formulae (6.6)–(6.9). Nevertheless it is important to understand the physics behind them, for we shall see that these singularities in Δ^2 can approach the physical region very closely.

The first and simplest singularity appears in the first term $F_1(k^2, \Delta^2)$ of the Born series (6.7) as a pole at $\Delta^2 = -\mu^2$. Evidently we are dealing with a dynamical singularity depending on the potential. The only reason why it is a simple pole is that we are considering, explicitly, a Yukawa potential. For instance, an exponential potential $e^{-\mu r}$ would give rise to a double pole at $\Delta^2 = -\mu^2$. The position of the pole is governed by the range of the potential; the longer the range, the smaller μ and the closer the pole to the physical region. The Coulomb potential α/r is a limiting case of the Yukawa potential as $\mu \to 0$; in this limit the singularity is just on the boundary of the physical region at $\Delta^2 = 0$, corresponding to forward scattering.

To understand the physics behind the higher terms of the Born series (6.6), one should recall that the term F_n can be interpreted intuitively as representing n successive elementary scatterings by the potential V. From this point on our argument is completely void of rigour, and is given only because it is simple, and because of the great complication of the more satisfactory arguments that might replace it.

The fact that the first term F_1 becomes infinite at $\Delta^2 = -\mu^2$ could be interpreted by saying that a great many particles are scattered with this unphysical value of the momentum transfer. Hence one would expect that on rescattering they will once again tend to acquire a momentum transfer of the same magnitude $\Delta = i\mu$, (Fig. 12.7.1.). In general this second transfer has a different orientation in space so that the directional singularity is smeared out, except where the two transferred momenta happen to combine in the same direction to give a resultant $2i\mu$. Accordingly one expects

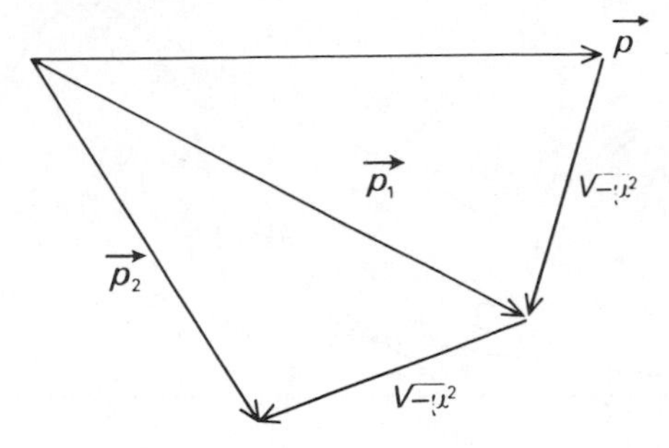

FIG. 12.7.1. Combination of singularities in momentum transfer.

F_2 to have a singularity at $\Delta^2 = -4\mu^2$, which should be a branch point. Similarly F_n will have a singularity at $\Delta^2 = -n^2\mu^2$.

But there exists also another mechanism for generating singularities, which we can see by considering the scattering angle instead of the momentum transfer. A singularity at $\Delta^2 = -\mu^2$ is simultaneously a singularity at the uniphysical angle θ_0 defined by $\cos\theta_0 = 1 - \Delta^2/2k^2 = 1 + \mu^2/2k^2$. Let us now pretend that this angle is physical. Then the scattered particles form a halo where the scattering is particularly strong, namely along a cone about the forward direction with half-opening-angle θ_0. If the particles in the halo rescatter, they will emerge once again with a preferred scattering angle θ_0, and will thus trace out a cone of half-opening-angle $2\theta_0$ (Fig 12.7.2). Accordingly we expect a specially high intensity of scattered particles along the surface of this cone, corresponding to a singularity at scattering angle $2\theta_0$. This mechanism is very similar to the previous one, but results in a singularity situated differently. One has

$$\cos\theta_0 = 1 + \frac{\mu^2}{2k^2} \quad \cos 2\theta_0 = 2\cos^2\theta_0 - 1 = 1 + \frac{2\mu^2}{k^2} + \frac{\mu^4}{2k^4}$$
$$\Delta^2 = 2k^2(1 - \cos 2\theta_0) = -4\mu^2 - \mu^4/k^2 \tag{7.1}$$

This last singularity, plus the corresponding one at $\Delta^2 = -4\mu^2$, are the only ones in F_2. The next term F_3 has three singularities: one at $\Delta^2 = -9\mu^2$, one at $\cos 3\theta_0$, and a third resulting from a combination of our two mechanisms, i.e. from a combination of enhancements at particular angles and momentum transfers. The term F_4 has five singularities, and the number of singularities of F_n grows faster than any power of n. The positions of all these singularities can be determined, and they all occur for real negative values of Δ^2 below $-4\mu^2$.

In this way one obtains an infinite number of singularities corresponding to *physical values of* k^2; they lie on curves called Landau curves sketched in figure (12.7.3).

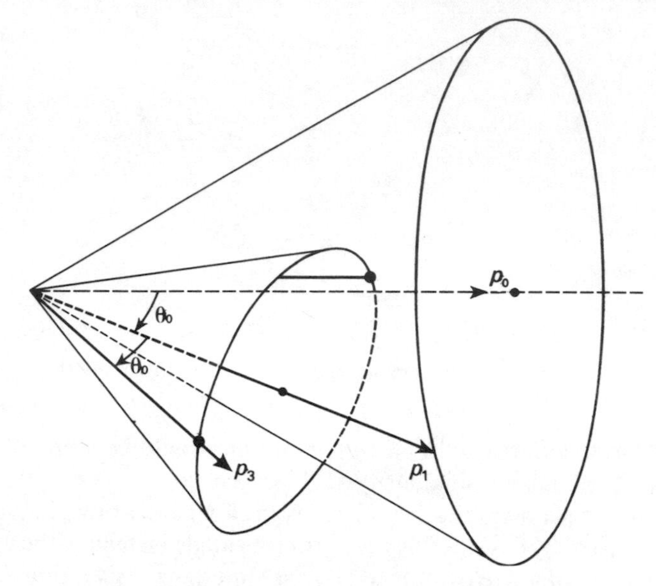

FIG. 12.7.2. Combination of singularities in $\cos\theta$.

It turns out that for large values of k^2 all these singularities tend asymptotically to $\Delta^2 = -n^2\mu^2$.

For unphysical values of k^2, i.e. for values that are not positive real, the only singularities are those not depending on k^2, namely those at $\Delta^2 = -n^2\mu^2$.

The end result is that $F(k^2, \Delta^2)$ is an analytic function of Δ^2 for all values of Δ^2 and k^2 except for real values of $\Delta^2 \leqslant -4\mu^2$, and for a pole at $\Delta^2 = -\mu^2$. If instead of a Yukawa potential we consider a superposition of such potentials expressible as

$$V(r) = \int_{\mu_0}^{\infty} \frac{e^{-\mu r}}{r} \rho(\mu)\, d\mu \tag{7.2}$$

then instead of a pole in Δ^2 there is a continuous line of singularities from $\Delta^2 = -\mu_0{}^2$ to $\Delta^2 = -\infty$. Note that one can avoid all the singularities in Δ^2 by making a cut along the real negative Δ^2 axis from $-\mu_0{}^2$ to infinity. Finally, a superposition of Yukawa potentials like

$$V(r) = \frac{Ge^{-\ {}_0 r}}{r} + \int_{\mu_1}^{\infty} \frac{e^{-\mu r}}{r} \rho(\mu)\, d\mu \tag{7.3}$$

$(\mu_0 < \mu_1)$, would lead to a pole at $\mu_0{}^2$ and to a branch cut strating from $\mu_1{}^2$.

In conclusion, we note that all the singularities, both in k^2 and in Δ^2, are real, and take advantage of this fact to represent them on one and the same diagram in figure (12.7.3).

The outcome is that we can treat the scattering amplitude as an analytic

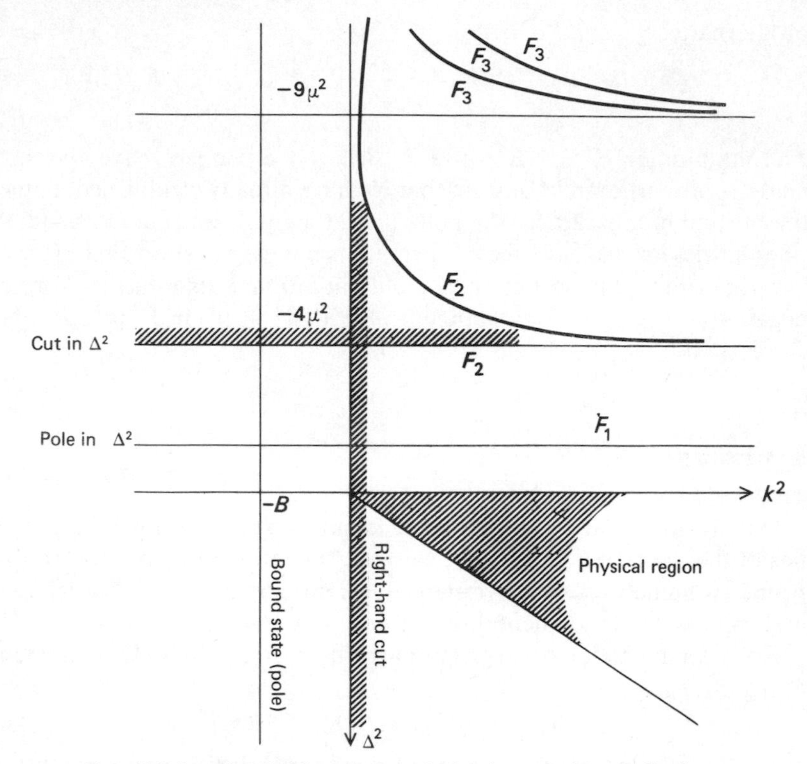

FIG. 12.7.3. Singularities of a nonrelativistic scattering amplitude.

function of its two arguments k^2 and Δ^2. In potential theory this is not a very useful result, because all the physically interesting information can be obtained by appeal to the Schroedinger equation. However, we shall soon be able to generalise these analytic properties to the relativistic case, thereby substituting for the specification of the dynamics by means of a potential, a specification by means of the corresponding singularities in Δ^2 plus the discontinuities across them.

One last remark about exchange potentials. By virtue of equations (1.4) and (1.5), they can be treated as follows. Let $f^+(k^2, \cos\theta)$ be the full scattering amplitude obtained by solving the Schroedinger equation with the potential $V^+(\mathbf{x})$ of equation (1.4), and let $f^-(k^2, \cos\theta)$ be the amplitude for the potential $V^-(\mathbf{x})$ of (1.5). The part of the true full amplitude $f(k^2, \cos\theta)$ which is even in $\cos\theta$ must evidently coincide with $f^+(k^2, \cos\theta)$, and its odd part must coincide with $f^-(k^2, \cos\theta)$, whence

$$f(k^2, \cos\theta) = \tfrac{1}{2}[f^+(k^2, \cos\theta) + f^+(k^2, -\cos\theta) + f^-(k^2, \cos\theta) - f^-(k^2, -\cos\theta)] \tag{7.4}$$

or alternatively

$$F(k^2, \Delta^2) = \tfrac{1}{2}[F^+(k^2, \Delta^2) + F^+(k^2, 4k^2 - \Delta^2) + F^-(k^2, \Delta^2) - F^-(k^2, 4k^2 - \Delta^2)] \tag{7.5}$$

The amplitudes $F^+(k^2, \Delta^2)$ and $F^-(k^2, \Delta^2)$ obviously have the same analytic properties in k^2 and Δ^2 that we have already established, namely they are analytic except for the poles in k^2, the right hand cut in k^2, and the singularities for real negative Δ^2. From this it is easy to show that $F(k^2, \Delta^2)$ has the same singularities in k^2, plus additional singularities for real negative $(4k^2 - \Delta^2)$. For instance, a pole $1/(\Delta^2 + \mu^2)$ in $F^+(k^2, \Delta^2)$ gives a pole $\frac{1}{2}/(4k^2 - \Delta^2 + \mu^2)$ in $F(k^2, \Delta^2)$.

8. Crossing

Having looked at potential theory for ideas about the analytic properties of the scattering amplitude, we shall turn to field theory for another property, namely crossing symmetry; it is intimately related to analyticity and will assume fundamental importance in the sequel.

Consider a reaction between the particles A, B, C, D with four-momenta p_1, p_2, p_3, p_4:

$$A(p_1) + B(p_2) \rightarrow C(p_3) + D(p_4) \tag{8.1}$$

and let $T_{AB\rightarrow CD}(p_1, p_2; p_3, p_4)$ be the Lorentz-invariant reaction amplitude, i.e. the collision matrix element between invariantly normalised states. This invariant amplitude is proportional to the usual amplitude $f(k^2, \cos\theta)$; for elastic scattering one has for instance

$$T = Wf \tag{8.2}$$

where W is the total energy in the centre of mass frame.

Consider also the so-called crossed reaction obtained from (8.1) by transposing two particles from one side of the reaction to the other, simultaneously changing them into their antiparticles:

$$A(q_1) + \bar{D}(q_2) \rightarrow C(q_3) + \bar{B}(q_4) \tag{8.3}$$

The corresponding invariant amplitude is $T_{A\bar{D}\rightarrow C\bar{B}}(q_1, q_2; q_3, q_4)$. From field theory one can derive the extremely simple crossing property expressed by the equation

$$T_{A\bar{D}\rightarrow C\bar{B}}(p_1, -p_4; p_3, -p_2) = T_{AB\rightarrow CD}(p_1, p_2; p_3, p_4). \tag{8.4}$$

It shows that two crossed reactions have the same invariant amplitude, provided only that the sign of the four-momentum of a particle is reversed when the particle is transposed from one side of the reaction to the other.

Obviously this is a purely mathematical condition inapplicable for physical values of the momenta. If p_1, p_2, p_3, and p_4 are physically admissible values of the four-momenta in reaction (8.1), then p_1, $-p_4, p_3$, and $-p_2$ are not physically admissible in reaction (8.3), since $-p_4$ and $-p_2$ correspond to negative energies. Crossing symmetry can be proved from field theory.

Note finally that many different reactions can be interrelated by repeated crossing. Thus one can deduce the equality of reaction amplitudes between particles and antiparticles:

$$\bar{A} + \bar{B} \to \bar{C} + \bar{D} \tag{8.5}$$

and

$$A + B \to C + D \tag{8.1}$$

(In field theory this follows also from the PCT theorem which we shall consider later on.)

Altogether there are essentially three distinct reactions related through crossing, for instance

$$p + n \to p + n \qquad p + \bar{n} \to p + \bar{n} \qquad p + \bar{p} \to \bar{n} + n \tag{8.6}$$

or

$$\pi + N \to \Lambda + K \qquad \pi + \bar{\Lambda} \to \bar{N} + K \qquad \pi + \bar{K} \to \bar{N} + \Lambda. \tag{8.7}$$

Spin modifies these relations only by introducing phase factors that are known explicitly.

Notice that crossing symmetry can be justified most simply by the remark that it is an immediate consequence of the Feynman rules.

9. Kinematics and crossed reactions

We have now gathered the essential tools for tackling the dynamics of strong interactions. It is convenient at this point to deal with the kinematics of reactions between relativistic particles, so that we need not return to these trivial questions later on.

Consider once again reaction (8.1) and let the masses of the four particles be m_1, m_2, m_3, and m_4. Energy-momentum conservation in the reaction imposes

$$p_1 + p_2 = p_3 + p_4 \, . \tag{9.1}$$

The reaction can be described in any arbitrary reference frame, for instance in the laboratory or in the centre of mass frame. But to take full advantage of relativistic invariance it is best not to confine oneself to any particular frame, and to describe the kinematics of the reaction through relativistic scalars.

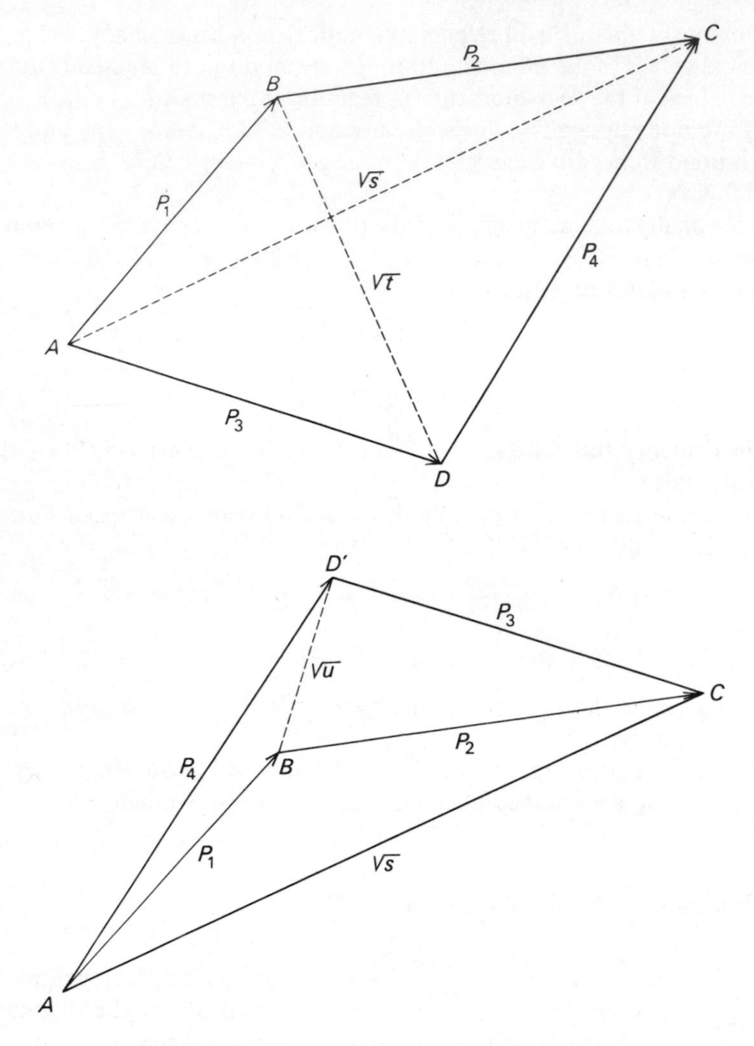

FIG. 12.9.1. The invariants of a two-body collision.

When one considers how such relativistic scalar kinematic variables should be defined, the first thing to note is that only three of the four momenta are independent, so that it sufficies to consider p_1, p_2, and p_3, the fourth being determined by (9.1). This equation has a simple geometric meaning; to see it, draw the four-dimensional vectors AB equal to p_1 and BC equal to p_2. If AD is drawn to equal p_3, then DC will equal p_4. A change of reference frame, i.e. a Lorentz transformation, corresponds to a displacement of the tetrahedron *ABCD* in four-dimensional space (Fig. 12.9.1.). Up to such transformations the tetrahedron is uniquely determined by the squared lengths of its sides, i.e. by

$$p_1^2 = m_1^2 \qquad p_2^2 = m_2^2 \qquad p_3^2 = m_3^2 \qquad p_4^2 = m_4^2 \tag{9.2}$$

and

$$s = AC^2 = (p_1 + p_2)^2 = (p_3 + p_4)^2 \tag{9.3}$$

$$t = BD^2 = (p_1 - p_3)^2 = (p_2 - p_4)^2. \tag{9.4}$$

Since the masses are fixed, we have two independent variables s and t; they will be used instead of the variables k^2 and $\cos\theta$ which depend on the reference frame.

Instead of taking the vectors in the above order, we could clearly have started by first drawing $AD' = p_4$ and $D'C = p_3$, which would have led to a different tetrahedron $ABCD'$. There is no reason for preferring either tetrahedron to the other, so that the new scalar variable

$$u = BD'^2 = (p_1 - p_4)^2 = (p_2 - p_3)^2 \tag{9.5}$$

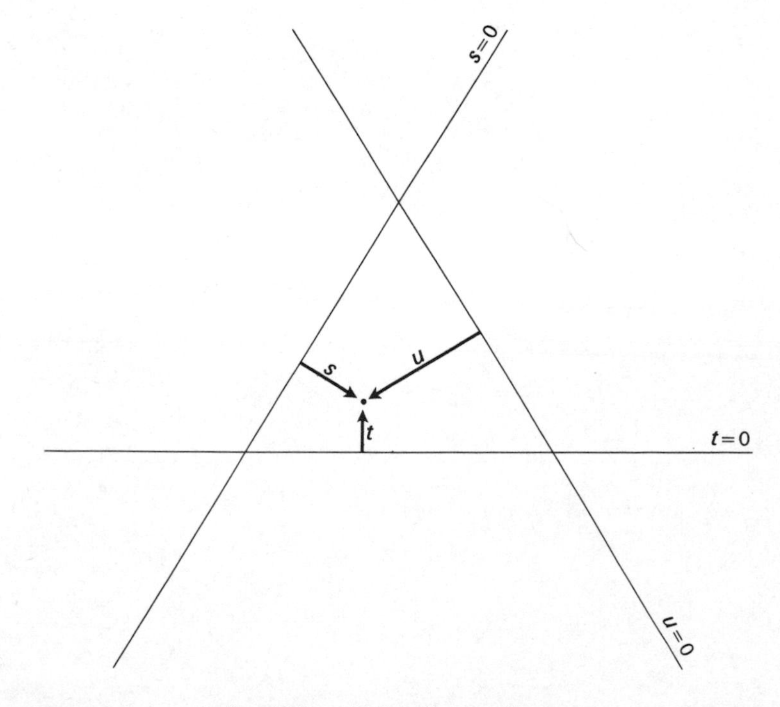

FIG. 12.9.2. Triangular coordinates.

must be treated on the same footing as t. Since there are only six scalar quantities, μ must be a function of s, t, and the masses; it is easy to check that

$$s + t + u = m_1^2 + m_2^2 + m_3^2 + m_4^2. \tag{9.6}$$

These variables admit a simple physical interpretation. In the centre of mass frame the momenta $\mathbf{p}_1$ and $\mathbf{p}_2$ are equal and opposite so that s reduces to the square of the total energy $(p_1^0 + p_2^0)^2$. When the masses are equal in pairs according to $m_1 = m_3, m_2 = m_4$, as in elastic scattering, the energies p_1^0 and p_3^0, p_2^0 and p_4^0 are also equal in pairs, so that t reduces to our former variable

$$t = -(\mathbf{p}_1 - \mathbf{p}_3)^2 = -\Delta^2. \tag{9.7}$$

If, further, $m_1 = m_2$, as for instance in nucleon-nucleon scattering, then the variable u can also be interpreted as the squared momentum transfer between particles 1 and 4. In the general case of unequal masses t and u retain their meaning as the scalar squared lengths of *four-dimensional* momentum transfers.

A final advantage of these variables lies in the simplicity of their behaviour under crossing. Thus, if particles B and D in reaction (8.1) are crossed

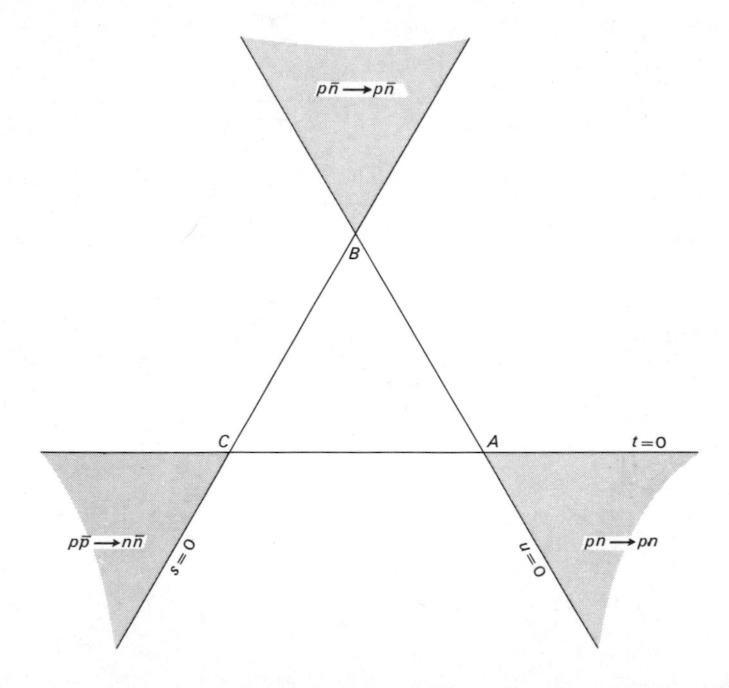

FIG. 12.9.3. Kinematics of proton-neutron scattering. The proton and neutron masses are both taken as m. The point A is at $s = 4\,m^2$, $t = 0$, $u = 0$. The physical regions are shaded.

to yield reaction (8.3), then the interchange of their momenta,

$$(p_1, p_2, p_3, p_4) \rightarrow (p_1, -p_4, p_3, -p_2) \tag{9.8}$$

corresponds simply to an interchange of the invariants:

$$(s, t, u) \rightarrow (u, t, s). \tag{9.9}$$

Accordingly it is natural to represent the physical regions of the three crossed reactions on a common plot with triangular coordinates s, t, and u, as shown in figure (12.9.2).

In each of these crossed reactions one of the scalar variables s, t, or u plays the role of the squared energy in the corresponding centre of mass frame, while the other two are momentum transfers.

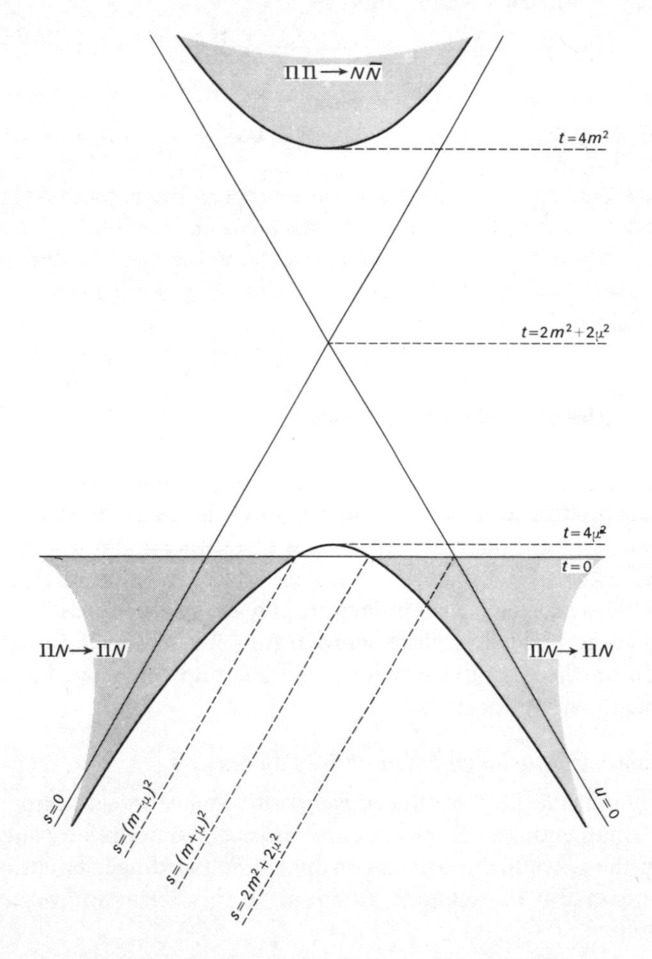

FIG. 12.9.4. Kinematics of pion-nucleon scattering.

The boundaries of the physical region are easily found from the following relations which govern the kinematics of reaction (8.1). The relative momentum of particles A and B in the centre of mass frame is given by

$$4q^2 = \frac{\lambda(s, m_1^2, m_2^2)}{s} \tag{9.11}$$

where

$$\lambda(a, b, c) \equiv a^2 + b^2 + c^2 - 2ab - 2bc - 2ca. \tag{9.12}$$

For the relative momentum of C and D we have similarly

$$4q'^2 = \frac{\lambda(s, m_3^2, m_4^2)}{s} \tag{9.13}$$

The scattering angle θ is determined by

$$4qq' \cos\theta = t - u + \frac{(m_1^2 - m_3^2)(m_2^2 - m_4^2)}{s} \tag{9.14}$$

The physical region corresponds to $s \geqslant (m_1 + m_2)^2$, $s \geqslant (m_3 + m_4)^2$, $-1 \leqslant \cos\theta \leqslant 1$.

Figures (12.9.3) and (12.9.4) show the shapes of the physical regions for the simplest reactions, namely for *nucleon-nucleon* and *pion-nucleon scattering*. *Proton-neutron scattering*: we take the proton and neutron masses to be equal to m. The point A is at $s = 4m^2$, $t = u = 0$. The physical regions are shown shaded.

10. Singularities of scattering amplitudes

We shall tackle the analytic properties of relativistic scattering amplitudes by exploiting crossing symmetry and the analytic properties of nonrelativistic amplitudes. What we aim to establish is not at all analyticity in energy and momentum transfer, but rather non-analyticity; in other words, we ask what singularities can be shown to be necessarily present by some rather straightforward arguments. Where appropriate we shall touch on the question of what can or cannot be proved at present from specific assumptions.

(a) *The nonrelativistic amplitude and its singularities*

Begin by considering once more the proton-neutron scattering amplitude. For small enough energies it can be described nonrelativistically by subjecting the wavefunction to an ordinary Schroedinger equation containing a direct and an exchange potential. In this discussion we continue to neglect spins.

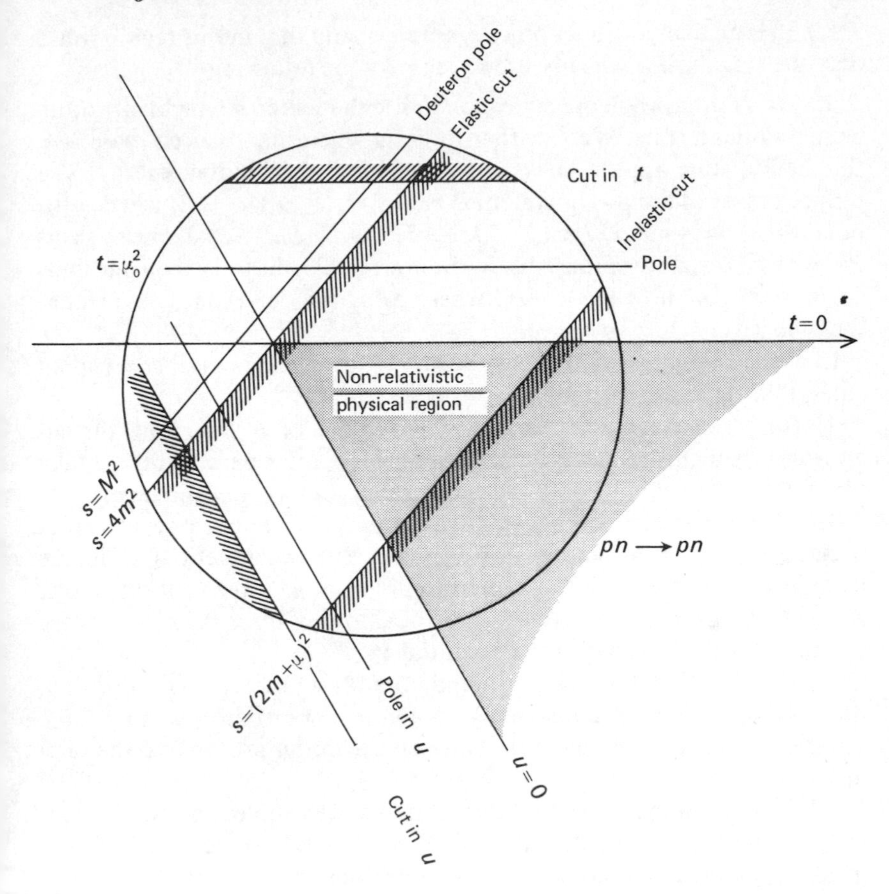

FIG. 12.10.1. Singularities of the nonrelativistic nucleon-nucleon scattering amplitude.

Assume that the potentials are not too singular, or more precisely that they can be written in the form (7.2) as superpositions of Yukawa potentials. This choice is suggested by the exchange of massive systems of mesons (one, two, three pions, etc.).

As long as the nonrelativistic approximation is valid, i.e. as long as the momenta are small compared to the nucleon mass m, the full scattering amplitude must have all the properties derivable from the Schroedinger equation. Under these conditions k^2, the squared momentum in the centre of mass frame, is small compared to m^2, and consequently the squared momentum transfer is also much smaller than m^2. Since we have $s = 4(k^2 + m^2)$ and $t = -\Delta^2$, we see that in a region of the (s, t, u) plane, roughly shown as a circle in figure (12.10.1), the invariant amplitude

$T(s, t, u)$ is an analytic function of s, t, and u; and that in this region it has only the singularities already found, namely the following.

1. $T(s, Jt, u)$ has a pole in s corresponding to the deuteron which is a proton-neutron bound state. We saw that in units where the reduced mass is $\frac{1}{2}$, this bound state appears at $k^2 = -B$, with B the binding energy. The reduced mass of the $p - n$ system is $m_p m_n/(m_p + m_n) = m/2$, whence the pole is at $k^2 = -mB$, i.e. at $s = 4(k^2 + m^2) = 4m(m - B)$. If one neglects B^2/m^2, as is certainly permissible in the nonrelativistic limit, then one finds $4m(m - B) \simeq (2m - B)^2 = M^2$, where M is the deuteron mass. Hence the pole is very close to $s = M^2$.

In a relativistic theory it is easy to show, for instance by the Feynman rules, that the pole is exactly at $s = M^2$.

2. $T(s, t, u)$ has a cut starting at $k^2 = 0$, i.e. at $s = 4m^2$. Thus the cut appears when the energy is high enough for scattering actually to take place. It must certainly continue to be present in a relativistic theory.

3. $T(s, t, u)$ has a cut for real positive t. It may even have a pole as well as a cut if the potential contains an isolated Yukawa potential as in the formula (7.3). In that case $T(s, t, u)$ has a pole $t = \mu_0{}^2$ and a cut starting at $t = \mu_1{}^2$. We have seen already that the positions of these poles or cuts are connected with the range of the potential V^+.

4. Similarly, $T(s, t, u)$ has a cut (and possibly a pole) *for real positive u*, due to the presence of an exchange potential; note that u is simply the squared difference in momentum between the initial proton and the final neutron.

It follows from the above that the positions of the poles and cuts in s are uniquely determined by the masses of the particles taking part in $p - n$ processes, while the positions of the singularities in t and u are determined by the ranges of the potentials.

(b) *Cuts in energy*

Let us look more closely at the cuts in energy. So far we have considered only elastic scattering. But for high enough energy inelastic processes will set in, pion production being the first to appear:

$$\begin{aligned} p + n &\to p + n + \pi^0 \\ &\to n + n + \pi^+ \\ &\to p + p + \pi^-. \end{aligned} \tag{10.1}$$

These reactions will begin to occur as soon as the total energy $\sqrt{s}$ in the centre of mass frame rises above the mass threshold $2m + \mu$ of the final state. One should ask whether these new reactions will affect the analytic properties of $T(s, t, u)$.

It is easy to see that they will. Below the inelastic threshold all the phase

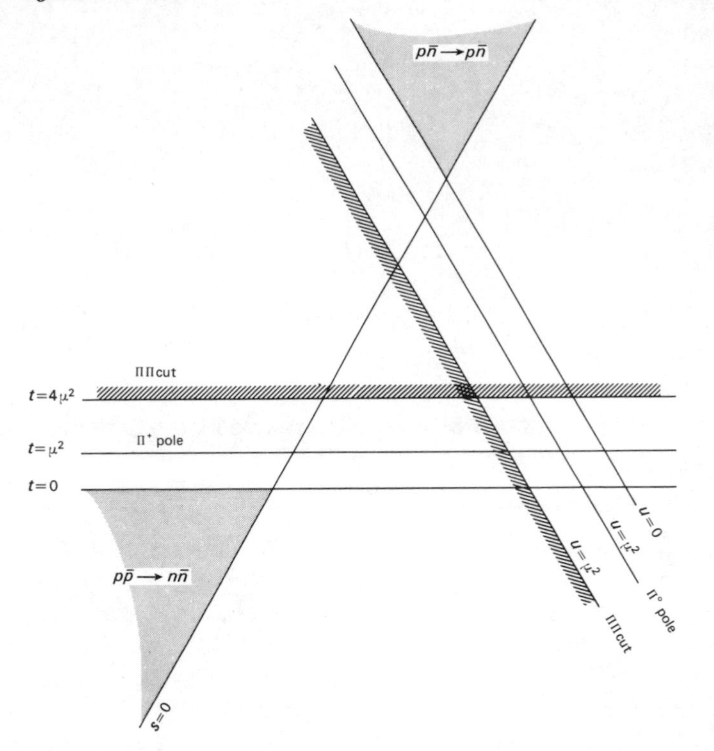

FIG. 12.10.2. The singularities of the proton-neutron scattering amplitude associated with the crossed channels. $= m^2$;

shifts $\delta_\ell(s)$ are trial, but they become complex for $s \geqslant (2m + \mu)^2$. Hence at $s = (2m + \mu)^2$ there must necessarily be a singularity which occurs both in the partial wave amplitude $a_\ell(s)$ and in the full amplitude $T(s, t, u)$; it is connected with the opening of the channel $2N + \pi$. More generally, we expect the amplitude to have a singularity at every energy corresponding to the threshold of a new channel. The existence of these singularities is verified in perturbation theory.

We are now in a position to formulate a general rule for all singularities in energy, including the deuteron pole as well as the elastic and inelastic cuts: there is a singularity in s at the opening of every channel, whether or not the channel leads to a physically realisable reaction. (In general the singularity is a branch point at the corresponding channel threshold.) Thus the deuteron pole is a singularity at the point $s = M^2$ which marks the physically unrealisable channel $p + n \rightarrow d$.

The cut can be taken to extend over all s corresponding to physically realisable values of the energy in the channel in question. Thus, the $p - n$ channel is associated with a cut from $s = 4m^2$ to infinity; the channel (10.1)

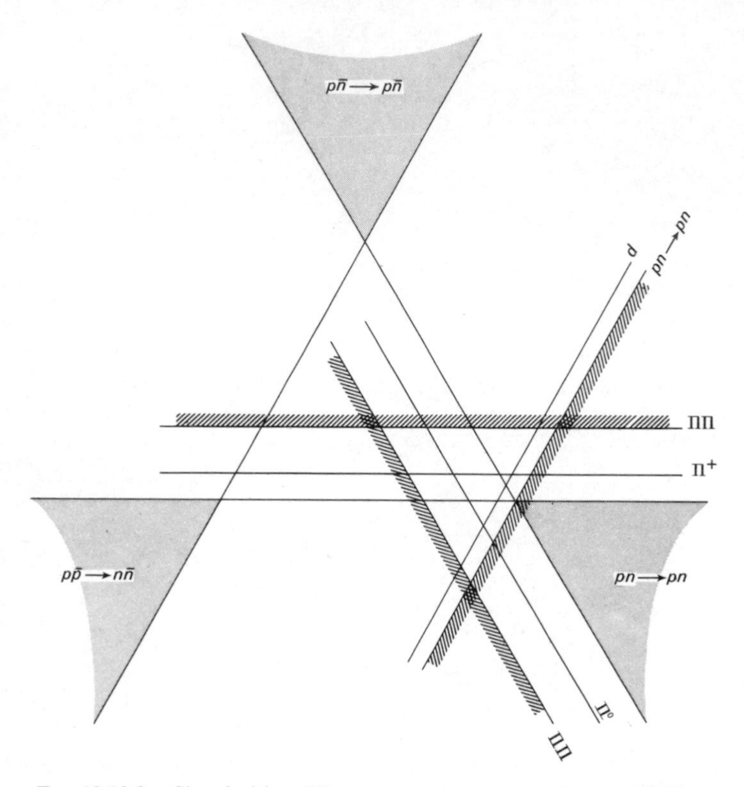

FIG. 12.10.3. Singularities of the proton-neutron scattering amplitude.

is associated with a cut from $s = (2m + \mu)^2$ to $+\infty$. By contrast, the mass M of the channel $p + n \rightarrow d$ is uniquely determined, and therefore the corresponding "cut" is completely localised at $s = M^2$. In this case the rule leads naturally to the existence of a pole.

(c) $p - \bar{n}$ *scattering*

Let us illustrate the rule by considering proton-antineutron scattering; this also happens to be a crossed reaction of proton-neutron scattering, so that we shall denote by $\sqrt{u}$ the energy of the system in its centre of mass frame. The corresponding amplitude is still denoted by $T(s, t, u)$, being identical to the proton-neutron amplitude by virtue of crossing.

The only single-particle state which, having the same quantum numbers as the $p - \bar{n}$ system, can be produced by it, is the π^+; hence by analogy with the deuteron pole expect $T(s, t, u)$ to have a pole at $u = \mu^2$.

Next comes the $\pi^+\pi^0$ channel opening at $u = (2\mu)^2$. It is obvious that the reaction $p + \bar{n} \rightarrow \pi^+\pi^0$ cannot be realised physically when u only just exceeds $4\mu^2$. Nevertheless, this channel gives rise to a branch cut extending from $4\mu^2$ to $+\infty$.

New singularities follow at $u = (3\mu)^2$ with the opening of the three-pion channel, at $(4\mu)^2$, etc. Similarly, there will be singularities at the openings of the $K - \bar{K}$, $K - \bar{K} - \pi$ channels, and so on; eventually one reaches $u = (2m)^2$ where the $p - \bar{n}$ channel enters into play, giving a new singularity which we shall call the elastic threshold. It is followed by the thresholds of $N\bar{N}\pi$, $N\bar{N}\pi\pi$, etc; (see figure (12.10.2). The same analysis for the other crossed reaction $p\bar{p} \rightarrow n\bar{n}$ is illustrated in figure (12.10.2); it shows the pole at $t = \mu^2$ due to the π^0.

All the singularities necessarily present in the proton-neutron scattering amplitude can be collected into a single diagram (figure 12.10.3).

Figure (12.10.4) shows the singularities which must enter the pion-nucleon scattering amplitude.

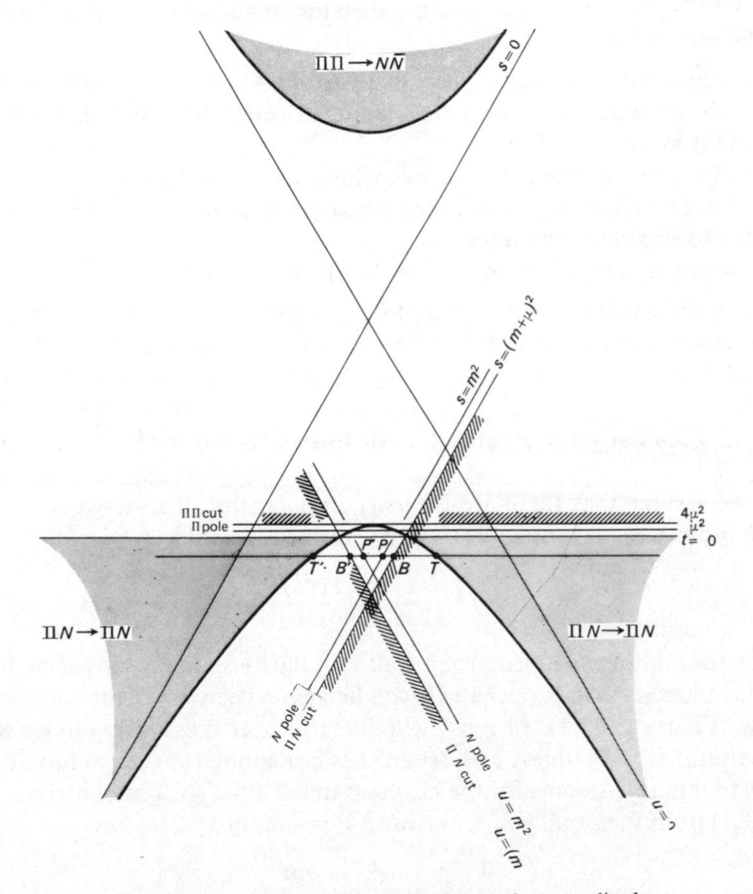

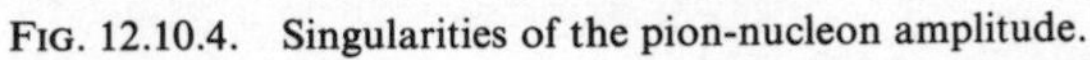

FIG. 12.10.4. Singularities of the pion-nucleon amplitude.

11. Pion-nucleon dispersion relations

We have discovered certain singularities of the scattering amplitude, the reason for whose presence lies in the physical processes themselves together with crossing symmetry. The crucial question to be answered next is whether these are the only possible singularities, or whether there are others. Before tackling this question theoretically in the later sections of this chapter, we shall set up straightaway an *experimental* test to check whether the singularities found so far are the only ones. Actually this test cannot verify the absence of all other singularities whatever, but only the absence of singularities in s at fixed $t = 0$ in the pion-nucleon system; in other words, it verifies that the forward scattering amplitude has only those singularities that we have already met.

If we consider a real negative, i.e. a physical value of the momentum transfer t in pion-nucleon scattering, then the singularities in s found so far can be read off the line $L$of figure (12.10.4); they are,

1. A cut from the point B at $s = (m + \mu)^2$ to $+\infty$; actually we have seen that one has many different cuts on top of each other, but this will be irrelevant in the following.
2. The nucleon pole P at $s = m^2$, whose residue we shall call g^2.
3. Another nucleon pole P' in the u-channel at $u = m^2$, having the same residue by crossing symmetry.
4. Another cut from B' at $u = (m + \mu)^2$ to $u = +\infty$.

Parts of the real s-axis lie in the physical region, either of the direct or of the crossed reaction, as can also be seen from the figure. One such part extends from a threshold T to $s = +\infty$, and the other from a threshold T' to $u = +\infty$. T is to the right of B unless $t = 0$, in which case T and B coincide. Similarly for B' and T'. All this is shown explicitly in figure (12.11.1).

If we assume that these are the *only* singularities, then for a complex value σ of s we can express $T(s, t)$ by Cauchy's formula

$$T(\sigma, t) = \frac{1}{2\pi i} \int_C \frac{T(s', t)\, ds'}{s' - \sigma} \tag{11.1}$$

where the contour C loops around all singularities, and is closed by two large semicircles in the upper and lower halves respectively of the complex s plane. (Figure 12.11.1). As a rule the integral over these semi-circles will contribute to (11.1), unless $T(s', t)$ vanishes fast enough as the modulus of s' tends to infinity. Assume for the moment that T does so. The contribution to (11.1) from the small circle γ around P is simply

$$-\frac{1}{2\pi i} \oint_\gamma \frac{g^2}{s' - m^2} \frac{ds'}{s' - \sigma} \tag{11.2}$$

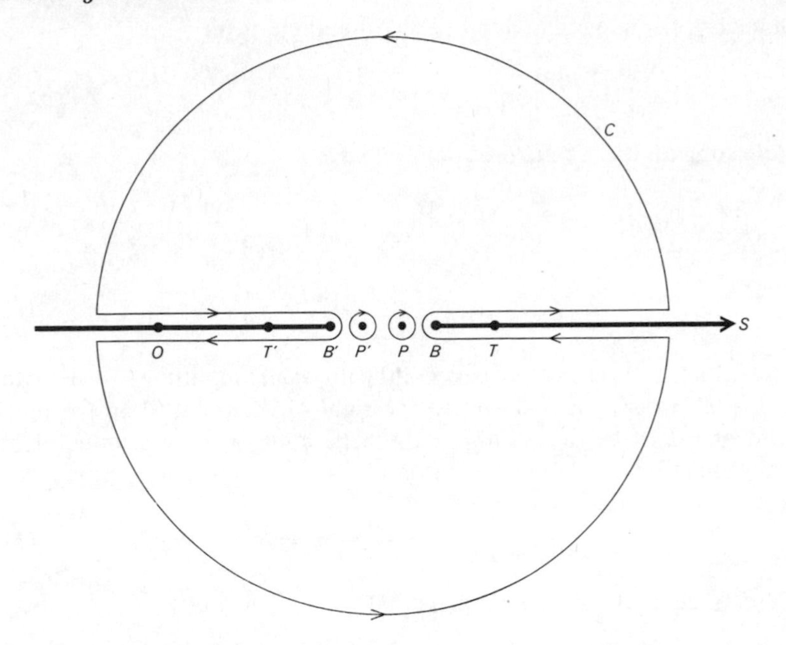

FIG. 12.11.1. The integration contour which leads to the dispersion relations. $s_p = m^2$; $s_B = (m + \mu)^2$.

where the integration is performed in the positive (anti-clockwise) direction. For this term Cauchy's formula yields

$$\frac{g^2}{\sigma - m^2}\,. \tag{11.3}$$

Similarly, the integral over γ' yields $g^2/(u - m^2)$, where $u = 2m^2 + 2\mu^2 - \sigma - t$. The integral over the contour around the right hand cut gives

$$\frac{1}{2\pi i}\int_{(m+\mu)^2}^{\infty} \frac{[T(s' + i0, t) - T(s' - i0, t)]}{s' - \sigma}\, ds'. \tag{11.4}$$

The amplitude $T(s, t)$ is real when t is negative and $(m + \mu)^2 - t < s < (m + \mu)^2$. We saw in chapter 8 that this is true nonrelativistically; it remains true also in the relativistic case. Hence,

$$\frac{1}{2i}[T(s' + i0, t) - T(s' - i0, t)] = \text{Im } T(s', t) \tag{11.5}$$

where the right hand side is the imaginary part of the *physical* amplitude. The expression (11.1) now becomes

$$\frac{1}{\pi}\int_{(m+\mu)^2}^{\infty} \frac{\text{Im } T(s', t)\, ds'}{s' - s}\,. \tag{11.6}$$

Similarly, the integral around the left-hand cut gives

$$-\frac{1}{\pi}\int_{-\infty}^{B'}\frac{\mathrm{Im}\,T(u',t)\,ds'}{s'-\sigma}=\frac{1}{\pi}\int_{(m+\mu)^2}^{\infty}\frac{\mathrm{Im}\,T(u',t)\,du'}{u'-u}. \tag{11.7}$$

Collecting all these contributions, we have

$$T(\sigma,t)=\frac{g^2}{\sigma-m^2}+\frac{g^2}{u-m^2}+\frac{1}{\pi}\int_{(m+\mu)^2}^{\infty}\frac{\mathrm{Im}\,T(s',t)\,ds'}{s'-s}$$
$$+\frac{1}{\pi}\int_{(m+\mu)^2}^{\infty}\frac{\mathrm{Im}\,T(u',t)\,du'}{u'-u} \tag{11.8}$$

Equation (11.8) yields a property of the physical amplitude if we let σ tend to $s + i0$, where s is real and to the right of the point T. In taking this limit one must be careful about the behaviour of the first integral; it is determined by the classical formula of distribution theory

$$\frac{1}{s'-s-i0}=\mathscr{P}\frac{1}{s'-s}+i\pi\,\delta(s'-s) \tag{11.9}$$

Here, $\mathscr{P}$ denotes the Cauchy principal value, defined by

$$\mathscr{P}\int_a^b\frac{f(s')}{s'-s}\,ds'=\lim_{\substack{\epsilon\to 0\\ \epsilon>0}}\left[\int_a^{s-\epsilon}\frac{f(s')\,ds'}{s'-s}+\int_{s+\epsilon}^{b}\frac{f(s')\,ds'}{s'-s}\right]. \tag{11.10}$$

Equating the real parts of the two sides of (11.8), one finds

$$\mathrm{Re}\,T(s,t)=\frac{g^2}{s-m^2}+\frac{g^2}{u-m^2}+\frac{1}{\pi}\mathscr{P}\int_{(m+\mu)^2}^{\infty}\frac{\mathrm{Im}\,T(s',t)\,ds'}{s'-s}$$
$$+\frac{1}{\pi}\int_{(m+\mu)^2}^{\infty}\frac{\mathrm{Im}\,T(u',t)\,du'}{u'-u} \tag{11.11}$$

This linear integral relation is called a *dispersion relation*.*

The dispersion relation assumes a particularly interesting form for zero momentum transfer, $t = 0$. Consider for instance π^--proton scattering. Then $\mathrm{Im}\,T(s, 0)$ can be expressed in terms of the total π^--p cross-section, $\sigma_t^{(-)}$, by virtue of the optical theorem (see chapter 7) and of the relation between the invariant amplitude T and the amplitude f (which differ by a factor $\sqrt{s}$). One finds

$$\mathrm{Im}\,T(s,0)=\frac{k\sqrt{s}\,\sigma_t^{(-)}}{4\pi}. \tag{11.12}$$

Similarly, $\mathrm{Im}\,T(u', 0)$ is the imaginary part of the forward amplitude for the crossed reaction $\pi^+p \to \pi^+p$. Let the cross-section for this be $\sigma_t^{(+)}$, and

* The name derives from optics where an analogous relation applies to the refractive index, with the frequency of the light as the variable.

let q be the momentum in the centre of mass frame; then one has

$$\text{Im } T(u, 0) = \frac{q\sqrt{u}\,\sigma_t^{(+)}(u)}{4\pi} \tag{11.13}$$

whence

$$\text{Re } T(s, 0) = \frac{g^2}{s - m^2} + \frac{g^2}{u - m^2} + \frac{1}{\pi}\mathscr{P}\int_{(m+\mu)^2}^{\infty} \frac{k'\sqrt{s'}\,\sigma_t^{(-)}(s')\,ds'}{s' - s}$$
$$+ \frac{1}{\pi}\int_{(m+\mu)^2}^{\infty} \frac{q'\sqrt{u'}\,\sigma_t^{(+)}\,du'}{u' - u} \tag{11.14}$$

In practice the measured cross-sections $\sigma_t^{(+)}$ and $\sigma_t^{(-)}$ appear to tend to nonzero constant values as the energy tends to infinity; then the integrals in (11.14) fail to converge. One can easily overcome this difficulty by writing a Cauchy formula not for $T(s, t)$ but for the function $T(s, t)/(s - s_0)$, where s_0 is real and fixed between B and B'. It is trivial to guess the final result from (11.14) without going through the whole calculation. Assume for the moment that the integrals converge, and subtract equation (11.14) as written for s_0 from its form as written for s. Setting $u_0 = 2m^2 + 2\mu^2 - s_0$, we get

$$\text{Re } T(s, 0) - \text{Re } T(s_0, 0) = \frac{g^2(s - s_0)}{(s - m^2)(s_0 - m^2)} + \frac{g^2(u_0 - u)}{(u - m^2)(u_0 - m^2)}$$
$$+ \frac{s - s_0}{\pi}\mathscr{P}\int_{(m+\mu)^2}^{\infty} \frac{k'\sqrt{s'}\,\sigma_t^{(-)}(s')\,ds'}{(s' - s)(s' - s_0)}$$
$$+ \frac{u_0 - u}{\pi}\int_{(m+\mu)}^{\infty} \frac{q'\sqrt{u'}\,\sigma_t^{(+)}(u')\,du'}{(u' - u)(u' - u_0)}. \tag{11.15}$$

In this last equation the integrals do converge; it is called a subtracted dispersion relation.

The right hand side of (11.15) can obviously be evaluated from the experimental results on pion-nucleon scattering. The same applies to the left hand side, since the forward elastic differential cross-section can be written as

$$\left.\frac{d\sigma_{\text{el}}}{d\Omega}\right|_{\theta=0} = \frac{1}{s}[(\text{Re } T(s, 0))^2 + (\text{Im } T(s, 0))^2] \tag{11.16}$$

where the contribution from Im T is known from the optical theorem.

Experiment confirms to very high precision the dispersion relation (11.15) as corrected (theoretically) to take into account the nucleon spin. Hence there is every reason to believe that at least in this well-defined instance the scattering amplitude has no singularities in energy other than those we have found. Moreover, we shall see later on that this dispersion relation has been proved in the framework of axiomatic field theory.

12. Mandelstam's method

In order to discover the full analyticity domain of the scattering amplitude, one needs a theory from which to derive it. By analogy with electrodynamics it might be tempting to exploit field theory and to investigate the analytic properties of the expressions contributed to the amplitude by the various Feynman diagrams. One could equally well try to develop an axiomatic method by abstracting some of the formal properties of field theory, and attempt to deduce therefrom the analytic properties of the full amplitude; we shall return to this method later on. Finally, one might try to rely on assumptions more closely related to experiment, and by starting from the theory of measurement to relate analyticity in energy to causality (no scattered wave can be produced before arrival of the incident wave); and to relate analyticity in momentum transfer to the finite range of strong interactions. All these methods lead to roughly the same domain of analyticity. But one cannot be sure that in actual fact the domain is not larger, since the limitations one encounters have no particular mathematical or physical significance. In general one assumes that the true domain of analyticity is bigger than that emerging from such calculations, and one blames the difference on one's lack of technical facility and on the complications inherent in this kind of investigation.

Under these conditions it is natural to adopt the working hypothesis that the only singularities are those derivable directly from the known spectrum of particles, from crossing symmetry, and from unitarity. This last condition expresses the fact that the total probability in outgoing waves equals the total probability in incoming waves; in elastic scattering for instance it entails that the phase shifts are real below the inelastic threshold. We have seen already how unitarity leads to branch points in energy at the various inelastic thresholds. Thus, by adopting this working hypothesis one exploits the most fundamental properties of scattering theory. In the special case of scattering between stable particles, like pion-pion, pion-nucleon, or nucleon-nucleon, the hypothesis implies that all the singularities occur at real values of the variables s, t, and u, being just those that we have already mentioned. This singularity structure was first proposed by S. Mandelstam in 1958.

Analyticity and dynamics

We must now come to the horses. So far, the reader has been led on a quest for the analytic properties of amplitudes, and at this stage he should be told how they can be used to formulate a theory of strong interactions. The crucial idea to which we now turn, and which derives from Mandelstam's hypothesis, is that analyticity has dynamical content by virtue of crossing symmetry.

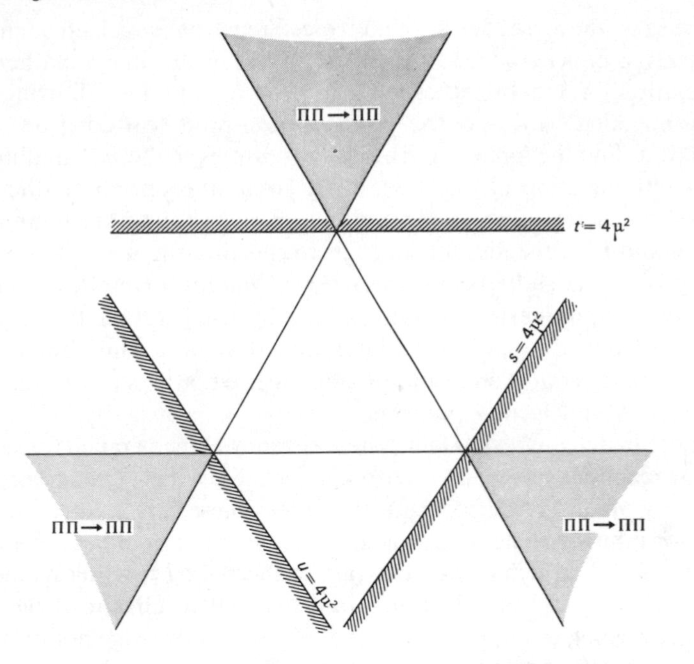

FIG. 12.12.1. Singularities of the pion-pion scattering amplitude.

This is best understood by considering a particular example; to avoid inessential complications we take the simplest case to be found in nature, namely pion-pion scattering. Then the three crossed reactions are basically the same, and according to Mandelstam's hypothesis all singularities are real and lie in regions of the (s, t, u) plane satisfying the inequalities: $s > 4\mu^2$, $t > 4\mu^2$, $u > 4\mu^2$ (see figure 12.12.1).

Consider one of these reactions, say that whose energy is denoted by s; we see that for fixed s the positions of the t and u singularities are determined by the properties of the crossed reactions. In practice the discontinuities across these singularities are also determined by the crossed reactions. But in our study of potential scattering we saw these t and u singularities were dynamical ones governed by the forces. We can now invert this correspondence by saying that the forces must be governed by the t and u singularities, which are themselves determined by the crossed reactions. To put it briefly, *crossing symmetry determines the forces*.

13. Bootstraps

For pion-pion scattering the dynamical problem which results from Mandelstam's hypothesis is basically one of self-consistency; this is clear from the fact the forces derive from crossed reactions which are identical to the reaction under study. More precisely, this means that a mathematical formulation of the hypothesis must proceed by the following steps. 1. Assume that we know the *physical* pion-pion scattering amplitude. 2. From it, find the forces, i.e. the discontinuities of the *full* amplitude (as an analytic function of t and u). 3. By some appropriate mathematical method, calculate the physical amplitude from its known discontinuities in t and u and in s, the discontinuity in s reflecting the unitarity condition (reality of phase shifts below inelastic threshold). Overall, the physical scattering amplitude is thus determined by itself, rather than by some prescribed outside quantities like a potential or a Hamiltonian. This selfconsistency condition results in nonlinear equations for the amplitude (the Chew-Mandelstam equations).

Unfortunately, in actual fact selfconsistency extends much further. The crossed reactions have cuts due to channels likes $\pi\pi \to \pi\pi\pi\pi$, $\pi\pi \to N\bar{N}$ and many others. The discontinuities across these cuts involve the amplitudes for multi-particle reactions and for the creation of heavier particles as in $\pi\pi \to N\bar{N}$, which in their turn are connected by crossing symmetry to pion-nucleon scattering. The end result is that a full knowledge of the forces governing pion-pion scattering involves knowledge not only of the pion-pion scattering amplitude, but also of practically every other strong reaction. Thus, what we actually have is not a selfconsistency condition on the pion-pion amplitude by itself, but a set of very complicated consistency conditions interconnecting all strong-interaction amplitudes.

How can one construct a theory of strong interactions under such conditions? From the logical point of view it would be best to proceed by analysing the overall structure of the consistency conditions, trying to deduce from them the greatest possible number of general properties of strong interactions. Unfortunately, the conditions rapidly become very complicated, and this programme has never been as much as formulated seriously. But it is not impossible that one day it will have to be faced, perhaps with the help of regularities suggested by experiment.

A less ambitious but far more precariously based method extracts from the general relations certain approximate ones which allow one to calculate the physical amplitude not everywhere but only at low energies. One starts by observing that in any calculation of strong interactions a discontinuity at a given point $z = z_0$ will always enter the dynamics through a Cauchy formula containing the denominator $(z - z_0)$; here, z represents s, t, or u, or some combination of these variables. Suppose that by confining oneself to low energies one can limit the range of z; then, faute de mieux, one can hope that the calculation will be dominated by small values of z_0 which give large values of $1/(z - z_0)$. This working hypothesis is called the "dominance of nearest singularities". Unfortunately, it is obvious from the start that such a hypothesis must be handled very carefully. To see this we need merely recall the pion-nucleon dispersion relations; in their simplest form they involved Cauchy integrals which diverge at high energy. Accordingly, this approach tends to divide the study of the structure of selfconsistency conditions into an analysis at low energies plus a separate analysis at high energies, for which special methods must be developed. In chapter 14 we shall take up the most promising of these high-energy theories, namely the Regge poles.

The selfconsistency basic to this theory is displayed most strikingly in the idea of the "bootstrap". If we accept dominance by nearest singularities, we see that only low energies are involved to any important extent in the crossed reactions determining the forces. But we saw in chapter 9 that at low energies the most salient features of an amplitude are the poles associated with bound states and resonances. In other words, the main characteristics of the forces depend on the particles appearing in the crossed channels. Thus, in pion-pion scattering the forces would be dominated by the ρ pole. On the other hand, given the forces one can calculate the physical amplitude and thereby in particular the resonances, the most important such resonance being the ρ itself. We see how the ρ regarded as a force creates the ρ regarded as a particle. This idea is called the bootstrap.*

Though we have formulated it only in the approximation of dominance by nearest singularities, it is obvious that basically the bootstrap idea expresses the selfconsistency of the theory. It is unfortunate that hitherto such selfconsistent structures have been studied only in too restricted a framework. However, this bootstrapping of particles is evidently intrinsic to the theory, independently of whatever approximations one may adopt. In this connection one could ask whether the basic internal symmetries (isotopic spin, $SU(3)$) might be necessary consequences of selfconsistency?

It is also conceivable that these equations, being nonlinear, determine the masses of all particles and the residues at all the corresponding poles, (i.e. all coupling constants), so that the theory contains no arbitrary parameters.

All such speculations are useless if one cannot calculate. To date, Mandelstam's hypothesis has been most useful precisely by enabling one to perform certain limited calculations. The bootstrap idea seems reasonable in that it does predict attractive forces in those channels where resonances and bound states exist, so that the forces do tend to create the particles. In this sense the bootstrap is qualitatively correct. Quantitatively, it has been tested only in calculations assuming dominance by

* The name reminds one that to extricate oneself from the mud one need only pull oneself up by one's bootstraps. This theory was first proposed in the eighteenth century by Baron Muenchhausen.

nearest singularities, which must be hedged with caution. To sum up, there is qualitative agreement, plus some quantitative results carrying little conviction as to detail.

By contrast, the basic idea that singularities exist, and that their discontinuities are connected with physical amplitudes, has been verified repeatedly in reactions with kinematics such that a singularity comes close to the physical region. At present our entire understanding of strong interaction phenomenology is organised around the concept of singularities, which does at least provide a framework well adapted to classifying the results of experiments.

14. Nuclear forces

To illustrate the use of analytic properties, let us return to the problem of nuclear forces, i.e. to nucleon-nucleon scattering. The kinematics and the singularities are shown in figure 12.10.3.

The pion pole at $t = \mu^2$ is seen to be the t-singularity nearest to the physical region; it is therefore expected to dominate the variation of the amplitude for small t. On the other hand this variation is determined basically by the phase shifts for large angular momenta, whence the pion pole governs the phase shifts for large J. This is just the result obtained earlier from our phenomenological discussion of nuclear forces. The pole has the form

$$\frac{g^2}{t - \mu^2}$$

where g^2 is the residue. When nucleon spin is taken into account, the contribution of the pion pole is found to be just the formula (3.9), which has been interpreted as the Born approximation for the Yukawa potential. Our present interpretation of this as a pole term dispels any remaining mystery as to why it should dominate the scattering at large angular momenta. Indeed, in this way one can evaluate g^2, which is found to be equal to G^2:

$$g^2/4\pi = 14{,}8 \tag{14.1}$$

Similarly, the contribution of the pion pole at $u = \mu^2$ can be associated with the Born term of a Yukawa *exchange* potential.

The other contributions to the forces derive from the discontinuities of the crossed reactions. Since both these are nucleon-antinucleon scattering, we see that direct and exchange potentials will be qualitatively similar, and shall not discuss them separately.

The main contributions to nuclear forces emerge once we equate the range of a singularity at t to $1/\sqrt{t}$.

1. The first cut starts from $t = 4\mu^2$. There are experimental reasons for

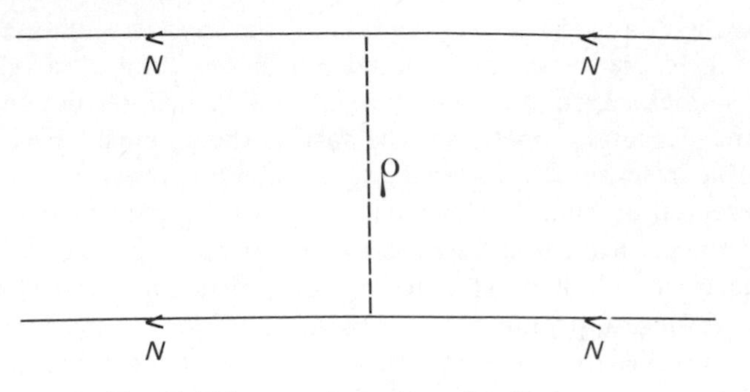

FIG. 12.14.1. ρ-exchange in nuclear forces.

believing in a strong pion-pion interaction at relatively low energy in the state with angular momentum 0 and isospin 0. This will be reflected in a contribution to nuclear forces independent of the spin and i-spin of the nucleons.

2. The two-pion resonance $\rho(I = 1, J = 1)$ leads to a nuclear force depending on spin and i-spin, and describable as *ρ-exchange* between the nucleons (see Fig. 12.14.1).

While pion exchange could be interpreted nonrelativistically as a Yukawa potential, the spin 1 of the ρ prevents at least one part of its contribution from being considered as a local potential.

3. Particularly important are the effects of the ω meson ($I = 0$, $J^P = 1^-$) which emerges from the 3π cut starting at $t = 9\mu^2$. The general features can be visualised by noting that the ω's quantum numbers $J^b = 1^-$ are the same as the photon's, and that its effects are analogous if the role of the charge (for photons) is assigned to baryon number in the case of the ω. Thus ω exchange leads to repulsion between baryon numbers of the same sign, and to attraction between baryon numbers of opposite signs. Therefore it gives repulsion between nucleons and attraction between nucleon and antinucleon. Since the discontinuity associated with the ω is very large, it results in a strong repulsion at short range ($1 \sim 1/m_\omega$), just the effect of a hard core.

4. Note finally that the η contributes via the four-pion cut.

5. The other singularities occur at larger t, and in practice their effects are masked by the ω.

One sees in this way how analytic methods allows nuclear forces to be discussed in concrete terms, even though the problem remains inaccessible by the techniques of perturbation theory.

Second part: Axiomatic field theory and S-matrix theory

In this second part of our study of strong interactions we aim to sketch the outlines of what is conventionally called axiomatic field theory. Here one deals with an abstraction containing the most essential features of ordinary field theory as encountered say in quantum electrodynamics. The aim is not to elaborate calculational techniques, but to determine what are the most general properties that a particle theory must have if it is to be describable in terms of quantised fields. Thus one tries to isolate clearly some essential assumptions, which are generally presented as axioms, along mathematical lines. Because the axioms are so clear-cut they are particularly well adapted to the derivation of certain properties, which can then be shown to apply under very general conditions. A discussion of this theory is justified by some exceptionally powerful results, namely the ΠCT theorem, the connection between spin and statistics, crossing symmetry, and proofs of dispersion relations.

15. The axioms

In presenting axiomatic field theory it is customary to confine oneself to the case where all stable particles are scalars and electrically neutral. The removal of these restrictions is a technical problem.

(a) *The S matrix*

In practice any experiment amounts ultimately to the production of initial particles and the detection of final ones. These initial and final particles are far enough separated in space to be considered as free and noninteracting. As such, we saw in chapter 11 that they are in states describable by means of creation operators acting on the vacuum. The particles which can be considered as free are the incoming ones (at time $-\infty$) and the outgoing ones (at time $+\infty$), rather than those present at intermediate times when the interaction is taking effect; hence one must actually introduce two vacua, one initial and one final, and two kinds of creation operators for incoming and outgoing particles respectively. We distinguish by indices *in* and *out* quantities associated with incoming and outgoing particles; then for the vacua we have

$$a^{in}(\mathbf{p}) \mid 0_{in}\rangle = 0 \qquad a^{out}(\mathbf{p}) \mid 0_{out}\rangle = 0 \tag{15.1}$$

For the creation operators we have the commutation relations

$$[a_{\substack{in\\out}}(\mathbf{p}), a^{+}_{\substack{in\\out}}(\mathbf{p}')] = \delta(\mathbf{p} - \mathbf{p}'). \tag{15.2}$$

These commutation relations have been studied in great detail. Consider two representations of (15.2) by operators acting on a Hilbert space *which contains a vacuum state*, i.e. such that (15.1) is obeyed; then one can prove the the representations are equivalent up to a unitary transformation.

From this it follows immediately that the operators a_{in} and a_{out} are connected by a unitary transformation which can be written as

$$S^+ a^{in}(\mathbf{k})\, S = a^{out}(\mathbf{k}) \tag{15.3}$$

$$SS^+ = I. \tag{15.4}$$

This transformation takes the initial into the final vacuum state

$$|\, 0_{out}\rangle = S^+ \,|\, 0_{in}\rangle$$

Note that initial and final states are associated with representations of the Poincaré group, and thereby with representations of the energy-momentum P_μ. Then the vacuum is an eigenstate of P_μ belonging to eigenvalue zero.

By a suitable choice of phase factors for the vacua, one can ensure

$$|\, 0_{out}\rangle = |\, 0_{in}\rangle \tag{15.5}$$

which defines what in future we shall call simply the vacuum.

The S operator establishes a correspondence between initial and final states having the same configuration. Consider two such states

$$|\, \mathbf{p}_1 \cdots \mathbf{p}_n \,\text{in}\rangle = a^{+in}(\mathbf{p}_1) \cdots a^{+in}(\mathbf{p}_n) \,|\, 0\rangle$$
$$|\, \mathbf{p}_1 \cdots \mathbf{p}_n \,\text{out}\rangle = a^{+out}(\mathbf{p}_1) \cdots a^{+out}(\mathbf{p}_n) \,|\, 0\rangle$$

Then by (15.3) one has

$$|\, \mathbf{p}_1 \cdots \mathbf{p}_n \,\text{out}\rangle = S^+ \,|\, \mathbf{p}_1 \cdots \mathbf{p}_n \,\text{in}\rangle \tag{15.6}$$

It is this equation which underlies the physical role of S as the collision operator. In a scattering experiment one first prepares a state, say $|\mathbf{p}_1 \cdots \mathbf{p}_n \text{ in}\rangle$, which at time $-\infty$ coincides with a state containing free particles of momenta $\mathbf{p}_1 \cdots \mathbf{p}_n$; then one allows this state to evolve; and finally, at time $+\infty$, one determines the projection of the resultant onto a state containing free particles of momenta $\mathbf{p}_1' \cdots \mathbf{p}_m'$. But just at time $+\infty$ such a state coincides with $|\mathbf{p}_1' \cdots \mathbf{p}_m' \text{ out}\rangle$; hence the projection, or in other words the collision amplitude, can be written as

$$\langle \mathbf{p}_1' \cdots \mathbf{p}_{m\,out}' \,|\, \mathbf{p}_1 \cdots \mathbf{p}_{n\,in}\rangle = \langle \mathbf{p}_1' \cdots \mathbf{p}_{m\,out}' \,|\, S \,|\, \mathbf{p}_1 \cdots \mathbf{p}_{n\,out}\rangle$$

$$= \langle \mathbf{p}_1' \cdots \mathbf{p}_{m\,in}' \,|\, S \,|\, \mathbf{p}_1 \cdots \mathbf{p}_{n\,in}\rangle \tag{15.7}$$

Therefore the matrix elements of S between the free *in* or *out* states are the probability amplitudes for collisions, and will be written simply as $\langle \mathbf{p}_1' \cdots \mathbf{p}_m' |\, S\, |\mathbf{p}_1 \cdots \mathbf{p}_n\rangle$. Note that very few assumptions are involved in this definition of S; in particular, it does not assume the existence of a unitary time evolution operator for the states, nor the existence of fields.

Hence S can turn up very early in presenting quantum mechanics; in this sense it is a basic object more elementary even than the Hamiltonian.

(b) *The axioms*

(i) *Asymptotic condition*

We shall now pursue the idea that there is a field theory underlying the models of particle theory discussed so far. Evidently one can always use the creation and annihilation operators to introduce *free* fields

$$\Phi_{\substack{in\\out}}(x) = \frac{1}{(2\pi)^{3/2}} \int \frac{d^3\mathbf{k}}{\sqrt{2k_0}} \left[a^{+\substack{in\\out}}(\mathbf{k})\, e^{-ikx} + a^{\substack{in\\out}}(\mathbf{k})\, e^{ikx}\right] \tag{15.8}$$

which are connected by $\Phi^{out}(x) = S^+\Phi^{in}(x)S$. A field theory results only if these free fields can be considered as the limits, at $t = \pm\infty$, of a field $\Phi(x)$ which is not free, often called the interpolating field of Φ^{in}_{out}. More precisely, we shall assume that the matrix elements of $\Phi^{in}_{out}(x)$ are limits of the corresponding matrix elements of $\Phi(x)$; or mathematically speaking, that the free fields $\Phi^{in}_{out}(x)$ are weak limits of $\Phi(x)$. In other words, for any states A and B, one has

$$\lim_{t\to\pm\infty} \langle A \mid \{\Phi(x) - \Phi_{\substack{out\\in}}(x)\} \mid B\rangle = 0. \tag{15.9}$$

This is called the asymptotic condition. Let us mention that there are also other formulations of field theory on much more abstract foundations; they allow a more leisurely deployment of mathematical ingenuity, with results that are not substantially different.

One must specify also what properties the field $\Phi(x)$ is assumed to have apart from the asymptotic condition. The most important are the following.

(ii) *Lorentz invariance*

Consider the Poincaré transformation $\{a, \Lambda\}$ where a is a translation and Λ a Lorentz transformation. This transformation acts on the state vectors according to a unitary representation $U\{a, \Lambda\}$, and naturally we shall assume that under it $\Phi(x)$ transforms in a normal way, like the free *in* and *out* fields; thus,

$$U^+\{a, \Lambda\}\, \Phi(x)\, U\{a, \Lambda\} = \Phi(\Lambda x + a). \tag{15.10}$$

(iii) *Causality*

We shall assume that the fields obey what is commonly known as the causality condition, demanding that measurements of the field at two space-time points x and y with space-like separation should commute, i.e.

$$[\Phi(x), \Phi(y)] = 0 \qquad \text{for} \quad (x-y)^2 < 0. \tag{15.11}$$

In quantum electrodynamics we saw that this condition ensures the measureability of the fields. Admittedly this is only a rough interpretation, and the causality condition remains the least explored part of field theory, even though it is the part ultimately responsible for the interesting results.

One can make other assumptions about the mathematical nature of matrix elements of the field; we shall not elaborate on them since we do not intend to enter into technical details.

Next, we consider what can be proved by starting from this clear-cut and disembodied version of field theory.

16. The reduction formula

The objects of physical interest are the S-matrix elements, which are the collision amplitudes. The objects which can be handled mathematically are the *Wightman functions*, which are vacuum expectation values of products of field operators

$$\langle 0 \mid \Phi(x_1)\, \Phi(x_2) \cdots \Phi(x_n) \mid 0 \rangle. \tag{16.1}$$

Hence it is especially important to be able to express the collision amplitudes in terms of Wightman functions. Their interrelation, together with the associated formulae, are called *reduction formulae*. They reduce collision amplitudes to matrix elements of the field.

To prove the reduction formulae one starts from the definition (15.7) of S matrix elements, and relies mainly on the causality condition (15.11). Here we shall not give the proof, but shall confine ourselves to quoting the most important results.

The reduction formulae involve the Klein–Gordon operator

$$K_x = \frac{\partial^2}{\partial x_0{}^2} - \nabla_{\mathbf{x}}{}^2 + m^2.$$

Let us express the scattering amplitude for two particles in terms of the fields. One begins by using the first reduction formula

$$\langle \mathbf{p}_1'\mathbf{p}_2' \mid S \mid \mathbf{p}_1'\mathbf{p}_2' \rangle = +\langle \mathbf{p}_1'\mathbf{p}_2' \mid \mathbf{p}_1\mathbf{p}_2 \rangle + i \int d^4x e^{-ip_2x} \langle \mathbf{p}_1'\mathbf{p}_2' \mid K_x\Phi(x) \mid \mathbf{p}_1 \rangle. \tag{16.2}$$

All states are taken to be *out* states. The first term on the right differs from zero only if initial and final states are the same. Note that the ket in the second term contains only one particle. This has been achieved by making explicit one of the creation operators in the initial state $|\mathbf{p}_1\mathbf{p}_2 \text{ in}\rangle$, and writing it as $a^{+in}(\mathbf{p}_2)|\mathbf{p}_1 \text{ in}\rangle$. Next, one uses the expansion of the Φ^{in} field in terms of the $a^{+in}(\mathbf{p})$, and the asymptotic condition, in order to express the creation operator in terms of the field $\Phi(x)$; eventually this leads to the formula (16.2).

This kind of manipulation can be continued by extracting a creation operator from the final state, say $a^{+out}(\mathbf{p}_2')$. In this way one is led to the new formula

$$\langle \mathbf{p}_1'\mathbf{p}_2' \mid S \mid \mathbf{p}_1\mathbf{p}_2\rangle = \langle \mathbf{p}_1'\mathbf{p}_2' \mid \mathbf{p}_1\mathbf{p}_2\rangle - \iint d^4x d^4y e^{ip_2'x - ip_2y} \times K_x K_y \langle \mathbf{p}_1' \mid [\Phi(x)\,\Phi(y)] \mid \mathbf{p}_1\rangle\, \theta(x^0 - y^0) \quad (16.3)$$

involving the commutator of the fields and the Heaviside step function

$$\theta(x^0) = \begin{cases} 1 & \text{for} \quad x^0 > 0 \\ 0 & \phantom{\text{for}} \quad x^0 < 0. \end{cases}$$

Finally one can extract all the creation operators and arrive at the following combination of Wightman functions:

$$\begin{aligned} &\langle \mathbf{p}_1'\mathbf{p}_2' \mid S \mid \mathbf{p}_1\mathbf{p}_2\rangle \\ &\quad = \langle \mathbf{p}_1'\mathbf{p}_2' \mid \mathbf{p}_1\mathbf{p}_2\rangle + \int d^4x_1 d^4x_2 d^4y_1 d^4y_2 e^{i(p_1'x_1 + p_2'x_2 - p_1y_1 - p_2y_2)} \\ &\quad \times \sum_P K_{x_1}K_{x_2}K_{y_1}K_{y_2} \times \langle 0 \mid \{[[\Phi(z_1), \Phi(z_2)], \Phi(z_3)], \Phi(z_4)]\} \mid 0\rangle \\ &\quad \times \theta(z_1{}^0 - z_2{}^0)\,\theta(z_2{}^0 - z_3{}^0)\,\theta(z_3{}^0 - z_4{}^0) \qquad (16.4) \end{aligned}$$

This involves multiple commutators which can be expressed in terms of Wightman functions only by writing them out explicitly. The sum $\sum_P$ runs over all permutations (z_1, z_2, z_3, z_4) of (x_1, y_1, x_2, y_2).

17. The methods of axiomatic field theory

Axiomatic field theory proceeds essentially by proving that scattering amplitudes are analytic functions of the momenta in domains which it is one's job to specify, and that the Wightman functions (16.1) are analytic functions of $(x_1 - x_2)$, $(x_2 - x_3)$, ... in a certain domain.

(a) *Analyticity of scattering amplitudes*

There are two basic reasons why scattering amplitudes are analytic: the presence of step functions in the reduction formulae ("retardation"), and the causality condition which limits the region where a commutator is nonzero. To understand this, consider the reduction formula (16.3). It shows that the scattering amplitude is the Fourier transform of a function of x and y, which vanishes for $x^0 - y^0 < 0$ because of retardation, and vanishes for $(x - y)^2 < 0$ by causality. Hence, if the amplitude is considered as a function of $(x - y)$ and $(x + y)$, it is seen to be nonzero only in the "future cone" of $(x - y)$ shown in figure (12.17.1).

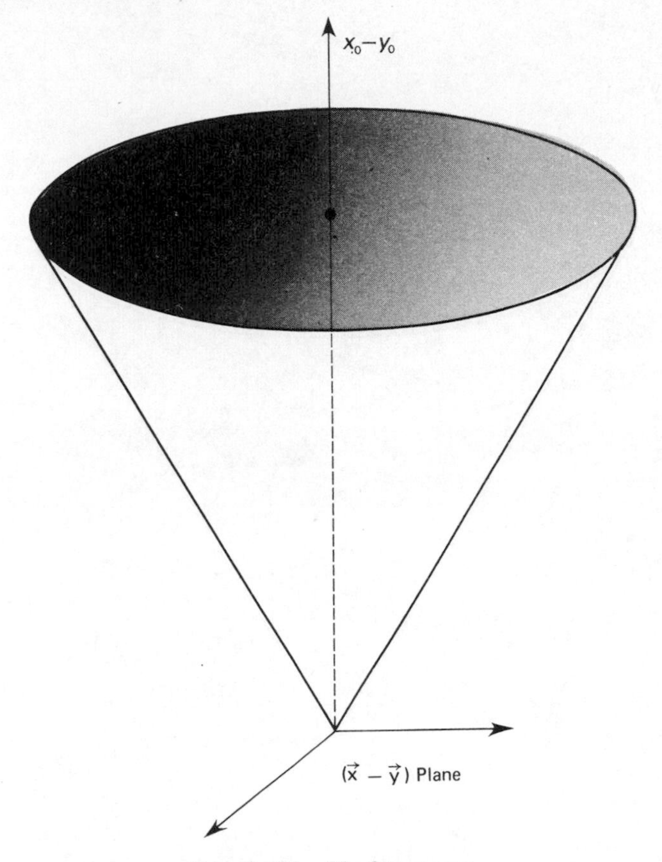

FIG. 12.17.1. The future cone.

This implies that the amplitude is an analytic function of the variable $\frac{1}{2}(p_2 + p_2')$, which is the Fourier-conjugate of $(x - y)$ according to the relation

$$p_2 x - p_2' y = \tfrac{1}{2}(p_2 + p_2')(x - y) + \tfrac{1}{2}(p_2 - p_2')(x + y). \qquad (17.1)$$

To see this, let us simplify things from the outset by assuming that $P = \frac{1}{2}(p_2 + p_2')$ has only a time-component P^0, while $\mathbf{P} = 0$. Under these conditions formula (16.3) is basically of the form

$$\int e^{iP^0(x^0 - y^0)}\, d^4(x - y)\, \theta(x^0 - y^0)\, A(x, y)$$

This is analytic in the upper-half complex P^0 plane,

$$\operatorname{Im} P^0 > 0 \qquad (17.2)$$

because $e^{iP^0(x^0 - y^0)}$ contains a rapidly decreasing convergence factor $e^{-\operatorname{Im} P^0(x^0 - y^0)}$, so that the integral exists and is differentiable in the half-plane $\operatorname{Im} P^0 > 0$.

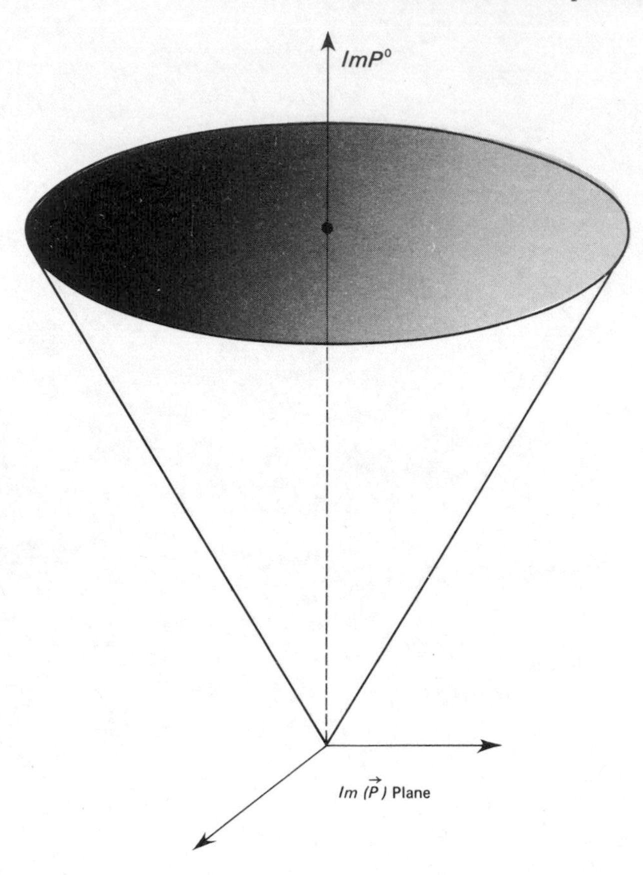

FIG. 12.17.2. The future cone in momentum space: domain of analyticity.

To discover for what value of P the amplitude is analytic in the general case, one takes Lorentz transforms of the domain $\text{Im}\ P^0 > 0$, thus generating the *future cone* for Im P, namely

$$\begin{gathered}\text{Im}\ P^0 > 0 \\ (\text{Im}\ P^0) - (\text{Im}\ \mathbf{P})^2 > 0.\end{gathered} \tag{17.3}$$

See figure (12.17.2).

(b) *Analyticity of the Wightman functions*

The reasons why Wightman functions are analytic functions of differences between their arguments depends on the fact that physical states have positive energy. Considering the simplest Wightman function

$$\langle 0 \mid \Phi(x)\, \Phi(y) \mid 0 \rangle \tag{17.4}$$

we note first that by virtue of the translation-invariance of the vacuum, the function remains unchanged if the same vector a is added to x and y; consequently one has

$$\langle 0 \mid \Phi(x)\, \Phi(y) \mid 0\rangle = \langle 0 \mid \Phi(x + a)\, \Phi(y + a) \mid 0\rangle \tag{17.5}$$

In particular, by taking

$$a = \frac{x + y}{2} \tag{17.6}$$

we obtain a function of the difference $\xi = x - y$, namely

$$W(\xi) = \langle 0 \mid \Phi(\xi/2)\, \Phi(-\xi/2) \mid 0\rangle \tag{17.7}$$

The same method shows quite generally that the Wightman functions depend only on the differences between their arguments.

To prove analyticity in ξ, insert a complete set of states $|n\rangle$, so that (17.6) becomes

$$\sum_n \langle 0 \mid \Phi(\xi/2) \mid n\rangle\langle n \mid \Phi(-\xi/2) \mid 0\rangle \tag{17.8}$$

Next, we appeal to the covariance property (15.10) of the fields and to the definition of the four-momentum, which implies that for a space-time translation T_a one has

$$T_a \mid n\rangle = e^{iP_n \cdot a} \mid n\rangle \tag{17.9}$$

Then it follows that

$$\langle 0 \mid \Phi(\xi/2) \mid n\rangle = e^{iP_n \xi/2} \langle 0 \mid \Phi(0) \mid n\rangle \tag{17.10}$$

$$\langle n \mid \Phi(-\xi/2) \mid 0\rangle = e^{iP_n \xi/2} \langle n \mid \Phi(0) \mid 0\rangle \tag{17.11}$$

whence we get a new expression for the Wightman function

$$\sum_n e^{iP_n \cdot \xi} \langle 0 \mid \Phi(0) \mid n\rangle\langle n \mid \Phi(0) \mid 0\rangle. \tag{17.12}$$

Since energy is positive, i.e. since the P_n are positive time-like, we see from (17.12) that the Wightman function $W(\xi)$ is the Fourier transform of a certain function of momentum which is nonzero only in the future cone. This situation is exactly complementary to our findings for scattering amplitudes, and implies therefore that $W(\xi)$ is an analytic function of ξ in the positive-imaginary future cone

$$\begin{gathered} \operatorname{Im} \xi^0 > 0 \\ (\operatorname{Im} \xi^0)^2 - (\operatorname{Im} \xi)^2 > 0. \end{gathered} \tag{17.13}$$

The programme of axiomatic field theory is to exploit such analytic properties. Typically, one obtains two kinds of results: consequences of the analyticity of Wightman functions in coordinate space (ΠCT theorem, connection between spin and statistics), and consequences of the analyticity of collision amplitudes in momentum space (crossing symmetry, dispersion relations). We proceed to a quick sketch of these results.

18. ΠCT theorem

We start by recalling the definitions of the transformations Π for space reflection, C for charge conjugation, and T for time reversal.

Π is unitary and transforms a state of a single particle A with momentum $\mathbf{p}$ and spin μ according to

$$\Pi \mid A\mathbf{p}\mu\rangle = \eta_P \mid A - \mathbf{p}\mu\rangle \tag{18.1}$$

where we have used Wigner's representation from chapter 5, and where the phase factor η_P is the intrinsic parity.

C is unitary and transforms a state of a single particle A into the antiparticle state with the same quantum numbers:

$$C \mid A\mathbf{p}\mu\rangle = \eta_C \mid \bar{A}\mathbf{p}\mu\rangle. \tag{18.2}$$

T is an *antiunitary* transformation as is shown by the complex-conjugation operator K in its definition

$$T \mid A\mathbf{p}\mu\rangle = K \mid A - \mathbf{p} - \mu\rangle. \tag{18.3}$$

We wish to discuss the fundamental result that the product ΠCT is conserved even in a theory which is not invariant under the transformations Π, C, and T individually. To show this one needs to assume only that it is possible to define quantised fields obeying the axioms of section 15. The proof proceeds by establishing invariance for the set of all Wightman functions, since it can be shown that any operation under which they are all invariant is necessarily a symmetry of the system and commutes with the S matrix. Note that ΠCT is an antinuitary operator.

We shall indicate only the *principle* on which the proof rests, and even that only for the case of neutral scalar spinless particles, where moreover the antiparticle $\bar{A}$ is identical with the particle A. From the definition of the state $|\mathbf{p}\rangle$ in terms of a creation operator acting on the vacuum, and from equations (18.1) to (18.3), we deduce immediately how Π, C, and T act on the *in* and *out* creation and annihilation operators; this then determines their effect on the fields Φ^{in} and Φ^{out}, and, by continuity, their action on the interpolating field $\Phi(x)$. One finds

$$\Pi\Phi(\mathbf{x}, t)\, \Pi^{-1} = \Phi(-\mathbf{x}, t) \tag{18.4}$$

$$C\Phi(\mathbf{x}, t)\, C^{-1} = \Phi(\mathbf{x}, t) \tag{18.5}$$

$$T\Phi(\mathbf{x}, t)\, T^{-1} = \Phi^{+}(\mathbf{x}, -t). \tag{18.6}$$

Defining $\Theta = \Pi CT$, it follows that

$$\Theta\Phi(\mathbf{x}, t)\, \Theta^{-1} = \Phi^{+}(-\mathbf{x}, -t). \tag{18.7}$$

Since Θ is antiunitary, the transformation law for the Wightman functions is

$$\langle 0 \mid \Phi(x_1)\,\Phi(x_2)\cdots\Phi(x_n)\mid 0\rangle \xrightarrow{\Theta} \langle 0 \mid \Phi^+(-x_1)\,\Phi^+(-x_2)\cdots\Phi^+(-x_n)\mid 0\rangle^*$$
$$= \langle 0 \mid \Phi(-x_n)\cdots\Phi(-x_2)\,\Phi(-x_1)\mid 0\rangle \qquad (18.8)$$

whence the ΠCT theorem is proved if one can establish the relation

$$\langle 0 \mid \Phi(x_1)\cdots\Phi(x_n)\mid 0\rangle = \langle 0 \mid \Phi(-x_n)\cdots\Phi(-x_1)\mid 0\rangle \qquad (18.9)$$

But the Wightman functions depend only on the differences $\xi_1 = x_1 - x_2$, $\xi_2 = x_2 - x_3, \ldots \xi_{n-1} = x_{n-1} - x_n$, and can be written as $W(\xi_1, \ldots \xi_{n-1})$. As we showed earlier, these functions can be generalised to complex values of $\xi_1 \ldots \xi_{n-1}$ in the positive-imaginary future cone (17.13). Then the invariance of the Wightman functions under Lorentz transformations entails that their extensions to complex ξ_i are actually analytic in a larger domain D, where they are invariant under proper but complex Lorentz transformations.* But it turns out that combined space and time reflection, though not a proper Lorentz transformation, is nevertheless connected to the identity by a sequence of complex Lorentz transformations; hence as a complex transformation it is proper. Thus in the domain D one has

$$W(\xi_1\cdots\xi_{n-1}) = W(-\xi_1\cdots-\xi_{n-1}) \qquad (18.10)$$

Now it so happens that the large domain D contains real values of $\xi_1 \ldots \xi_n$ such that the separations between the corresponding points $x_1 \ldots x_n$ are all spacelike. Then the causality condition implies that the fields at these points commute, whence one can write

$$W(\xi_1\cdots\xi_{n-1}) = W(\xi_{n-1},\ldots,\xi_1) \qquad (18.11)$$

This relation holds for the real points in D. Next, it can be shown that these real points form a large enough subset of D for (18.11) to be extended by analytic continuation to all points of D. Thus by using (18.10) and (18.11) one finally obtains

$$W(\xi_1\cdots\xi_{n-1}) = W(-\xi_{n-1},\ldots,-\xi_1) \qquad (18.12)$$

which is just (18.9).

The same line of argument can be extended to particles with spin, and which are not their own antiparticles. The proof is due to R. Jost.

19. The connection between spin and statistics

It is very remarkable that in every case that can be checked, particles of integer spin obey Bose–Einstein statistics, while particles of half-integer

* In the notation of chapter 4 a complex Lorentz transformation is defined by $\exp(i\boldsymbol{\alpha}\cdot\mathbf{J} + i\boldsymbol{\beta}\cdot\mathbf{K})$, where $\boldsymbol{\alpha}$ and $\boldsymbol{\beta}$ are complex.

spin obey Fermi–Dirac statistics. We saw in electrodynamics that this condition can be satisfied by ensuring that the asymptotic fields $\Phi_{in}(x)$ and $\Phi_{out}(x)$ for bosons obey commutation relations, while those for fermions obey anticommutation relations. This entails that the causality condition must involve either the commutator or the anticommutator of two fields, depending on the statistics. In other words, for bosons we must assume

$$[\Phi(x), \Phi(y)] = 0 \tag{19.1}$$

when $(x - y)$ is spacelike, whereas for fermions one has

$$\{\Phi(x), \Phi(y)\} = 0 \qquad \text{for} \qquad x - y \text{ spacelike.} \tag{19.2}$$

Although we shall not go into the details, it can be proved by combining algebraic and analytic considerations that for integer spin particles the "wrong" choice (19.2) of causality condition leads to $\Phi(x)|0\rangle = 0$, in which case the theory does not exist; and vice versa for half-integer spins. The essential difference between the two cases is that a half-integer spin field is multiplied by -1 under a rotation through 2π. This entails that the Wightman functions for fermion and boson fields behave differently under the complex Lorentz transformation $x \to -x$ which, as we saw, leads to the condition (18.10). It is from this difference that the connection between spin and statistics is proved.

20. The proof of dispersion relations

The reduction formula (16.2) provides the starting point for proving dispersion relations for fixed momentum transfer $t = (p_1' - p_1)^2$, i.e. for investigating the analytic properties of the scattering amplitude in s for fixed t. For certain amplitudes and for certain negative values of the momentum transfer, one can prove from the causality condition that the dispersion relations in s and u are direct consequences of the axioms of field theory. In this way they have been proved for pion-pion scattering for $-28\mu^2 < t < 0$, and for pion-nucleon scattering for $t_0 < t < 0$, but not for nucleon-nucleon scattering. Whether or not a dispersion relation can be proved by such methods, and the exact value of t_0, depend on the mass values of the particles in the reaction and on the thresholds. The consequent limitations seem to lack any practical significance, and are due simply to the fact the proof uses only causality and the mass spectrum. The proof, which is very delicate, was given by N. N. Bogoliubov in 1956.

In 1965 A. Martin succeeded in extending the proof by introducing a new property derived from the unitarity of the S matrix, namely that the imaginary part of any partial wave $e^{i\delta} \sin \delta$ (δ complex) is necessarily positive; this follows from the fact that basically it is the total reaction probability in that partial wave. In this way Martin could derive analytic properties of the amplitude in s *and* t simultaneously. He obtains a domain of analyticity which is large, but not large enough to justify Mandelstam's hypothesis. It seems likely that once again one could go further by introducing more information. But unfortunately it appears that in order to extend the analyticity domain of the two-particle scattering amplitude, one needs to know something about amplitudes involving more than two final state particles; and thereby the problem becomes very complicated.

In the same general field one should mention also the proof of crossing symmetry by J. Bros, H. Epstein and V. Glaser. Here it is important to realise that the very formulation of the crossing property involves assumptions about analyticity. Indeed, for two reaction amplitudes related by crossing, it is impossible that the invariants s, t, and u should be in their respective physical regions for both reactions simultaneously, since the two physical regions are disjoint. Hence it is possible to equate the two amplitudes only if at least one of them can be defined outside its physical region, i.e. if it can be continued analytically along some path in the complex s, t, u space. The last-mentioned authors have shown explicitly that this can be done in a complex neighbourhood of each of the three physical regions related by crossing; thereby they have constructed a rigorous proof of crossing symmetry from field theory.

Third part

21. *S*-matrix theory

No satisfactory theory of strong interactions exists at the moment. Analyticity properties and the dispersion relations reflecting them enable us to relate different processes; for instance, the two-pion cross-section which is not directly accessible to experiment can be extracted from analysis of pion production in the reaction

$$\pi + N \rightarrow N + \pi + \pi \tag{21.1}$$

In this sense, analytic properties are very useful.

In our discussion of axiomatic field theory we saw that its techniques can prove the existence of large analyticity domains of amplitudes. These domains are often limited by the method of proof and by the fact that the proof does not make use of all the information that we have about the amplitudes. From this viewpoint it seems reasonable to explore what consequences would follow from analyticity domains even larger than those proven. If these consequences have much physical interest, then one should persevere with the mathematical analysis of field theory; otherwise the investigation is likely to prove academic.

From the way we have discussed analytic properties at the beginning of this chapter it is clear that some singularities of the amplitude, including at least those that were mentioned, are intimately connected with obvious physical features of the system, like the mass spectrum in a given channel. The discontinuities on the other hand can be approached by judicious use of the unitarity condition. The so-called 'analytic S-matrix theory' rests on the assumption that these are the only singularities present, whatever the number of particles involved. Actually it is difficult to formulate this theory precisely, because one knows of no way to specify in a finite number of steps all the singularities that might be generated by the mass spectrum and by unitarity. But as we have already said it is in any case interesting to exploit such analyticity assumptions, since in some sense they specify the most stringent properties of the amplitude that we can hope for.

At the present time there is an unhappy confusion on this point. Some physicists studying properties of the S matrix have proposed that analyticity should no longer be justified from field theory, but that it should be regarded as a basic axiom subject only to those limitations that are inevitable. In a way, the convenient working hypothesis we have just mentioned has been transformed into an axiom, namely 'all amplitudes are maximally analytic, i.e. as analytic as they can be without contradicting unitarity and the mass spectrum'. This change of viewpoint entails drastic consequences, such as the proposal to abandon field theory altogether when constructing the dynamics. So unequivocal a position has occasionally elicited an equally extreme response from some specialists in field theory, who refuse to examine the consequences of any analytic properties that have not been proved rigorously. Such entrenched positions seem clearly inappropriate on either side, since neither axiomatic field theory nor S-matrix theory are much more as yet then research programmes; only an analysis of their internal consistency and of their consequences could put one in a position to be dogmatic. At the present time it seems best to work on the assumption that the true analytic properties of the S matrix are extensions of those derived from field theory, generalised far enough to allow contact with experiment.

But from another point of view, the mainly artificial confrontation between S-matrix and field theory does conceal a consistency problem well worth studying. We know that field theory is a dynamical theory, at least in the case of quantum electrodynamics. Nevertheless, the method runs into difficulties connected with renormalisation and with the divergence of the renormalisation constants; these cancel much of its charm and cast doubts on its foundations. Thus it seems interesting to ask whether one could avoid these difficulties while still remaining in the framework of field theory, by adopting as calculational methods the assumptions of *S*-matrix theory formulated as dispersion relations. In this way one could in principle calculate the Lamb shift by dispersion methods without needing to renormalise nor, a fortiori, to handle divergent quantities. Whether one formulates it within or outside of field theory, there arises the fundamental question: do analytic properties amount to a formulation of dynamics? And if they do not, what additional assumptions are needed for dynamical calculations? Recently there has been a small advance in this direction. Considering electron-proton scattering in nonrelativistic approximation, and taking into account the zero mass of the photon, it has been shown that the analytic properties of the *S* matrix lead uniquely to the classical Coulomb amplitude, and thereby determine all the properties of the hydrogen atom; there is no need to appeal to any kind of Schroedinger equation. In this instance there are reasons for thinking that analyticity could indeed have dynamical content.

In any case, complementary as they are, both field theory and *S*-matrix theory must produce more new and concrete results to justify further study. Much remains to be done in this direction, and it is impossible to assess the chances of success for either programme while progress in each continues to be reported regularly.

CHAPTER 13

THE REGULARITIES OF THE STRONG INTERACTIONS. *SU*(3)

We give a quick introduction to a "higher symmetry" of the strong interactions, namely to invariance under the group $SU(3)$.

1. Formulation of the problem

We know already how an invariance property like isotopic spin leads to particle multiplets and to a reduction of the number of independent transition amplitudes needed to describe various reactions. Another equally striking feature of invariance properties is that they distinguish between the different interactions through the quantum numbers which these interactions conserve. One can thus draw up a list of interactions and of the quantum numbers associated with them; it shows that isospin conservation distinguishes the strong from the electromagnetic interactions, while strangeness conservation distinguishes the strong and electromagnetic interactions from the weak ones.

It is fair to ask whether, as it now stands, our rough categorisation of interactions by their strength really amounts to a complete classification. The evident correlation between interaction type and conservation laws suggest a closer look. Thus we ask whether the strong interactions themselves might not be subdivided into several distinct types by extending the traditional structure, which regularly subjects the stronger interaction to the bigger invariance group. It could happen that there are two such classes of strong interactions, say A and B, differing by their conservation laws, A conserving more than B. Note at once that such a distinction has meaning only by reference to the interaction strength: we can always split the Hamiltonian into one part H_A invariant under any arbitrary group, plus another part H_B which is simply $\mathrm{H} - \mathrm{H}_A$. For the separation to be significant, it is useful and indeed necessary that the interactions A and B should differ in strength. Thus we shall envisage the possibility that the strong interactions subdivide into the super-strong and the medium-strong.

If this were true, it would imply that the medium-strong interactions conserve isotopic spin and strangeness, while the super-strong interactions might be invariant under a larger group. The larger group would of course have to include the isospin transformations as a subgroup, since isospin is conserved by all strong interactions. The larger group could embrace strangeness in two different ways. Either strangeness conservation remains unconnected with the other conservation laws, as hitherto; or it is included in the invariance group of strong interactions, as charge conservation is included in isotopic spin, whose third component takes on the basic role of electric charge. A priori the second is the more interesting case, since it would make for an analogy between the conservation laws for strangeness and electric charge; this would lead to a higher degree of unity and give room for hopes for more simplicity. Hence we investigate the second alternative.

These ideas may be formalised by assuming that the strong interaction Hamiltonian H is separable into two terms,

$$\mathrm{H}_F = \mathrm{H}_A + \mathrm{H}_B \tag{1.1}$$

where the Hamiltonian H_A of the superstrong interactions is invariant under the transformations of a certain group G, while the Hamiltonian H_B of the medium-strong interactions conserves only strangeness and isospin, and may be regarded as a perturbation on H_A. If one neglects H_B in first approximation, then particles states must be classifiable under unitary irreducible representations of the group G.

From the outset of such an investigation one must face a basic question by deciding whether to look for a group whose irreducible representations have finite or infinite multiplicity; or in other words, whether the classification is to be into multiplets containing finite or infinite numbers of particles. The two cases correspond to very different kinds of groups, whose unitary representations are of finite or infinite dimensionality, respectively. Mathematicians call such groups compact or noncompact, and distinguish them also by their topological properties. The difference can be shown by an example. The rotation group is compact, which means mathematically that any infinite sequence of rotations contains a subsequence converging to some definite limiting rotation; the group space contains no point at infinity, there is no 'infinite rotation', and the unitary representations are finite-dimensional. By contrast, the Poincaré group, and the homogeneous Lorentz group more especially, are noncompact in the sense that there exist 'infinite' Lorentz transformations corresponding to reference frames with relative velocities tending to c. It can be shown that the unitary representations of the Lorentz group are infinite-dimensional. Though the classification of infinite numbers of particles is worth studying, in the following we shall investigate only the case of finite multiplets, i.e. of compact groups.

The group G must admit at least two additive quantities that are measurable simultaneously, corresponding to strangeness and to charge respectively. It could happen that G includes other quantities conserved by H_A, like baryon number, or other quantum numbers whose conservation by H_A is masked by H_B, and which have not yet been observed. Once again we confine ourselves to the simplest case where G includes only two simultaneously measurable additive quantities. If a larger group does exist then this method would reach only one of its subgroups. Two of the generators of G must therefore correspond to I_3 and to Y (Y being linearly connected to strangeness), and must therefore commute with each other, while no other generator may commute with I_3 and Y simultaneously. Mathematically speaking such a group is said to be of rank 2. The states within a given unitary irreducible representation are distinguished from each other by two additive quantum numbers. By comparison, the three-dimensional rotation group is of rank 1; the states of a given irreducible representation are labeled by a single additive quantum number, namely by the z-component of the angular momentum.

Luckily it turns out that Elie Cartan has classified all compact groups, and that only a limited number satisfy our conditions. Thus one can survey these groups to decide whether any of them play a role in physics. A successful candidate must have irreducible unitary representations to which one can assign particles sharing the same kinematic quantum numbers (spin and parity) and the same baryon number. The only acceptable group emerging from such a survey is the one called $SU(3)$ by mathematicians. It was first proposed by M. Gell-Mann and by Y. Ne'eman.

2. The group $SU(3)$

We proceed to specify the properties of the group $SU(3)$.

It consists of the group of three-dimensional unitary unimodular matrices, i.e. of matrices α with three rows and three columns, satisfying the two conditions

$$\alpha^+ = \alpha^{-1} \qquad (2.1) \qquad \text{unitarity}$$

$$\det \alpha = 1 \qquad (2.2) \qquad \text{unimodularity}$$

The group is named after these properties: U means unitary, S means special, i.e. unimodular, and the number 3 specifies the dimensionality of the matrices.

To investigate the Lie algebra of this group we must first discover on how many parameters the matrices depend. A priori, a 3×3 matrix

TABLE SHOWING INTERACTIONS AND CONSERVED QUANTUM NUMBERS

Interaction	Isotopic spin	Strangeness	Electric charge	Baryon number	Lepton number	Electron number	Parity	Charge Conjuation	Time reversal
Strong	Yes	Yes	Yes	Yes	Yes	Yes	Yes	Yes	Yes
Electromagnetic	No	Yes	Yes	Yes	Yes	Yes	Yes	Yes	Yes
Weak	No	No	Yes	Yes	Yes	Yes	No	No	No (by a few %)

depends on 9 complex coefficients, i.e. on 18 real parameters. The unitarity condition (2.1) equates two Hermitean matrices $\alpha\alpha^+$ and 1. Since the diagonal elements of a Hermitean matrix are real, we obtain three real conditions by equating them; the three complex elements above the main diagonal yield three complex conditions, i.e. we have nine real conditions altogether. Equation (2.2) amounts only to one further real condition, since the unitarity of α entails automatically that its determinant has unit *modulus*. In conclusion, the matrices α depend on $18 - 9 - 1 = 8$ real parameters.

We have seen already that one basic step in studying a group is to introduce its Lie algebra, i.e. the family of matrices $(1/i) \log \alpha$. We saw likewise that the Lie algebra of $SU(3)$, which we shall call A, consists of Hermitean 3×3 matrices with zero trace, i.e. of matrices obeying the two conditions

$$h = h^+ \tag{2.3}$$

$$\text{Trace } h = 0 \tag{2.4}$$

These conditions follow directly from (2.1) and (2.2). The matrices h themselves form an 8-parameter family, which is particularly simple in this case, being an 8-dimensional vector space. If we define a set of 8 such linearly independent matrices $\lambda_1, \lambda_2, \ldots, \lambda_8$, then every matrix h of A can be expressed uniquely in the form

$$h = \sum_{i=1}^{8} y_i \lambda_i \tag{2.5}$$

where the y_i are real coefficients. One example of such a matrix basis is the set

$$\lambda_1 = \begin{pmatrix} 0 & 1 & 0 \\ 1 & 0 & 0 \\ 0 & 0 & 0 \end{pmatrix} \quad \lambda_2 = \begin{pmatrix} 0 & -i & 0 \\ i & 0 & 0 \\ 0 & 0 & 0 \end{pmatrix} \quad \lambda_3 = \begin{pmatrix} 1 & 0 & 0 \\ 0 & -1 & 0 \\ 0 & 0 & 0 \end{pmatrix}$$

$$\lambda_4 = \begin{pmatrix} 0 & 0 & 1 \\ 0 & 0 & 0 \\ 1 & 0 & 0 \end{pmatrix} \quad \lambda_5 = \begin{pmatrix} 0 & 0 & -i \\ 0 & 0 & 0 \\ i & 0 & 0 \end{pmatrix} \quad \lambda_6 = \begin{pmatrix} 0 & 0 & 0 \\ 0 & 0 & 1 \\ 0 & 1 & 0 \end{pmatrix} \tag{2.6}$$

$$\lambda_7 = \begin{pmatrix} 0 & 0 & 0 \\ 0 & 0 & -i \\ 0 & i & 0 \end{pmatrix} \quad \lambda_8 = \frac{1}{\sqrt{3}} \begin{pmatrix} 1 & 0 & 0 \\ 0 & 1 & 0 \\ 0 & 0 & -2 \end{pmatrix}$$

The matrices $\frac{1}{2}\lambda_i$ are an explicit representation of the infinitesimal generators of the group $SU(3)$, and we shall denote them by $F_i (i = 1, \ldots, 8)$. There is an alternative notation which is more convenient for exhibiting certain features of their algebra, namely

$$\begin{aligned} F_1 &= I_1 \qquad F_2 = I_2 \qquad F_3 = I_3 \\ F_4 &= K_1 \qquad F_5 = K_2 \\ F_6 &= L_1 \qquad F_7 = L_2 \\ F_8 &= M \end{aligned} \tag{2.7}$$

It is useful to introduce also the notation

$$\begin{aligned} I_\pm &= I_1 \pm iI_2 \\ K_\pm &= K_1 \pm iK_2 \\ L_\pm &= L_1 \pm iL_2 \,. \end{aligned} \tag{2.8}$$

Let us now study the commutation rules for the infinitesimal generators F_i; for this it suffices to evaluate explicitly the commutators of the matrices λ_i, which can be expressed in the general form

$$[\lambda_i\,, \lambda_j] = \sum_{k=1}^{8} 2if_{ijk}\lambda_k \tag{2.9a}$$

Accordingly, the commutation relations for the generators are given by

$$[F_i\,, F_j] = i\sum_k f_{ijk}F_k \tag{2.9b}$$

involving the same coefficients f_{ijk}.

To develop some feeling for the information contained in these commutation relations it is best to use the notation (2.7) for the generators. One notes from the outset that the generators I_3 and M commute, just like their representative matrices λ_3 and λ_8:

$$[I_3\,, M] = 0. \tag{2.10}$$

Similarly one can convince oneself by explicit calculation that no other combination of generators commutes with I_3 and M simultaneously.

The matrices (λ_1, λ_2, λ_3) are simply Pauli matrices bordered by zeros to make them into 3×3 matrices; this shows immediately that their associated generators I_1, I_2, and I_3 satisfy the commutation rules for the generators of $SU(2)$, which we can write as

$$\begin{aligned} [I_1\,, I_2] &= iI_3 \\ [I_3\,, I_\pm] &= \pm I_\pm \end{aligned} \tag{2.11}$$

Hence they can be considered as generators of the isospin group which thus appears as a subgroup of $SU(3)$.

Note that this is not the only way in which a group isomorphic to $SU(2)$ can appear as a subgroup of $SU(3)$. Indeed, the matrices λ_4 and λ_5 on one

TABLE 13.1

The f_{ijk} are completely antisymmetric in their three indices:

$$f_{ijk} = f_{jki} = f_{kij} = -f_{jik} = -f_{ikj} = -f_{kji}$$

They are given by the table

ijk	123	147	156	246	257	345	367	458	678
f_{ijk}	1	$\frac{1}{2}$	$-\frac{1}{2}$	$\frac{1}{2}$	$\frac{1}{2}$	$\frac{1}{2}$	$-\frac{1}{2}$	$\sqrt{\frac{3}{2}}$	$\sqrt{\frac{3}{2}}$

Index combinations (ijk) not mentioned in the table correspond to vanishing f_{ijk}.

hand, and λ_6 and λ_7 on the other, are also obtainable by supplementing Pauli matrices with one row and one column of zeros. Let us define the matrices

$$\mu = \begin{bmatrix} 1 & 0 & 0 \\ 0 & 0 & 0 \\ 0 & 0 & -1 \end{bmatrix} \qquad \nu = \begin{bmatrix} 0 & 0 & 0 \\ 0 & 1 & 0 \\ 0 & 0 & -1 \end{bmatrix}$$

and the generators corresponding to $\mu/2$ and $\nu/2$, which will be denoted by K_3 and L_3. Then we see that the sets of three generators

$$(K_1, K_2, K_3) \qquad (L_1, L_2, L_3)$$

also have the same internal commutation relations as the Lie algebra of $SU(2)$; hence they generate new subgroups of $SU(3)$, also isomorphic to $SU(2)$. Actually the generators K_3 and L_3 are linear combinations of I_3 and M:

$$\begin{aligned} K_3 &= (1/2)I_3 + \frac{\sqrt{3}}{2} M \\ L_3 &= (-1/2)I_3 + \frac{\sqrt{3}}{2} M \end{aligned} \tag{2.12}$$

The remaining commutation relations are specified by Table (13.1), which gives the coefficients f_{ijk}.

3. The representations of $SU(3)$

A unitary representation of $SU(3)$ in Hilbert space automatically provides one with a representation of the generators $F_1, \ldots, F_8$ by Hermitean operators. The states in such a representation are labeled by the eigenvalues of a complete commuting set of operators. To have something definite in mind we can think of the rotation group, with generators J_1, J_2, J_3, whose irreducible representation are labeled by the eigenvalues of the operator $\mathbf{J}^2 = J_1{}^2 + J_2{}^2 + J_3{}^2$. In the case of $SU(3)$, two operators are needed to label an irreducible representation, one of the two being

$$F^2 = \sum_{i=1}^{8} F_i^2$$

and the other a polynomial G^3 of third order in the F_i. Such invariants are much less interesting in $SU(3)$ than in the rotation group, and we shall not need to use them. Instead of labeling an irreducible representation by the eigenvalues of F^2 and G^3, we shall therefore confine ourselves for the moment to a label γ which will be made more explicit presently.

In the case of the rotation group, once we have fixed on an irreducible representation, i.e. once we have chosen $\mathbf{J}^2$, the different states belonging to this representation can be labeled further by the eigenvalue of J_3. In the case of $SU(3)$ it will evidently be necessary to give the eigenvalues i_3 and m of the generators I_3 and M, which are simultaneously measurable by virtue of (2.10). Since M moreover commutes with I_1 and I_2 as well, one can diagonalise simultaneously I_3, M, and $\mathbf{I}^2 = I_1{}^2 + I_2{}^2 + I_3{}^2$, the last

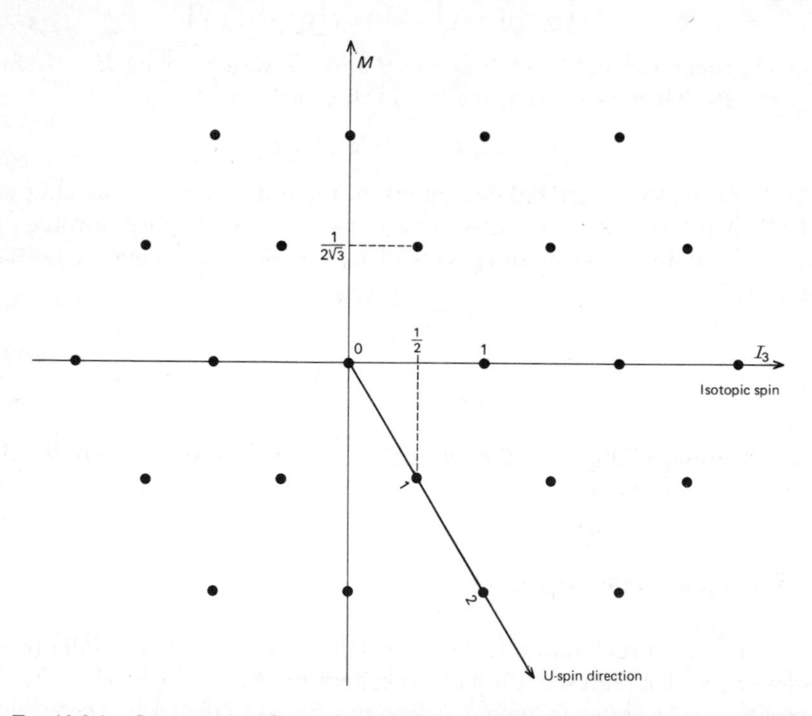

FIG. 13.3.1. System of coordinates for describing representations of $SU(3)$. It is customary to represent the elements of an irreducible representation of $SU(3)$ as points, with coordinates M and I_3. Such points lie on a hexagonal lattice. The contents of various representations of $SU(3)$ are given in the succeeding figures. Points represent single states; crosses represent two states with identical values of Y and I_3, but differing in their values of I; squares represent three states.

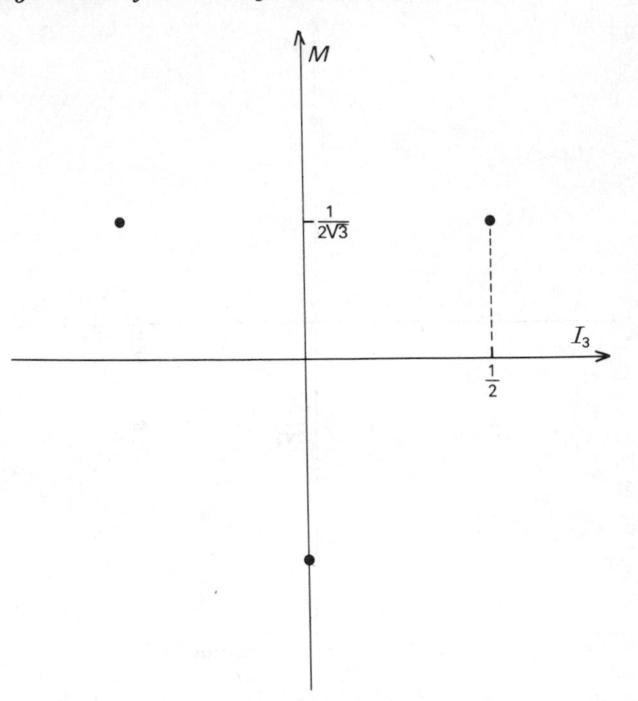

FIG. 13.3.2. The representation {3}.

operator being simply the isotopic spin. Most remarkably, it turns out the eigenvalues of these three quantum numbers suffice to determine uniquely a state in an irreducible representation of $SU(3)$; such a state can therefore be labeled in the notation

$$|\gamma, I. i_3, m\rangle$$

where $I(I + 1)$ is the eigenvalue of the total isotopic spin operator $\mathbf{I}^2$.

We proceed to investigate the most important irreducible representations. The simplest is clearly the one-dimensional identity representation which we denote by {1}; it is called the singlet representation. The corresponding quantum numbers are necessarily $I = 0, i_3 = 0, m = 0$.

Since the group was defined by means of unitary 3×3 matrices, there exists a 3-dimensional representation which we denote by {3} (Fig. 13.3.2). In this representation of the group by matrices α, the generators are represented by the matrices $\frac{1}{2}\lambda_i$. Then the expressions (2.6) for the λ_i show directly that the three basis vectors correspond, respectively, to the following eigenvalues of I_3 and M:

$$(i_3, m) = \left(\frac{1}{2}, \frac{1}{2\sqrt{3}}\right), \left(-\frac{1}{2}, \frac{1}{2\sqrt{3}}\right), \left(0, -\frac{1}{\sqrt{3}}\right)$$

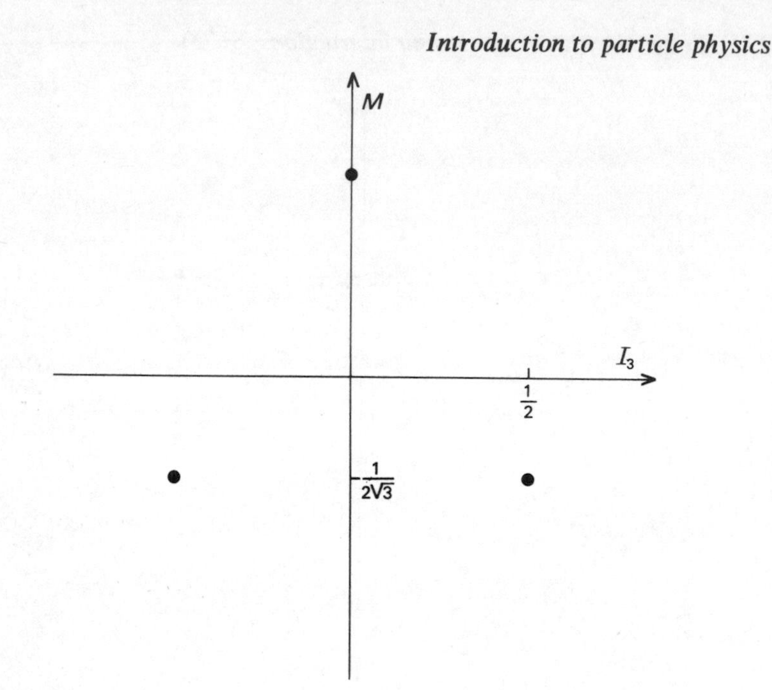

FIG. 13.3.3. The representation $\{3^*\}$.

To avoid the ubiquitous factors $\sqrt{3}$, we introduce the new quantum number $Y = 2M/\sqrt{3}$; this operator is called the hypercharge. Consider the state vectors corresponding to the first two columns of the matrices $\frac{1}{2}\lambda_i$; we see that on these vectors the matrices $\frac{1}{2}\lambda_1$, $\frac{1}{2}\lambda_2$, $\frac{1}{2}\lambda_3$ act just like the matrices $\frac{1}{2}\boldsymbol{\sigma}$, while they give zero when acting on the state vector corresponding to the third column. Therefore these three states have the following quantum numbers:

$$\begin{aligned} |\{3\}I, I_3, Y\rangle &= |\{3\}\tfrac{1}{2}, \tfrac{1}{2}, \tfrac{1}{3}\rangle \\ &\quad |\{3\}\tfrac{1}{2}, \tfrac{1}{2}, \tfrac{1}{3}\rangle \\ &\quad |\{3\}0, 0, -\tfrac{2}{3}\rangle \end{aligned} \tag{3.1}$$

Accordingly, the representation $\{3\}$ contains an isospinor with hypercharge $Y = \frac{1}{3}$ and an isoscalar with $Y = -\frac{2}{3}$.

There exists another representation closely related to $\{3\}$, given by the complex conjugates α^* of the matrices α belonging to $\{3\}$ (Fig. 13.3.3). By definition, its infinitesimal generators are represented by the matrices $-\frac{1}{2}\lambda_i^*$, whence the quantum numbers of the three basis states are

$$\begin{aligned} |\{3^*\}I\,I_3Y\rangle &= |\{3^*\}\tfrac{1}{2}, \tfrac{1}{2}, -\tfrac{1}{2}\rangle \\ &\quad |\{3^*\}\tfrac{1}{2}, -\tfrac{1}{2}, -\tfrac{1}{2}\rangle \\ &\quad |\{3^*\}0, 0, \tfrac{2}{3}\rangle \end{aligned} \tag{3.2}$$

Denote by x_i the three components of a vector transforming under $SU(3)$ according to the representation $\{3\}$, and by y_i^* the components of another vector transforming according to $\{3^*\}$; then an $SU(3)$ transformation acts according to

$$x_i \to x_i' = \alpha_{ij} x_j$$

$$y_i^* \to y_i'^* = \alpha_{ij}^* y_j^*.$$

The products $x_i y_j^*$ form the basis for a 9-dimensional representation of $SU(3)$. This representation is reducible, because the scalar product $x_1 y_1^* + x_2 y_2^* + x_3 y_3^*$ is an invariant by virtue of the fact that the matrices α are unitary. If this invariant is removed from the basis $x_i y_j^*$, one obtains an 8-dimensional representation of $SU(3)$. This representation is unitary; it is denoted by $\{8\}$ and called the *octet*.

The isospin and hypercharge content of the octet representation $\{8\}$ is easily found by considering the *symmetric* direct products of the two bases (3.1) and (3.2) for $\{3\}$ and $\{3^*\}$ respectively. The hypercharges add; likewise the values of I_3. The values of $\mathbf{I}^2$ are obtained from the rules for adding isospins, which yield

$$|\{3\}I = \tfrac{1}{2}, Y = \tfrac{1}{3}\rangle \otimes |\{3^*\}, I = \tfrac{1}{2}, Y = -\tfrac{1}{3}\rangle$$
$$= |I = 1, Y = 0\rangle + |I = 0, Y = 0\rangle$$

$$|\{3\}I = \tfrac{1}{2}, Y = \tfrac{1}{3}\rangle \otimes |\{3^*\}I = 0, Y = +\tfrac{2}{3}\rangle = |I = \tfrac{1}{2}, Y = +1\rangle$$

$$|\{3\}I = 0, Y = \tfrac{2}{3}\rangle \otimes |\{3^*\}I = \tfrac{1}{2}, Y = -\tfrac{1}{3}\rangle = |I = \tfrac{1}{2}, Y = -1\rangle$$

$$|\{3\}I = 0, Y = -\tfrac{2}{3}\rangle \otimes |\{3^*\}I = 0\ \ Y = \tfrac{2}{3}\rangle = |I = 0, Y = 0\rangle.$$

Since one of the states $|I = 0, Y = 0\rangle$ corresponds to the representation $\{1\}$ having the basis $x_1 y_1^* + x_2 y_2^* + x_3 y_3^*$, we see that the representation $\{8\}$ contains (Fig. 13.3.4):

$$\begin{array}{lll} \text{one isoscalar} & I = 0, & Y = 0, \\ \text{two isospinors} & I = \tfrac{1}{2}, & Y = \pm 1, \\ \text{one isovector} & I = 1, & Y = 0. \end{array} \tag{3.3}$$

Figures (13.3.5) and (13.3.6) show how the most important baryons and mesons can be assigned to this representation.

Just as the three Pauli matrices transform like the components of an isovector under the similarity transformation $A\sigma_i A^{-1}$, where A is a matrix belonging to $SU(2)$, the eight Gell-Mann matrices λ_i transform according to the octet representation under the transformation $\alpha\lambda_i\alpha^{-1}$, where α belongs to $SU(3)$.

Another important representation can be obtained by considering the tensor products $x_i x_j x_k$ formed from the components x_i of a vector transforming like $\{3\}$ under $SU(3)$. How many of these products are independent can be worked out easily after noting the following.

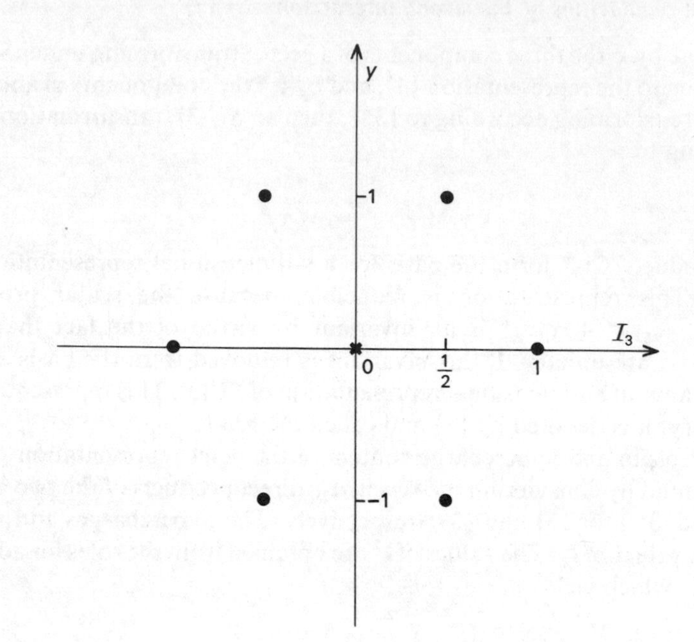

FIG. 13.3.4. The representation {8}.

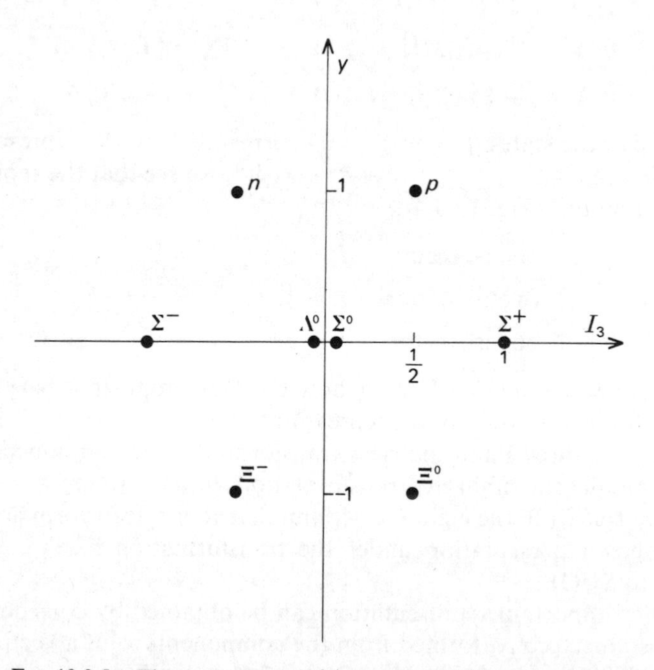

FIG. 13.3.5. The representation {8} for baryons (The baryon octet).

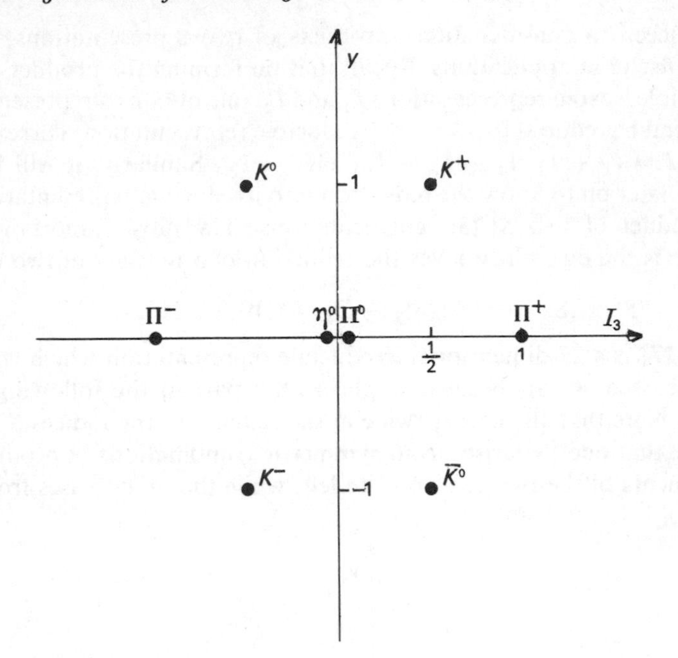

FIG. 13.3.6. The representation {8} for mesons. (The meson octet.)

Let $i \neq j, j \neq k, i \neq k$; then (i, j, k) is a permutation of (1, 2, 3), leading to a single product $x_1x_2x_3$.

Let $i = j$, $i \neq k$; then i can take three values ($i = 1, 2, 3$) for each of which k can take two values, leading to six independent products $x_1^2x_2$, $x_2^2x_3$, $x_2^2x_1$, $x_2^2x_3$, $x_3^2x_1$, $x_3^2x_2$.

Let $i = j = k$; this gives three independent products x_1^3, x_2^3, x_3^3.

Hence this is a 10-dimensional representation. It can be shown to be irreducible, is denoted by {10}, and is called a *decuplet*.† Its isospin and hypercharge content is easily found from the rules for adding isospin.

It is written as

$$\{10\} : (I, Y) = (\tfrac{3}{2}, 1), (1, 0), (\tfrac{1}{2}, -1), (0, -2) \tag{3.4}$$

(See figure (13.3.7).)

By starting with the products $x_i^*x_j^*x_k^*$ instead of $x_ix_jx_k$, one obtains another irreducible representation denoted by {10*}, whose isospin and hypercharge content is also given by (3.4), up to the sign of the hypercharge; thus,

$$(I, Y) = (\tfrac{3}{2}, -1), (1, 0), (\tfrac{1}{2}, 1), (0, 2) \tag{3.5}$$

† Translator's note: Though etymologically inconsistent, the nomenclature triplet, octet, decuplet has established itself in common usage. Purists speak of an octuplet, but are in a small minority.

We proceed to consider direct products of two representations, which will be useful in applications. Recall that on forming the product of two irreducible isospin representations I_1 and I_2, one obtains a representation which can be reduced to a sum of irreducible representations corresponding to $I = I_1 + I_2, I_1 + I_2 - 1, ..., |I_1 - I_2|$. Similarly, it will be important later on to know the reduction into irreducible representations of the product of two $SU(3)$ representations. The most important such formula is the one which gives the reduction of a product of two octets:

$$\{8\} \times \{8\} = \{1\} \oplus \{8\}_S \oplus \{8\}_A \oplus \{10\} \oplus \{10^*\} \oplus \{27\} \tag{3.6}$$

Here, $\{27\}$ is a 27-dimensional irreducible representation which we have not discussed so far, because it plays little part in the following (Fig. 13.3.8). Note that $\{8\}$ occurs twice in the reduction; the indices S and A indicate that one $\{8\}$ arises from symmetric combinations of products of components of the two $\{8\}$'s on the left, while the other arises from the

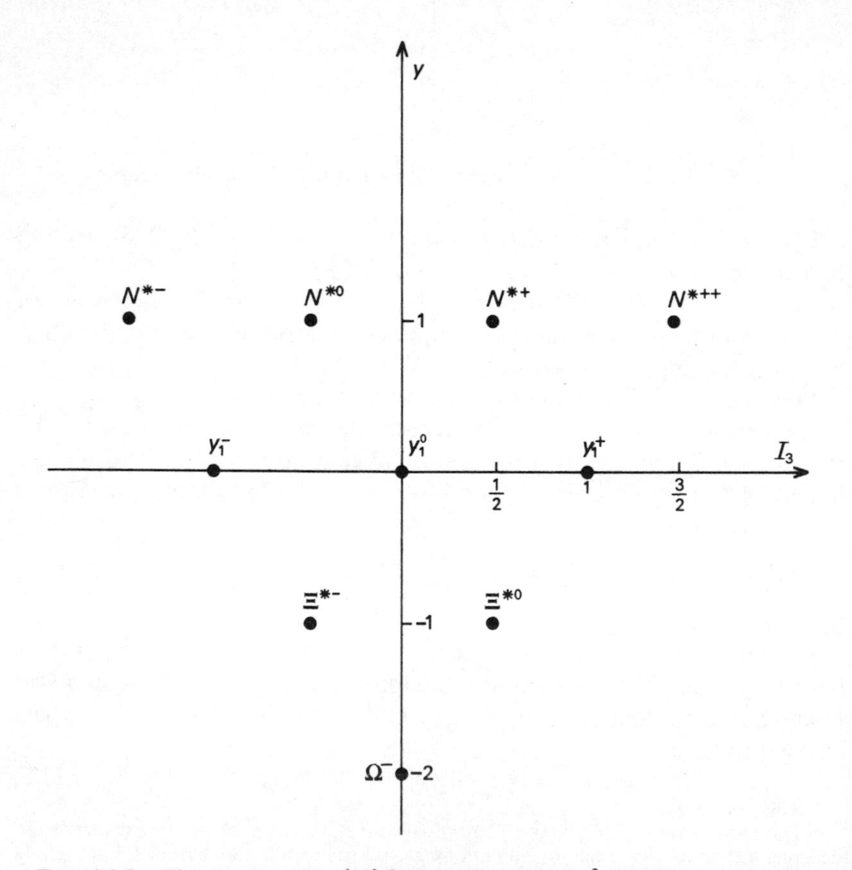

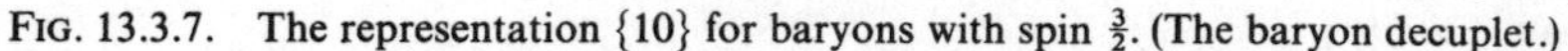

FIG. 13.3.7. The representation $\{10\}$ for baryons with spin $\frac{3}{2}$. (The baryon decuplet.)

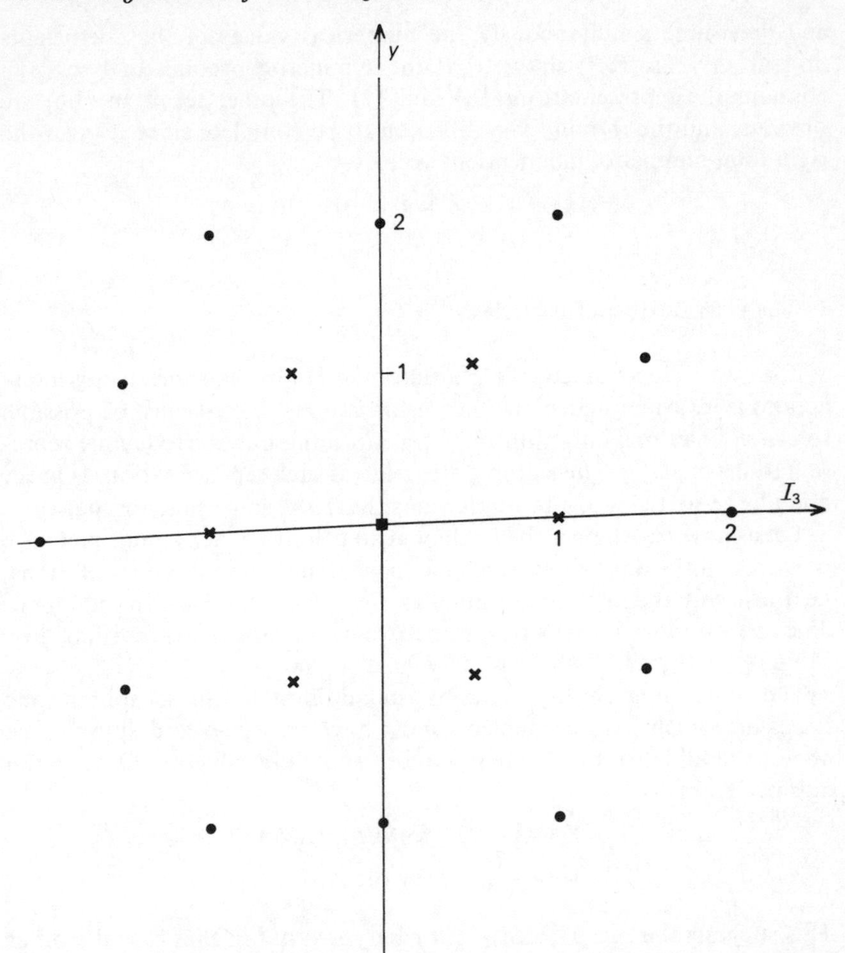

FIG. 13.3.8. The representation {27}.

antisymmetric combinations.* To see this, note that by virtue of the commutation rules (2.9a) for the λ_i, the commutator $[\lambda_i, \lambda_j]$, which is antisymmetric in i and j, is a sum of matrices λ_i and therefore it itself transforms according to {8}; hence an octet must be present in the antisymmetric product of two {8}'s. Similarly, we can derive by explicit calculation the anticommutation relations

$$\{\lambda_i \lambda_j\} = 2d_{ijk}\lambda_k + \tfrac{4}{3}\delta_{ij} \tag{3.7}$$

* Translator's note: $\{8\}_S$ is often denoted by $\{8\}_D$, and $\{8\}_A$ by $\{8\}_F$.

and determine simultaneously the numerical values of the coefficients d_{ijk}; in any case (3.7) shows that the symmetric product of two $\{8\}$'s contains the representations $\{8\}$ and $\{1\}$. The other terms in (3.6) are obvious, and the formula is easily seen to be complete since it gives the right total number of independent vectors:

$$8 \times 8 = 1 + 8 + 8 + 10 + 10 + 27$$

4. The classification of particles

If, as we have assumed, the Hamiltonian H_A of the superstrong interactions is invariant under $SU(3)$ transformations, then it must be possible to classify hadrons into multiplets transforming under irreducible representations of $SU(3)$. The isotopic structure of such representations is fixed, and all the particles of a multiplet must have the same spin and parity.

Let us now see whether the nucleons can belong to such a multiplet. The most straightforward procedure is to inspect the simplest representations, i.e. those with the lowest multiplicities, for candidates that can accommodate an isospinor. Actually this is possible with all the representations $\{3\}$, $\{3^*\}$, $\{8\}$, $\{10\}$, $\{10^*\}$ which we have met so far.

The representation $\{3\}$ contains, in addition to the isospinor, one isoscalar, which like the nucleon must have parity + and spin $\frac{1}{2}$. The obvious candidate is the Λ. Then the corresponding values of the quantum number Y are

$$Y = \tfrac{1}{3} \qquad \text{for the nucleon}$$

$$Y = -\tfrac{2}{3} \qquad \text{for the } \Lambda$$

This suggests the rule $Y = S + \frac{1}{3}$ for baryons; but in that case the other known hyperons with spin $\frac{1}{2}$ and parity +, namely the Σ and the Ξ, cannot be accommodated in any irreducible representation without making room for many other states not yet discovered. Therefore we abandon this line of approach.

The next simplest possibility is to assign the nucleons to an octet. This looks more promising because the $\{8\}$ consists of two isospinors, one isoscalar, and one isovector, so that it is just filled by the N, the Ξ, the Λ, and the Σ. Figure (13.3.4) suggests that the corresponding values of Y be taken as

$$Y_N = +1, \qquad Y_\Sigma = 0, \qquad Y_\Lambda = 0, \qquad Y_\Xi = -1,$$

whence we have for baryons

$$Y = 1 + S$$

The terms in this multiplet differ considerably in mass; the differences are ascribed to effects of the medium-strong interactions.

Next we consider the pseudoscalar mesons. Eight of them, K, $\bar{K}$, η, and π, once again provide a set of two isospinors, one isoscalar, and one isovector, characteristic of an octet. Comparison between the strangeness and the values of Y given by (3.3) shows that for these mesons one has

$$Y = S$$

This suggests the following general relation between hypercharge, baryon number B, and strangeness S:

$$Y = B + S. \tag{4.1}$$

The resonances with spin $\frac{3}{2}$ and parity + can be assigned to a decuplet, which then consists of

$$N^* \qquad Y_1{}^* \qquad \Xi^* \qquad \Omega^-$$

Note the presence of Ω^- which is stable under strong interactions. Comparing the values of the hypercharge Y as imposed by $SU(3)$ with the other quantum numbers of these particles, we again verify the relation (4.1) between Y, B, and S, which strengthens our confidence in the classification.

The case of the 1^- vector mesons is more delicate. There are nine such mesons, K^*, $\bar{K}^*$, ρ, ω, and φ, which can be assigned to an $\{8\}$ and a $\{1\}$. The quantum numbers make it clear that K^*, $\bar{K}^*$, and ρ must belong to the $\{8\}$. The problem is to discover which of the two i-spin 0 and hypercharge 0 mesons, ω and φ, belongs to the $\{8\}$ and which belongs to the the $\{1\}$. We shall return to this later.

5. The Gell-Mann–Okubo formula†

Consider next the experimental support for $SU(3)$. It must be admitted as a first qualitative success that we have been able to arrange particles into irreducible representations of $SU(3)$, with correct values of isospin and hypercharge; but more detailed confirmation of the assumptions is desirable. The main difficulty of principle is the existence of medium-strong interactions which are not invariant under $SU(3)$; judged by the particle masses these interactions are far from negligible. Indeed if $SU(3)$ symmetry were exact, as for the Hamiltonian H_A, then all particles belonging to the same multiplet would have the same mass. But if one compares for instance the masses of the pseudoscalar mesons assigned to the octet representation, one notes that the pion mass is 135 MeV while

† Translator's note: the customary English hyphenation hides the fact that Gell-Mann is one person and Okubo another.

the η mass is 548 MeV, which shows that the medium-strong interactions are very appreciable.

Since the Hamiltonian H_B of the medium-strong interactions is so important, there is no reason a priori why one should be able to treat it by perturbation theory. Gell-Mann and Okubo did so nevertheless, in order to find out how $SU(3)$ violations tend to change masses. The first problem is to discover the form of the Hamiltonian H_B, which must not be invariant under $SU(3)$ but which must conserve isospin and strangeness. The only $SU(3)$ generator to satisfy these conditions is F_8, whence one can try a Hamiltonian of the form

$$H_8 = fF_8 \tag{5.1}$$

where f is a coupling constant. Of course this is not the only possible choice, and by combining higher powers of the $SU(3)$ generators one could construct very complicated forms of H_B.

If one uses first order perturbation theory to calculate the mass shifts, due to the Hamiltonian (5.1), of the particles in a multiplet, one finds that the contribution of H_B to the mass of a particle α is given by

$$\Delta m_\alpha = \langle \alpha \mid H_B \mid \alpha \rangle.$$

Within a given $SU(3)$ representation this matrix element depends only on a few parameters. To see this, consider the case where the particle $|\alpha\rangle$ belongs to an octet. Then, by (3.6), the vector $H_B |\alpha\rangle$ is a superposition of components transforming according to the representations $\{1\}$, $\{8\}_S$, $\{8\}_A$, $\{10\}$, $\{10^*\}$, and $\{27\}$, in view of the fact that H_B itself transforms like an octet component. But Δm_α is calculated by taking the projection of $H_B |\alpha\rangle$ onto the vector $|\alpha\rangle$ belonging to $\{8\}$; therefore only the octet components of $H_B |\alpha\rangle$ can give a nonvanishing contribution to Δm_α. Since there are two such octet components in $H_B |\alpha\rangle$, we see that Δm_α will depend on only two parameters. This holds subject only to the condition that H_B transform under $SU(3)$ like F_8; it need not actually be a numerical multiple of the operator F_8. An explicit calculation shows that, to first order, the masses of particles in an octet are determined by their isotopic spin and by their hypercharge according to the Gell-Mann–Okubo formula

$$m = m_0 + m_1 Y + m_2[I(I+1) - Y^2/4]. \tag{5.2}$$

where m_0 is the mass unperturbed by the medium strong interactions. This formula applies quite accurately to the baryon masses, and to squared masses of the pseudoscalar bosons. Indeed its accuracy is much too high to be explicable in a satisfactory way by first order perturbation theory, and constitutes an unsolved problem in its own right.

For states $|\alpha\rangle$ belonging to a decuplet, like the $\frac{3}{2}^+$ resonances, the formula (5.2) is replaced by

$$m = m_0 + m_1 Y \tag{5.3}$$

When $SU(3)$ was first proposed the baryon Ω^- was unknown, and from (5.3) its mass was predicted to be 1679 MeV. The experimental value is 1672 MeV.

Apart from the mass formulae whose very success still remains to be understood, we can mention some other experimental tests of $SU(3)$ involving baryon-meson coupling constants; in principle these constants can be obtained by analysing dispersion relations, and from the relations between the masses and decay rates of particles. For details we refer to the bibliography. The cleanest test is one exploiting electromagnetic mass differences, since they can justifiably be treated by perturbation theory. The test is based on the observation that the electromagnetic interaction Hamiltonian is proportional to the charge, which can be expressed in terms of the isotopic spin and the hypercharge by

$$Q = I_3 + Y/2 \tag{5.4}$$

Recalling the generator $M = F_8$ this becomes

$$Q = I_3 + \frac{1}{\sqrt{3}} M = 2K_3 \tag{5.5}$$

Here we have used the definition (13.13) of the generator K_3, which bears the same relation to K_1 and K_2 as I_3 bears to I_1 and I_2. The three operators K_1, K_2, K_3 generate a group which is isomorphic to $SU(2)$, but which is not the isospin group, and which is not an invariance group of the strong interactions since K_1 and K_2 do not commute with the medium-strong interaction Hamiltonian H_B. This analogue of isotopic spin is called U-spin. From the symmetry of the $SU(3)$ commutation rules between the two sets of operators (I_1, I_2, I_3) and (K_1, K_2, K_3), it is obvious that we can classify particles according to their U-spin. The electromagnetic contribution to the mass depends only on the U-spin, and a perturbation calculation yields the Coleman–Glashow formula for the electromagnetic mass splittings:

$$m_{\Xi^-} - m_{\Xi^0} = (m_{\Sigma^-} - m_{\Sigma^+}) - (m_n - m_p)$$

The measured values are

$$m_{\Xi^-} - m_{\Xi^0} = 6{,}5 \pm 1{,}0 \text{ Mev}$$

$$m_{\Sigma^-} - m_{\Sigma^+} = 7{,}67 \pm 0{,}25 \text{ Mev}$$

$$m_n - m_p = 1{,}29 \text{ Mev}$$

in good agreement with the theoretical formula.

The magnetic moments can be analysed similarly; this leads one to predict the following relations between them:

$$\mu_\Lambda = -\mu_{\Sigma^0} = \mu_n/2$$

$$\mu_{\Sigma^-} = -(\mu_p + \mu_n).$$

These relations can be checked experimentally, though so far the results are not precise enough for a significant test.

6. Outlook

At the time of writing it seems that invariance under $SU(3)$ can be considered as a well-established property of strong interactions.

There have been attempts to extend the invariance group of strong interactions in various ways, for instance by including $SU(3)$ into a larger and less well conserved compact group, or by including it into a noncompact group whereby infinite numbers of particles can be classified simultaneously. But no truly convincing results have been obtained along such lines.

One of the boldest and most interesting assumptions posits the existence of particles called 'quarks', transforming under $SU(3)$ like members of a triplet $\{3\}$. From equation (3.1), giving the properties of this representation, we see that the three quarks subdivide into an isospinor doublet and an isoscalar singlet. We denote them by p, n, and λ, by analogy with the nucleons and the Λ; thus we have

$$|p\rangle = |\{3\}\, \tfrac{1}{2}, \tfrac{1}{2}, \tfrac{1}{3}\rangle$$

$$|n\rangle = |\{3\}\, \tfrac{1}{2}, -\tfrac{1}{2}, \tfrac{1}{3}\rangle$$

$$|\lambda\rangle = |\{3\}\, 0, 0, -\tfrac{2}{3}\rangle$$

Since every representation of $SU(3)$ can be constructed from direct products of the representations $\{3\}$ and $\{3^*\}$, we can picture all particles as bound states of quarks or antiquarks. Thus the pseudoscalar and vector bosons could be bound states of a quark q and an antiquark $\bar{q}$; the baryon octet and decuplet could be bound states of three quarks. This implies that the baryon number of the quark is $B = \frac{1}{3}$. Since the electric charge is given by (5.4), we see that the quark charges are fractions of the electron charge.

Very thorough experimental searches have failed to find any evidence for quarks with a mass below 3 GeV. Nevertheless many results in surprisingly good agreement with experiment can be derived from crude analyses of models which picture particles as bound states of quarks, more or less along atomic lines. It does not follow from this that quarks exist, because even if they do it is hard to see how they could be described by such simple-minded models; rather it seems likely that there exist regularities in particle physics which have not been formulated as yet.

Problems

1. Derive the reduction rules (3.5) and (3.6).
2. Derive the Coleman–Glashow formula.
3. Use the formula (3.7) to study the representation $\{27\}$.

CHAPTER 14

REGGE POLES

We give a brief introduction to the basic ideas about Regge poles, both for interpolating between families of resonances, and in the theory of interactions at high energies.

1. Resonances and internal quantum numbers

In the last few years phase shift analyses of pion-nucleon scattering have been extended to high energies, over 1 GeV, and have revealed a fair number of resonances whose spins and squared masses are shown in figure 14.1.1. The data display a certain regularity when, as we have done, one considers separately sets of resonances with a common value of isotopic spin and parity. We see that the resonances, at least those high on the diagram, tend to lie on straight lines whose slopes are of the order of 1 GeV^{-2} for all isospins and parities. If on the other hand one considers experiments at higher energies, say up to 3 GeV, then both in the differential and in the total cross-sections one notices peaks suggestive of new resonances, whose spins are unfortunately difficult to measure; but as we shall see later, they could quite naturally extend the linear progressions already in evidence on figure 1. As regards the resonances lower down on the diagram, they correspond to small spins at high energies and are consequently difficult to detect; hence it would not be incompatible with the evidence to assume that they too are subject to regularities.

As often before, we shall let ourselves be guided by an analogy in order to isolate these regularities and to find ideas linking them with other phenomena. For orientation we have drawn two diagrams analogous to figure (14.1.1) and appropriate to the two simplest models in nonrelativistic physics: figure (14.1.2) for the hydrogen atom and figure (14.1.3) for the harmonic oscillator. In both cases one is dealing with bound states rather than resonances, which is not an essential difference for the purpose in hand. Accordingly, figures 2 and 3 show the angular momenta of the hydrogen and oscillator bound states as functions of the energy. They are

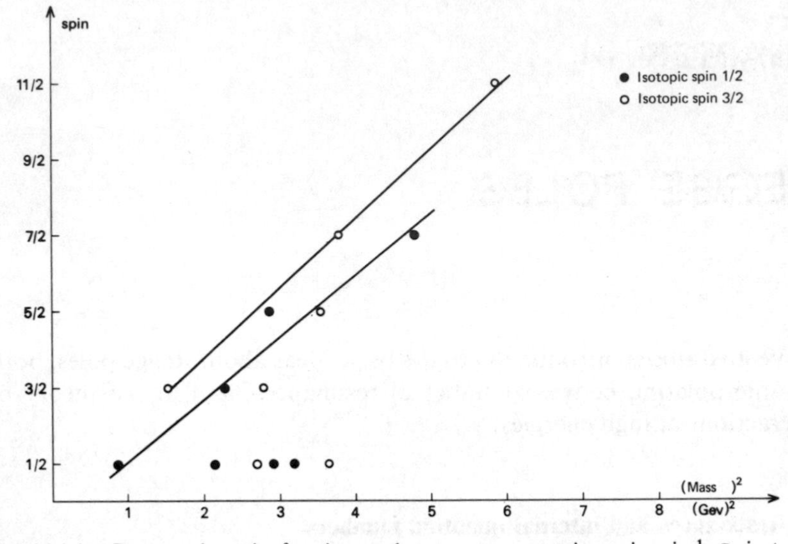

FIG. 14.1.1. Regge trajectories for pion-nucleon resonances: ● isotopic spin $\frac{1}{2}$; ○: isotopic spin $\frac{3}{2}$.

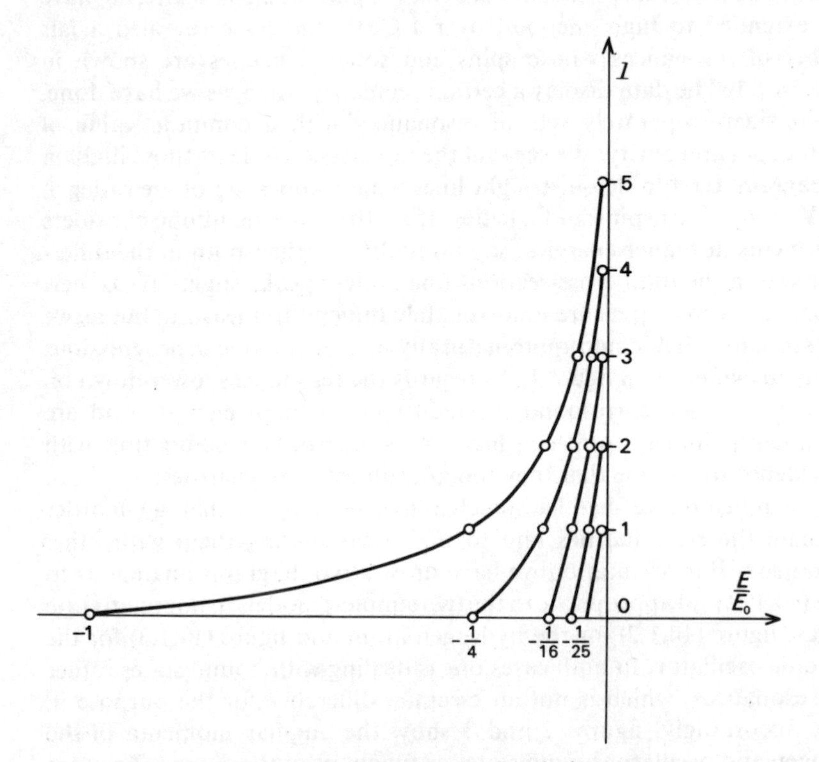

FIG. 14.1.2. Bound states of the hydrogen atom.

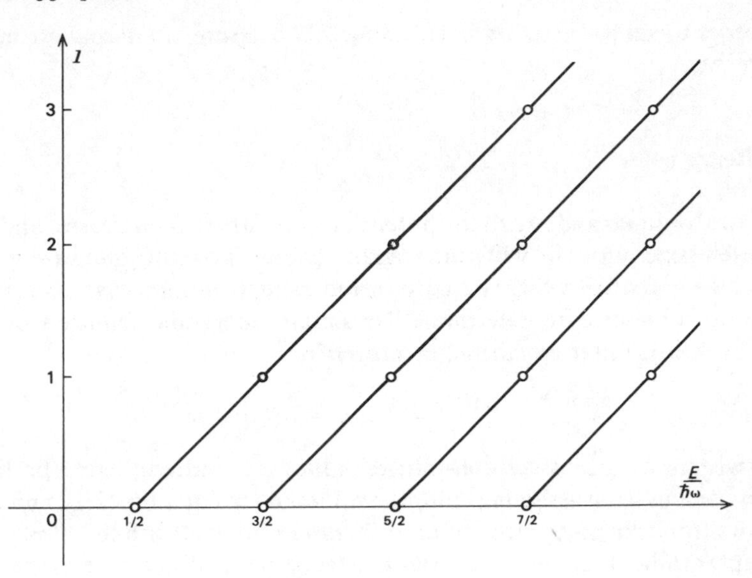

FIG. 14.1.3. Bound states of the harmonic oscillator.

seen to fall neatly into families, just like the pion-nucleon resonances; but here we know what is the common feature for each family. All the hydrogenic levels on one hyperbola, and all the oscillator levels on one straight line, have wavefunctions with the same number of radial nodes (zeros of the radial wavefunction). From this point of view the bound states highest on the diagram are the simplest ones, since their radial wavefunctions have no nodes.

For the hydrogen atom the number of radial nodes is well known, being given in terms of the principal quantum number n and of the orbital angular momentum ℓ by the relation

$$\text{number of radial nodes} = n - \ell.$$

It is an *internal quantum number*, completely different in kind from ℓ and m, which are connected with rotational invariance. The number of radial nodes has no such simple group-theory interpretation;* it represents rather a common topological property of the wavefunctions, suggestive of the possibility of passing continuously from one to the other. In other words such families of bound states invite us to look more closely at what they may have in common; thus we might take more seriously the connecting lines between the points on the figures, i.e. *interpolate* between the points, forgetting for the moment the quantisation of angular momentum

* This need not prevent a topologist from treating it in terms of group theory, but so far no one has succeeded in exploiting this possibility.

in order to gain a better understanding of the nature of internal quantum numbers.

2. Regge poles

The Coulomb and oscillator potentials are rather special cases, and for greater generality we will turn again to our favourite guinea-pig the Yukawa potential, which actually includes the Coulomb case as a limit. The bound states are determined by square integrable solutions of the radial Schroedinger equation for integer ℓ:

$$\frac{d^2u_\ell(r)}{dr^2} + \left[\frac{\ell(\ell+1)}{r^2} + k^2 - V(r)\right] u_\ell(r) = 0 \tag{2.1}$$

We wish now to consider non-integer values of ℓ. Indeed there is nothing to prevent us from assigning complex values to ℓ in equation (2.1); and one knows from experience that often it is simpler to deal with functions of a complex rather than of a real variable. Hence we shall study the properties of solutions of equation (2.1) for complex ℓ.

In their essentials, the considerations of chapter 6 about the Schroedinger equation apply just as well for complex as for real ℓ, as we shall see presently.

First we specify the boundary conditions defining the interesting solution of the Schroedinger equation. The point $r = 0$ is a singular point of the equation because of the centrifugal term. Near $r = 0$, the most singular terms of (2.1) are the first and the second, so that one has asymptotically

$$\frac{d^2u_\ell}{dr^2} + \frac{\ell(\ell+1)}{r^2}u_\ell \simeq 0,$$

This is solved by

$$u_\ell(r) = r^{\ell+1} \times \text{constant}$$

$$u_\ell(r) = r^{-\ell} \times \text{constant}$$

Hence equation (2.1) has two independent solutions behaving respectively like $r^{-\ell}$ and $r^{\ell+1}$ as r tends to zero. For integer ℓ, the solutions corresponding to bound or scattering states behave at the origin like r^ℓ. Therefore we define our interpolation by the boundary condition

$$\lim_{r\to 0} r^{-\ell} u_\ell(r) = 1 \tag{2.2}$$

which determines the solution $u_\ell(r)$ uniquely.

A fundamental property of $u_\ell(r)$ as a function of ℓ can now be derived from Poincaré's theorem. From the fact that equation (2.1) and the boundary condition (2.2) are analytic in ℓ, it follows that $u_\ell(r)$ must be an *entire* function of ℓ for all r.

We can rewrite scattering theory for complex ℓ. To this end one notes that as r tends to infinity, the Schroedinger equation always has two independent solutions behaving like $e^{\pm ikr}$. Hence, as r tends to infinity, $u_\ell(r)$ tends to a linear combination of these two solutions, and can be written as

$$u_\ell(r) \xrightarrow[r\to\infty]{} \frac{1}{2ik}\left[f^{(-)}(\ell, k)\, e^{ikr} - f^{(+)}(\ell, k)\, e^{-ikr}\right] \tag{2.3}$$

which defines the Jost functions $f^{(\pm)}(\ell, k)$ for complex ℓ. The asymptotic behaviour could equally well have been expressed in terms of a phase shift $\delta(\ell, k)$ by writing

$$u_\ell(r) \xrightarrow[r\to\infty]{} \text{constant} \times \sin\left(kr - \frac{\ell\pi}{2} + \delta(\ell, k)\right). \tag{2.4}$$

By comparing the last two equations we can express the collision matrix element $S(\ell, k) = e^{2i\delta\ell(k)}$ as the ratio of the two Jost functions, just as in the physical case:

$$S(\ell, k) = \frac{f^{(-)}(\ell, k)}{f^{(+)}(\ell, k)} \tag{2.5}$$

The analytic properties of $S(\ell, k)$ as a function of ℓ follow immediately from those of the wavefunction $u_\ell(r)$. By (2.3) both Jost functions must be entire functions of ℓ, exactly like $u_\ell(r)$. Hence singularities of $S(\ell, k)$ can be due only to zeros of the denominator, and we get the following very simple result:

The collision matrix element $S(\ell, k)$ can be continued to complex values of ℓ; the only singularities of this continuation are poles in ℓ, (it is a meromorphic function of ℓ), and the poles occur at zeros of the Jost function $f^{(+)}(\ell, k)$. These properties are obviously shared by the partial wave scattering amplitude:

$$a(\ell, k) = \frac{e^{i\delta(\ell,k)} \sin\delta(\ell, k)}{k} = \frac{1}{2ik}\left[S(\ell, k) - 1\right] \tag{2.6}$$

Such poles of $S(\ell, k)$ or zeros of $f^{(+)}(\ell, k)$ are called Regge poles.

3. Regge trajectories

As k^2 varies, each zero of $f^{(+)}(\ell, k)$, i.e. each Regge pole, moves in the complex ℓ-plane, thus defining a function

$$\ell = \alpha(k^2)$$

As k^2 varies from $-\infty$ to $+\infty$, the function $\alpha(k^2)$ is called a *Regge trajectory*. Examples of trajectories calculated for a Yukawa potential are given in figures (14.3.1) and (14.3.2). As long as k^2 is negative, $\alpha(k^2)$ is real.

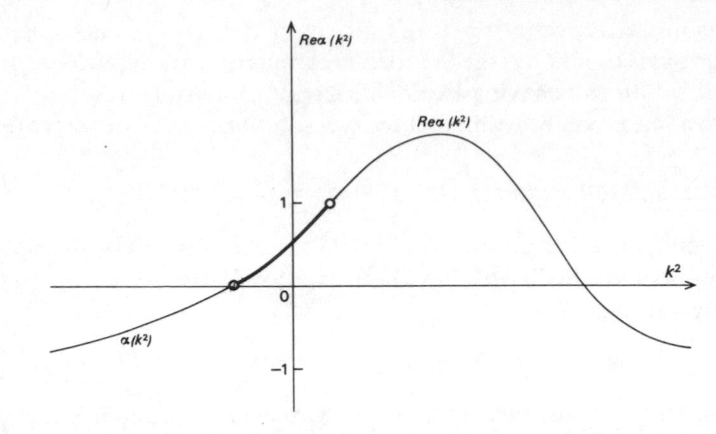

FIG. 14.3.1. Regge trajectory for a Yukawa potential.

In that case $u_\ell(r)$ as well as $e^{\pm ikr}$ are real functions (k being pure imaginary). Hence the Jost function $f^{(+)}(\ell, k)$ is real, and its zeros with it. (Under these conditions two trajectories could in principle become complex conjugates, but this has neven been known to happen.) When k^2 is real positive, the Jost function becomes complex for real ℓ, and $\alpha(k^2)$ has both real and imaginary parts.

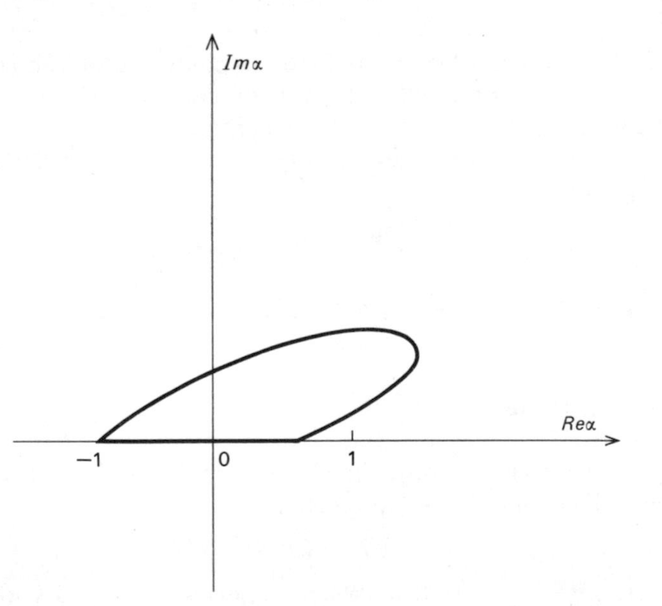

FIG. 14.3.2. Regge trajectory for a Yukawa potential, shown in the complex angular momentum plane.

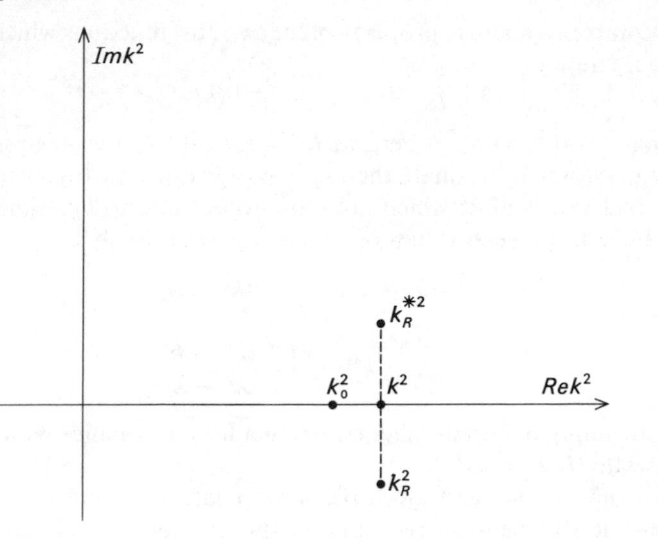

FIG. 14.3.3.

It is easy to see that when $\alpha(k^2)$ goes through a physical value for negative k^2, i.e. through a nonnegative integer value of ℓ, the value of k^2 for which it does so corresponds to the energy of a bound state. For in that case $f^{(+)}(\ell, k^2) = 0$, so that in view of (2.3) the wavefunction behaves for large r like

$$\frac{1}{2ik} f^{(-)}(\ell, k)\, e^{ikr}$$

But, k being imaginary, $(k = i\kappa)$, this is a decreasing exponential

$$-\frac{1}{2\kappa} f^{(-)}(\ell, i\kappa)\, e^{-\kappa r}$$

and the wavefunction is square integrable; and since ℓ is physical, we certainly have a bound state for this value of k^2. This shows that the Regge trajectories are indeed the desired interpolating functions between bound states.

Next, let us consider resonances. For positive k^2, $\alpha(k^2)$ is complex, but it can happen that Re $\alpha(k^2)$ goes through an integer ℓ_0 for non-negative $k^2 = k_0{}^2$, while Im $\alpha(k^2)$ still remains small (Fig. 14.3.3). By continuity there must exist a complex value $k_R{}^2$ of k^2 where $\alpha(k^2)$ becomes exactly equal to the physical value l_0. Let us write

$$k_R{}^2 = k_1{}^2 + ik_2{}^2$$

In the vicinity of $k_R{}^2$ we can write

$$f^{(+)}(k^2, \ell_0) = C(k^2 - k_R{}^2)$$

The complex conjugate property of the two Jost functions which is obvious from (3) implies

$$f^{(-)}(k^{*2}, \ell_0) = [f^{(+)}(k^2, \ell_0)]^*$$

whence $f^{(-)}(k^2, \ell_0)$ has a zero at $k^2 = k_R^{2*}$. If, as we have assumed, the imaginary part $k_2{}^2$ is small, then $k_R{}^2$ and k_R^{*2} are close together, and there exist real values of k^2 which are close to $k_R{}^2$ and to k_R^{*2} simultaneously (fig. 14.3.3). For such values of k^2 one has accordingly

$$f^{(-)}(k^2, \ell_0) = C^*(k^2 - k_R^{*2})$$

and

$$S(\ell_0, k^2) = \frac{C^*}{C}\frac{(k^2 - k_R^{*2})}{(k^2 - k_R{}^2)} \tag{3.1}$$

This is simply the Breit–Wigner formula for a resonance with energy $k_1{}^2$ and width $(\Gamma/2) = k_2{}^2$.

To sum up, Regge trajectories interpolate not only between bound states but also between resonances; this cannot but encourage us even further to use them for interpolating between objects (resonances and bound states) having identical internal quantum numbers.

Before comparing this situation with the relativistic case we must say a word about the effects of exchange potentials. When we were studying the strong interactions we saw that the two crossed reactions corresponding to a given reaction gave rise to both direct and exchange forces; in our nonrelativistic model this must be allowed for by adding an exchange potential to the direct potential V.

When it acts on a spatial wavefunction $\psi(\mathbf{x})$, the exchange potential is an operator $\mathscr{V}$ such that

$$\mathscr{V}\psi(\mathbf{x}) = V_E(r)\,\psi(-\mathbf{x})$$

where $V_E(r)$ is an ordinary function. For physical values of the orbital angular momentum ℓ this gives

$$\mathscr{V}u_\ell(r) = V_E(r)(-1)^\ell\, u_\ell(r)$$

Hence in the presence of an exchange potential the Schroedinger equation takes the form

$$\frac{d^2u_\ell}{dr^2} + \left[\frac{\ell(\ell+1)}{r^2} + k^2 - V(r) - V_E(r)\right]u_\ell(r) = 0 \qquad \text{for } \ell \text{ even} \tag{3.2a}$$

$$\frac{d^2u_\ell}{dr^2} + \left[\frac{\ell(\ell+1)}{r^2} + k^2 - V(r) + V_E(r)\right]u_\ell(r) = 0 \qquad \text{for } \ell \text{ odd} \tag{3.2b}$$

Each of these Schroedinger equations, involving the potentials $V_1 = V + V_E$ and $V_2 = V - V_E$ respectively, can be continued to complex values of ℓ, and they will give rise to two systems of Regge trajectories

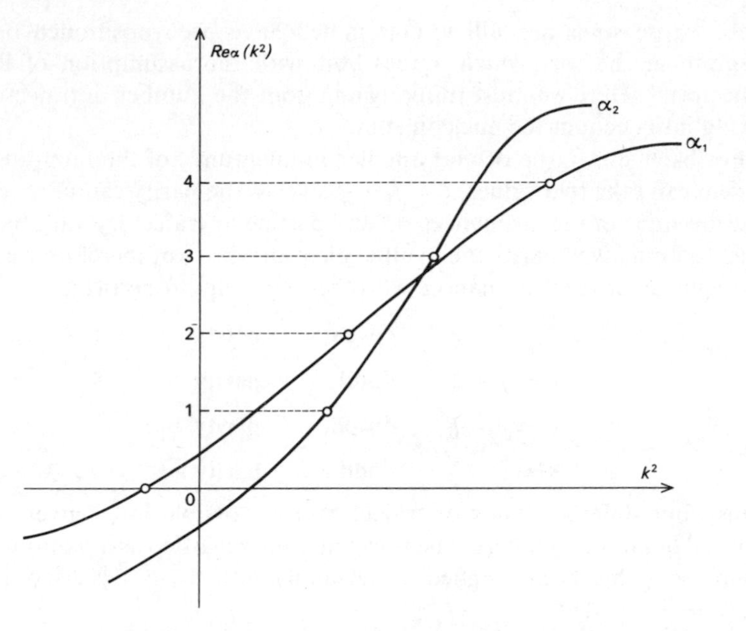

FIG. 14.3.4. Regge trajectories in presence of an exchange potential. The points on the trajectories show the locations of bound states and resonances.

$\alpha_1(k^2)$ and $\alpha_2(k^2)$. The only difference from the previous case is the following. When $\alpha_1(k^2)$ goes through an even integer it gives rise to a bound state or to a resonance; but when it goes through an odd integer, it gives rise only to a bound state of equation (3.2a) which is *not* the Schroedinger equation appropriate to this value of ℓ. Accordingly, $\alpha_1(k^2)$ interpolates only between bound states or resonances with even angular momenta, while $\alpha_2(k^2)$ interpolates only between states of odd angular momenta. This is shown in figure (14.3.4).

4. The relativistic case

If Regge poles exist in the real world, they could manifest themselves through regular families of resonances like those we have just described. All the resonances belonging to one family would share the same internal quantum numbers (baryon number, isotopic spin, charge, strangeness, and charge-conjugation or G parity if applicable). Since exchange forces are expected, all resonances within a family will have the same parity, and successive spin values will differ by two units.

On this subject diagrams speak better than words. The families of pion-nucleon resonances are shown in figure (14.1.1). Resonances of known mass whose spins are still undetermined have been positioned on the diagram in the way which agrees best with the assumption of Regge trajectories. Here we must think again about the number of trajectories, taking into account the nucleon spin.

For each spin j, the orbital angular momentum ℓ of the pion-nucleon system can take two values, $\ell = j \pm \frac{1}{2}$, so that the parity can be $+$ or $-$. But the spins of the resonances on any particular trajectory vary by two, whence for a given parity the ℓ values also vary by two; therefore we must distinguish between resonances with the following properties:

$$\begin{array}{lll} \ell = j + \frac{1}{2} & \ell \text{ even}, & \text{parity} -, \\ \ell = j + \frac{1}{2} & \ell \text{ odd}, & \text{parity} +, \\ \ell = j - \frac{1}{2} & \ell \text{ even}, & \text{parity} -, \\ \ell = j - \frac{1}{2} & \ell \text{ odd}, & \text{parity} +. \end{array}$$

Thus, four different kinds of trajectory are possible for a given set of internal quantum numbers. The nucleon itself, regarded as a pion-nucleon bound state, has been assigned to the family with $\ell = j - \frac{1}{2}$, ℓ odd, parity $+$, $(j = \frac{1}{2}, \ell = 1)$.

The diagrams show that there is very satisfactory agreement between the assumptions and experiment. Indeed, confirmation is limited only by the experimental difficulty in identifying high energy resonances and their spins. It is a most remarkable fact that these plots of j against s are practically straight lines, and that these lines are all practically parallel to one another with a universal slope of $1(\text{GeV})^{-2}$. This is an extremely simple property for which the Yukawa potential has left us unprepared, and which recalls rather the harmonic oscillator.

By entering the hyperon resonances ($\Lambda - \pi$, $\Sigma - \pi$, and $K - n$) on figure 14.1.1, one could further confirm the universality of the slopes; $SU(3)$ breaking then shows up as a slight relative (parallel) displacement of the different trajectories corresponding to the same multiplet.

Comparison with experiment is much more difficult for bosons, because high energy boson resonances are very hard to find and their spins even more so. But even here one can detect signs of linearly rising trajectories with the same universal slope of $1(\text{GeV})^{-2}$.

It seems that here we are offered a genuine glimpse of the laws of nature. These quasi-linear trajectories with their universal slope recall by their simplicity the spectrum of the hydrogen atom, which provided the handle for solving the problems of atomic physics. Most of all they seem to suggest that beneath the inevitable complications due to the multiplicity of particles and to the constraints of unitarity, the structure of particles involves very simple and elegant laws which will probably open the door to a simple and elegant interpretation of their properties.

One is faced with the problem whether the trajectories continue to rise linearly to infinite values of j, the resonances becoming indistinct because of their strong inelasticity and because of their widths overlapping; or whether the trajectories stop somewhere, and if so where?

5. The crossed reactions

The abscissa s in the figures we have just seen is the squared mass of the resonance, and at first sight it seems impossible to continue the trajectories to negative s. Nevertheless it turns out that in actual fact they can be continued, and that crossing symmetry not only gives meaning to these continuations, but also opens up new and fascinating applications and unifications of the theory.

Consider for instance the familiar figure (12.9.3) which shows the physical regions of three crossed reactions, and concentrate for convenience on the variable t rather than on s. Positive values of t represent the total squared energy of the system in its centre of mass frame for reaction II, while negative values of t represent the squared momentum transfer in reaction I. Vice versa, s represents the squared energy for reaction I, but a momentum transfer for reaction II; hence it is simply related to the scattering angle in reaction II. Now one can ask the following question. Consider the amplitude $A(t, \cos\theta_t)$ for reaction II *as a function of* $\cos\theta_t$, rather than as a function primarily of t, (as in the treatment of families of resonances). Does this function have any properties whereby Regge poles could make themselves felt in this channel II? Eventually we shall proceed to verify these properties in the physical region of reaction I, so that there is no reason to confine $\cos\theta_t$ to its physical range between -1 and $+1$.

The function $A(t, \cos\theta_t)$ does indeed have such a property, namely, as we are going to see in detail, its asymptotic behaviour as $\cos\theta_t$ tends to infinity. We shall consider later how such an asymptotic property can be investigated experimentally.

The problem is therefore the following: find the asymptotic behaviour of the amplitude $A(t, \cos\theta)$ as $\cos\theta$ tends to infinity, given that the amplitude has Regge poles.

Since Regge poles are properties of partial waves, we start from the partial wave expansion of $A(t, \cos\theta)$:

$$A(t, \cos\theta) = \sum_{\ell=0}^{\infty} (2\ell + 1)\, a_\ell(t)\, P_\ell(\cos\theta) \qquad (5.1)$$

Our assumption is that $a_\ell(t)$ can be generalized into an analytic function of ℓ, whose only singularities are poles. Even if other singularities in ℓ do exist, we ignore them for the moment and confine ourselves exclusively to the detection of Regge poles. To display explicitly the analyticity of $a(\ell, t)$ in ℓ, we replace the sum (5.1) by an integral over ℓ, using a very simple idea from the calculus of residues. An individual term of the sum can always be written as

$$(2\ell + 1)\, a_\ell(t)\, P_\ell(\cos\theta) = \frac{1}{2\pi i} \oint_{C_\ell} (2\lambda + 1) \frac{a(\lambda, t)}{\lambda - \ell} P_\lambda(\cos\theta)\, d\lambda \tag{5.2}$$

where the contour C_ℓ encircles the point $\lambda = \ell$ (Fig. 14.5.1). This formula applies by virtue of the fact that the Legendre polynomials $P_\ell(\cos\theta)$ are special cases of the Legendre functions $P_\lambda(\cos\theta)$; the latter solve the same differential equation as do the $P_\ell(\cos\theta)$, but for complex values of ℓ:

$$(1 - z^2) \frac{d^2 P_\lambda(z)}{dz^2} - 2z \frac{dP_\lambda(z)}{dz} + \lambda(\lambda + 1) P_\lambda(z) = 0 \tag{5.3}$$

They obey the boundary condition

$$P_\lambda(1) = 1 \tag{5.4}$$

Poincaré's theorem shows at once that $P_\lambda(z)$ is an entire function of λ, which allows us to insert it into the formula (5.2) without any further precautions.

One can try to extend formula (5.2) to the entire sum (5.1), instead of applying it only to an individual term. To this end, the function $(\lambda - \ell)$ in

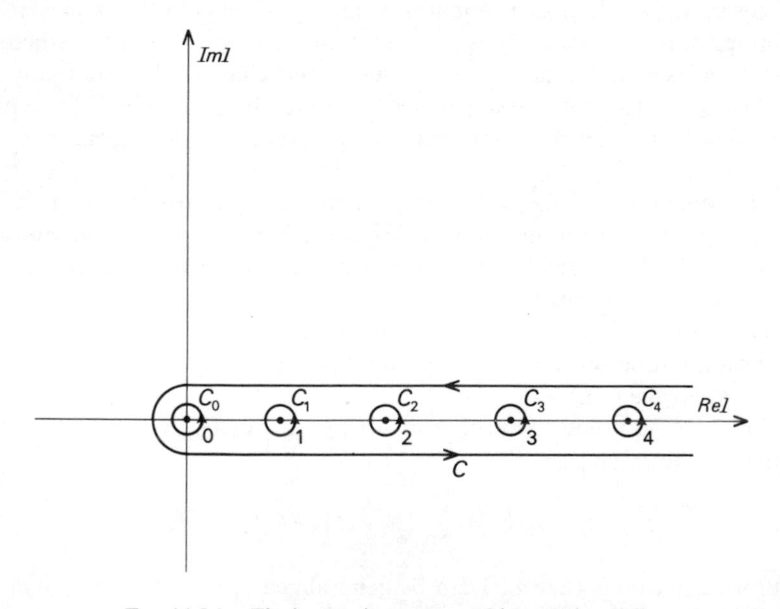

FIG. 14.5.1. The integration contour C in equation (5.5).

the denominator, which vanishes only for a single value of ℓ, must be replaced by another function which vanishes whenever ℓ is a positive integer. One obvious candidate for this role is $\sin \pi\lambda$. The function $1/\sin \pi\lambda$ has poles at $\lambda = 0, 1, 2, 3, \ldots$ with residues $\pm\pi$.

To allow for the alternating signs of the residues, we note that

$$P_\ell(-\cos\theta) = (-1)^\ell P_\ell(\cos\theta)$$

By appeal to the residue formula this allows us to write

$$A(t, -\cos\theta) = \sum_\ell (2\ell + 1)\, a_\ell(t)(-1)^\ell P_\ell(\cos\theta)$$

$$= \frac{1}{2i}\int_C (2\lambda + 1)\frac{a(\lambda, t)}{\sin \pi\lambda} P_\lambda(\cos\theta)\, d\lambda \tag{5.5}$$

where the contour C now loops around the real positive axis.

In order to relate this formula to the asymptotic behaviour of the amplitude in $\cos\theta$, we shall need the behaviour of $P_\lambda(z)$ as z tends to infinity; it can be written as

$$P_\lambda(z) \underset{z\to\infty}{\simeq} \frac{2^\lambda}{\sqrt{\pi}}\frac{\Gamma(\lambda + \frac{1}{2})}{\Gamma(\lambda + 1)} z^\lambda \qquad \text{for} \qquad \operatorname{Re}\lambda > -\tfrac{1}{2} \tag{5.6}$$

where Γ is Euler's gamma function. This shows that the rate of growth in z increases with the real part of λ, and we are led to display the dominant values of λ in (5.5). To do this, still for the physical values of $\cos\theta$ for which the foregoing calculations converge, we complete the contour C by a circle at infinity (which is explicitly possible at least in potential scattering). Then the contour can be deformed. Following the suggestion of equation (5.6) we shift the contour towards the left until it runs along the line L parallel to the imaginary axis and defined by $\operatorname{Re}\lambda = -\frac{1}{2}$; in doing so we also include a small loop around every Regge pole swept over as the original contour is deformed into L, (Figure 14.5.2). We denote the positions of such Regge poles by $\alpha_n(t)$, $(n = 1, 2, \ldots)$, and their residues by $\beta_n(t)$; then we can evaluate their contributions explicitly, and find

$$A(t, \cos\theta) = -\pi \sum_n \frac{(2\alpha_n(t) + 1)}{\sin \pi\alpha_n(t)} \beta_n(t)\, P_{\alpha_n(t)}(-\cos\theta)$$

$$+ \frac{1}{2i}\int_L (2\lambda + 1)\frac{a(\lambda, t)}{\sin \pi\lambda} P_\lambda(-\cos\theta)\, d\lambda \tag{5.7}$$

The $-$ sign in the Regge terms stems from the direction in which the Regge poles have been encircled.

The asymptotic behaviour of $A(t, \cos\theta)$ in $\cos\theta$ is given explicitly by (5.7). By virtue of (5.6) the integral along L behaves like $|\cos\theta|^{-1/2}$ and one has

$$A(t, \cos\theta) \underset{\cos\theta\to\infty}{\simeq} -\pi \sum_n \frac{(2\alpha_n(t) + 1)}{\sin \pi\alpha_n(t)} \beta_n(t)\, P_{\alpha_n(t)}(-\cos\theta) \tag{5.8}$$

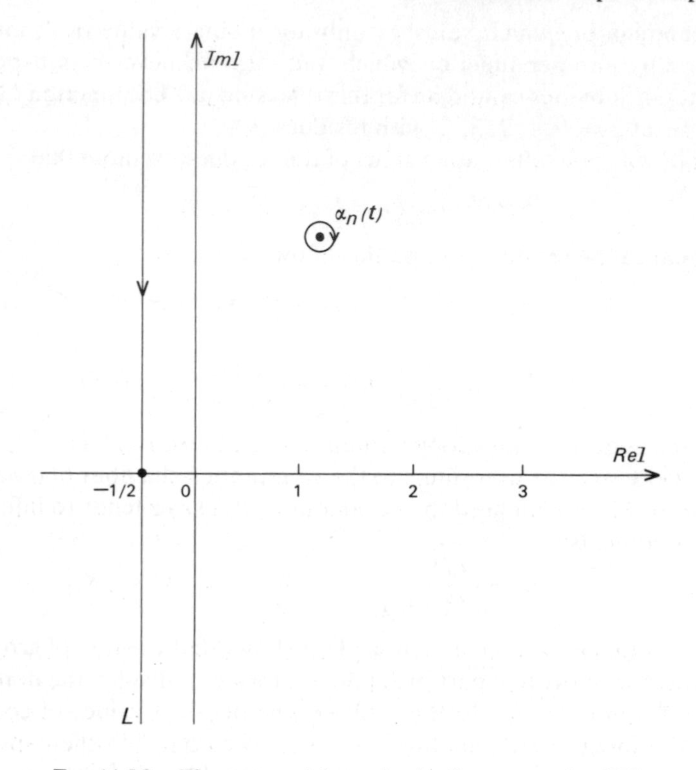

FIG. 14.5.2. The contour deformation leading to equation (5.7).

This formula must be modified because of exchange forces. Along the lines of section 3, we denote by $A_1(t, \cos\theta)$ the amplitude obtained by solving (3.2a), and by $A_2(t, \cos\theta)$ the amplitude obtained by solving (3.2b). In view of the fact the $A(t, \cos\theta)$ has the same $\cos\theta$-even part as A_1, and the same $\cos\theta$-odd part as A_2, we can write

$$A(t, \cos\theta) = \tfrac{1}{2}[A_1(t, \cos\theta) + A_1(t_1 - \cos\theta) + A_2(t, \cos\theta) - A_2(t, -\cos\theta)] \tag{5.9}$$

It is easy to check from the partial wave expansion that this is compatible with unitarity. Introducing the Regge poles $\alpha_n{}^1(t)$ and residues $\beta_n{}^1(t)$ of A_1, and similarly for A_2, using the same method as above, one gets

$$A(t, \cos\theta) \simeq -\frac{\pi}{2}\sum_n \frac{(2\alpha_n{}^1(t) + 1)}{\sin \pi\alpha_n{}^1(t)} \beta_n{}^1(t)[P_{\alpha_n{}^1(t)}(\cos\theta) + P_{\alpha_n{}^1(t)}(-\cos\theta)]$$

$$-\frac{\pi}{2}\sum_m \frac{(2\alpha_m{}^2(t) + 1)}{\sin \pi\alpha_m{}^2(t)} \beta_n{}^2(t)[P_{\alpha_n{}^2(t)}(-\cos\theta) - P_{\alpha_n{}^2(t)}(\cos\theta)] \tag{5.10}$$

Finally, the Legendre functions can be replaced by their asymptotic form (5.6). At the same time we redefine the residue $\beta_n(t)$, which will be written as $\gamma_n(t)$ after absorbing some sign factors, $\sqrt{\pi}$'s, and gamma functions. Note that the asymptotic behaviour of $P_\lambda(-z)$ involves $(-z)^\lambda$, which for positive z may be written as $e^{i\pi\lambda}z^\lambda$. In this way, the brackets in (5.10) yield the asymptotic terms

$$(\cos\theta)^{\alpha(t)}\,[1 \pm e^{i\pi\alpha(t)}]$$

with a + (−) sign for the Regge poles of $A_1(A_2)$. These poles are said to have the *signature* + or − respectively. A more compact expression for the asymptotic behaviour results from dealing jointly with the poles $\alpha_n^1(t)$ and $\alpha_n^2(t)$, denoting their signature by $\xi = \pm 1$. Thus one obtains

$$A(t, \cos\theta_t) \underset{\cos\theta\to\infty}{\simeq} \sum_n \frac{1 + \xi_n e^{i\pi\alpha_n(t)}}{\sin \pi\alpha_n(t)}\,\gamma_n(t)(\cos\theta_t)^{\alpha_n(t)} \tag{5.11}$$

By using the relation

$$4p_t p_t' \cos\theta_t = s - u + \frac{(m^2 - \mu^2)(m'^2 - \mu^{2'})}{t} \tag{5.12}$$

one can rewrite (5.11) in terms of the scalar variables s and t; here, p_t and p_t' are the initial and final momenta in the centre of mass frame, and m, μ, m', μ', are the masses of the reacting particles. If we are interested in the case where s tends to infinity, with t fixed, then by using the relation

$$s + t + u = m^2 + m'^2 + \mu^2 + \mu'^2$$

we can write

$$s - u = 2s + t + m^2 + m'^2 + \mu^2 + \mu'^2.$$

Therefore one has asymptotically,

$$4p_t p_t' \cos\theta_t = 2s$$

In order to display the behaviour of (5.12) as s tends to infinity, it is convenient to introduce a fixed value s_0 of s, and to absorb a factor $(s_0/2p_t p_t')\alpha(t)$ into the residue; then one has

$$A(s, t) \underset{s\to\infty}{\simeq} \sum_n \frac{1 + \xi_n e^{i\pi\alpha_n(t)}}{\sin \pi\alpha_n(t)}\,\gamma_n'(t)\left(\frac{s}{s_0}\right)^{\alpha_n(t)}. \tag{5.13}$$

Recall that this behaviour obtains when s tends to infinity for a given fixed value of t.

6. Experimental tests

The expression (5.13) is one of the predictions which follow from assuming Regge poles; let us now see how it can be tested. Consider for instance a reaction which is one of the simplest from this point of view, namely pion-nucleon scattering with charge exchange:

$$\pi^- + p \rightarrow \pi^0 + n$$

at energy $\sqrt{s}$. The formula (5.13) tell us the asymptotic behaviour of the amplitude as the energy tends to infinity for a fixed value of the momentum transfer. By 'infinity' one means the highest energy attainable by a pion, i.e. about 20 GeV, corresponding to a value of s of about $400(\text{GeV})^2$.

In order to discover which trajectories can enter the expression (5.13), recall that the poles $\alpha_n(t)$ are those of the crossed reaction where t is the squared energy, i.e. of the reaction

$$\pi^- + \pi^0 \rightarrow \bar{p} + n.$$

The poles $\alpha_n(t)$ must have the quantum numbers of this reaction, namely isotopic spin 1, charge -1, G-parity $+$, and zero baryon number and strangeness. We know of one particle, the ρ^-, with these quantum numbers. If the ρ^- belongs to a Regge trajectory, we can expect that one of the $\alpha_n(t)$ goes through 1 at $t = m_\rho^2$. In terms of the potential model, for which signature has been discussed, the ρ with its negative parity would be a Regge pole of A_2, i.e. it has signature $\xi = -1$.

One knows of no other $\pi - \pi$ resonances with the same quantum numbers, which leads to the belief that the Regge trajectories other than that of the ρ must lie considerably lower: $\alpha_n(t) < \alpha_\rho(t)$; then we need keep only the ρ contribution to the asymptotic formula (5.13). Accordingly we expect

$$A(s, t) \underset{s\to\infty}{\simeq} \frac{1 - e^{i\pi\alpha_\rho(t)}}{\sin \pi\alpha_\rho(t)} \gamma(t) \left(\frac{s}{s_0}\right)^{\alpha_\rho(t)}. \tag{6.1}$$

This formula is compared with experiment in figure (14.6.1), which shows that there is remarkable agreement. One obtains for $\alpha_\rho(t)$ a form which extrapolates naturally, and with the universal value $1(\text{GeV})^{-2}$ of the slope, to 1 at $t = m_\rho^2$.

To apply this analysis to other reactions, one goes through the same arguments, and investigates which poles can enter the corresponding crossed reactions. In general, several poles occur in a given reaction, and their contributions interfere. Hence it is inadvisable as a rule to consider one reaction on its own, as we succeeded in doing for the pion-nucleon charge-exchange; instead, one analyses jointly a whole set of reactions involving the same trajectories.

In such a joint analysis there is one important condition on the residues, called *factorisability*, which we shall now derive.

Consider for instance the ρ trajectory. As we saw, it can occur in pion-nucleon scattering,

$$\pi + N \rightarrow \pi + N$$

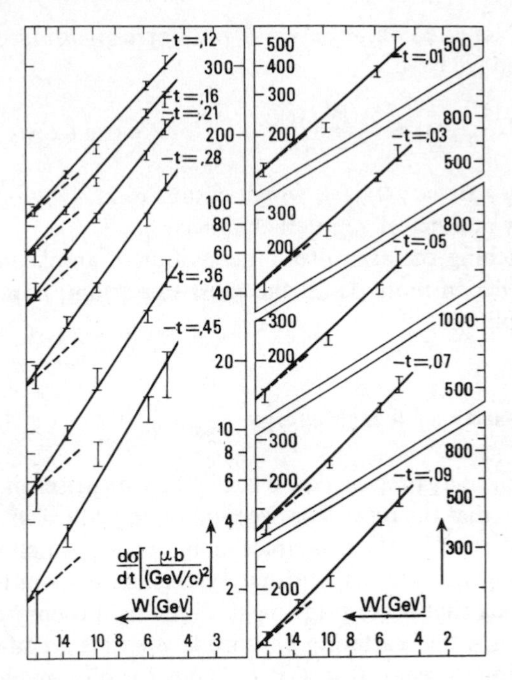

FIG. 14.6.1. Differential cross-section $d\sigma/dt$ for charge exchange scattering $\pi^- p \to \pi^0 n$. Logarithmic scales. (From G. Hoehler, S. Baacke, H. Schlaile, and P. Sonderegger, Physics Letters, 20, 79 (1966).) The curves represent $\ln d\sigma/dt$ as a function of $\ln W$, where $W = \sqrt{s}$. Regge pole theory predicts straight lines.

where we will denote its residue by $\gamma_{\pi\pi, N\bar{N}}$. It can occur also in elastic $\pi - \pi$ or $N - N$ scattering, where its residues will be denoted by $\gamma_{\pi\pi,\pi\pi}$ and $\gamma_{N\bar{N},N\bar{N}}$ respectively. It can occur in other reactions as well which we omit for simplicity. Symbolising the $\pi\pi$ channel by a, and the $N\bar{N}$ channel by b, we have three residues γ_{aa}, γ_{ab}, and γ_{bb}. The unitarity of the S-matrix implies that they are related by the factorisability condition

$$\gamma_{ab}^2 = \gamma_{aa}\gamma_{bb} \tag{6.2}$$

The matrix $S(t, \ell)$ for the two channels $a \to a, b$ and $b \to a, b$ has two rows and two columns. Denote by $\Sigma(t, \ell)$ the analytic continuation of $[S(t, \ell)]^+$ in t; then the unitarity condition can be expressed by the relation

$$S(t, \ell) = \frac{1}{\Sigma(t, \ell)}$$

This shows that $S(t, \ell)$ has a pole where the determinant of $\Sigma(t, \ell)$ vanishes. In general the minors of $\Sigma(t, \ell)$ do not vanish simultaneously with the determinant itself, so that $\Sigma(t, \ell)$ has only a single eigenvector corresponding to eigenvalue zero. Let $\alpha(t)$ be the value of ℓ for which $\det \Sigma(t, \ell)$

vanishes, and let the components of the corresponding eigenvector be $\gamma_a(t)$, $\gamma_b(t)$. Then we have

$$S_{ij}(t, \ell) \simeq \frac{\gamma_i(t)\,\gamma_j(t)}{\ell - \alpha(t)} + \cdots \text{ for } \ell \text{ close to } \alpha(t)$$

It follows that $\gamma_{ij}(t) = \gamma_i(t)\gamma_j(t)$, which entails (6.2). Obviously this proof applies for any number of coupled channels.

Before broaching the asymptotic behaviour of amplitudes in general, we must consider in more detail the most important, namely the elastic scattering amplitude.

7. Total cross-sections at high energy

The very high energy region could hold many surprises; it is all the more useful to realise that the total cross-section of any collision at high energy tends to be restricted by the principles of analyticity which we have taken as basic in strong interactions. Ultimately, this result stems from unitarity, and from the fact that according to axiomatic field theory or to Mandelstam's hypothesis, the scattering amplitude $A(s, t)$ is an analytic function of t in the region of real t, $0 < t < t_0$. Thus for pion-nucleon scattering one has $t_0 = 4m_\pi{}^2$. We shall not give a rigorous proof but only the main points of the argument.

One starts by translating the assumed analyticity in t into a property of the partial waves. To this end, the elastic scattering amplitude is written as

$$A(s, t) = \Sigma(2\ell + 1)\, a_\ell(s)\, P_\ell(\cos\theta). \tag{7.1}$$

We note that for large values of s, the order of magnitude of the momentum in the centre of mass frame is asymptotically given by

$$4p^2 \simeq s$$

whence

$$t = -4p^2(1 - \cos\theta) \simeq -s(1 - \cos\theta)$$

Therefore analyticity in t implies analyticity of $A(s, t)$ in $\cos\theta$ in the unphysical region

$$1 < \cos\theta < 1 + t_0/s. \tag{7.2}$$

In general, the domain of analyticity of a series depends on the asymptotic behaviour of its terms, in this case on the behaviour of $a_\ell(s)$ as ℓ tends to infinity. Since $P_\ell(\cos\theta)$ is a polynomial, (7.1) can fail to be analytic only because of a failure of the series to converge, and not because of singularities in any finite term.

As ℓ tends to infinity with cos θ real and greater than 1, we have

$$P_\ell \simeq \frac{1}{\sqrt{2\pi\ell}} \frac{(z+\sqrt{z^2-1})^{\ell+1/2}}{(z^2-1)^{1/4}}. \tag{7.3}$$

The series (7.1) converges up to $z = z_0 = 1 + t_0/s$; it can do so only if for all smaller values of z the exponential increase of $P_\ell(\cos\theta)$ with increasing ℓ is compensated by a decrease of $a_\ell(s)$. For large ℓ, $a_\ell(s)$ consequently behaves like

$$a_\ell(s) \sim (z_0 + \sqrt{z_0^2-1})^{-\ell}. \tag{7.4}$$

To sum up, analyticity in t implies an exponential decrease of $a_\ell(s)$ with increasing ℓ.

In order to see how this bears on the total cross-section, note that at high energies it is useful to replace the angular momentum ℓ by the impact parameter b, where $\ell = pb$. Thus, the statement that $a_\ell(s)$ decreases exponentially in ℓ implies that the interaction decreases exponentially with increasing impact parameter b. This will obviously tend to restrict the total cross-section. A detailed mathematical analysis exploiting the unitarity of partial waves yields the explicit restriction

$$\sigma_T < C(\text{Log } s)^2. \tag{7.5}$$

This is *Froissart's theorem*. Experimental results seem to suggest that total cross-sections tend to constant values at high energy, though a logarithmic increase or decrease would clearly be difficult to detect.

Another important result bearing on high energy processes is the Pomeranchuk theorem. Consider the scattering of a particle B, and also of its antiparticle $\bar{B}$, from one and the same target particle A, and assume that in both cases the total cross-sections tend to constant values at infinity; for instance

$$\sigma_T(p+\pi^+) \to C_1$$
$$\sigma_T(p+\pi^-) \to C_2 .$$

It can then be shown that the dispersion relations imply $C_1 = C_2$. We shall not embark on the proof.

8. Elastic scattering and Regge poles

Consider next how the theory of Regge poles can be applied to elastic scattering, say to

$$p+\pi^+ \to p+\pi^+ \qquad \pi^+ + \pi^+ \to \pi^+ + \pi^+ \qquad p+\bar{p} \to p+\bar{p}.$$

The corresponding Regge trajectories are those of the crossed reaction, namely

$$p + \bar{p} \rightarrow \pi^- + \pi^+ \qquad \pi^+ + \pi^- \rightarrow \pi^+ + \pi^- \qquad p + \bar{p} \rightarrow p + \bar{p}.$$

in the three cases respectively. In all three cases the quantum numbers of the crossed reaction are those of a boson with strangeness 0 and charge 0. Some of these trajectories support particles; this is the case for $(\rho^0, \omega^0, f^0, f^{*0}, \pi^0)$. Conversely, the observation that these particles exist implies the existence of corresponding Regge trajectories.

The asymptotic high energy behaviour of the total cross-section can be connected with the values of these Regge trajectories at $t = 0$. We know that the total cross-section is connected with the imaginary part of the elastic scattering amplitude through the optical theorem

$$\sigma_T = \frac{4\pi \operatorname{Im} f(s, 0)}{p}.$$

The amplitude $A(s, t)$ which we have been discussing hitherto is related to the conventional amplitude $f(s, t)$ by $A(s, t) = \sqrt{s}f(s, t)$, whence

$$\sigma_T = \frac{4\pi \operatorname{Im} A(s, 0)}{p\sqrt{s}}. \tag{8.1}$$

The Regge pole assumption makes it easy to write down the asymptotic expression for $\operatorname{Im} A(s, 0)$. For negative t the trajectories $\alpha_n(t)$ and residues $\gamma_n(t)$ must be real; hence the imaginary part is prescribed uniquely by the coefficient $1 + \xi\, e^{i\pi\alpha_n(t)}$, and for the total cross-section we obtain

$$\sigma_T \simeq \text{constant} \sum_n \xi_n \gamma_n(t) \left(\frac{s}{s_0}\right)^{\alpha_n(0)-1}. \tag{8.2}$$

Thus, Froissart's theorem demands

$$\alpha_n(0) \leqslant 1 \tag{8.3}$$

If total cross-sections really do tend to constant values, then for the largest of the $\alpha(0)$, which we shall call $\alpha_p(0)$, one must have

$$\alpha_P(0) = 1$$

On analysing the experimental data one sees that this trajectory $\alpha_P(t)$ does not appear to support any particles. It has signature + (the coefficient ξ_1 in (8.2) must be positive since σ_T is) and isotopic spin zero, and it automatically validates the Pomeranchuk theorem. Note that all its quantum numbers are those of the vacuum; hence it is called the vacuum trajectory or the Pomeranchuk trajectory, depending on the context.

At the present time the precise nature of this vacuum singularity is highly controversial. It has been suggested that one is dealing with a trajectory having a slope much smaller than the universal slope of 1 $(\text{GeV})^{-2}$; or with a fixed pole at $\alpha = 1$; or with an essential singularity rather than a pole. Such a discussion would take us too far, and would in any case be liable to be quickly overtaken by experimental developments.

9. Current work

The comparison of Regge pole theory with experiment, and the development of the theory itself, are both changing so rapidly at present that it seems unwise to review them in an introduction such as this.

CHAPTER 15

WEAK INTERACTIONS

We give a brief introduction to the most important aspects of weak interactions, namely the Hamiltonian and the conservation or non-conservation laws.

What is most remarkable about the weak interactions is the fact that they break most of the conservation laws, whether of isotopic spin, strangeness, parity or charge conjugation, and up to a point even time reversal invariance. As far as one can tell at the moment, only the conservation laws for some of the charge-type quantum numbers are obeyed, like those of the electric charge itself, of baryon number, and of its analogues the electron and muon numbers.

Perhaps equally interesting is the observation that violation of the conservation laws is not haphazard, but that it obeys detailed rules of its own. Thus, parity and charge-conjugation invariance are not only broken, but in a certain sense they are broken maximally. Non-conservation of strangeness and of isotopic spin obey subtil rules with an amazing resemblance to conservation laws; we shall see for instance that the difference $\Delta I = I_f - I_i$ in isotopic spin between the initial and final states of a decay process often obeys conditions like $|\Delta I| = \frac{1}{2}$.

Following the discovery of parity violation, the most important task in recent years has been to analyse the structure of weak interactions. Since they are weak enough to be treated, in principle, by first order perturbation theory, the problem is to discover a Hamiltonian which embodies and correlates all the experimental results. This work is creditably far advanced, even though the recently discovered violation of time-reversal invariance is a reminder that it is not yet complete. The outcome is a Hamiltonian which in many respects is strongly reminiscent of the electromagnetic interaction.

In this chapter we shall try to exhibit the most important characteristics of weak interactions without embarking on detailed calculations, which are generally long and uninspiring. Thus we shall ignore one of the most interesting aspects, which would be to reenact, by analysing the experiments, the historical emergence of the Hamiltonian as it is known to-day.

Fortunately, this aspect is treated in exemplary fashion in Källen's book listed in the bibliography.

1. β-decay and parity violation

(a) *Beta decay*

Historically, weak interactions first appeared in nuclear β-decay. The simplest such process is neutron decay

$$n \to p + e + \bar{\nu}_e . \tag{1.1}$$

The complementary reaction

$$p \to n + \bar{e} + \nu_e \tag{1.2}$$

cannot be observed directly because the proton is less massive than the neutron; but it can occur inside a nucleus where the proton can acquire the necessary energy from the rest of the nucleus. When the electron-producing reaction (1.1) takes place inside a nucleus, the nucleus is said to have undergone β-decay; the positron-producing reaction (1.2) is called β^+-decay. Note also the reaction

$$p + e \to n + \nu_e \tag{1.3}$$

which can occur if a proton in a nucleus absorbs one of the atomic electrons. It requires a nonzero probability for the electron to be found at the nucleus, which because of the centrifugal barrier happens only to S-electrons; in spectroscopic notation, they belong to the K shell. Hence reaction (1.3) is called K-capture.

(b) *Parity violation in beta decay*

As we saw in chapter 6, the fact that parity is not conserved can be established by investigating the β-decay of a polarised nucleus. One needs an unstable nucleus with high spin like Co^{60}, which is then polarised so that its spin points in a known direction (Fig. 15.1.1).

The nucleus decays, emitting an electron with some momentum $\mathbf{p}$; the experiment consists simply in measuring the directional distribution of this momentum. The result one obtains has the form

$$\frac{dP}{d\Omega} = A + B\mathbf{s} \cdot \mathbf{p} \tag{1.4}$$

where $dP/d\Omega$ is the emission probability per unit solid angle. This shows that more electrons are emitted into one than into the other of the two hemispheres situated on opposite sides of the plane Oxy.

One can see immediately that this implies parity violation. If one inverts

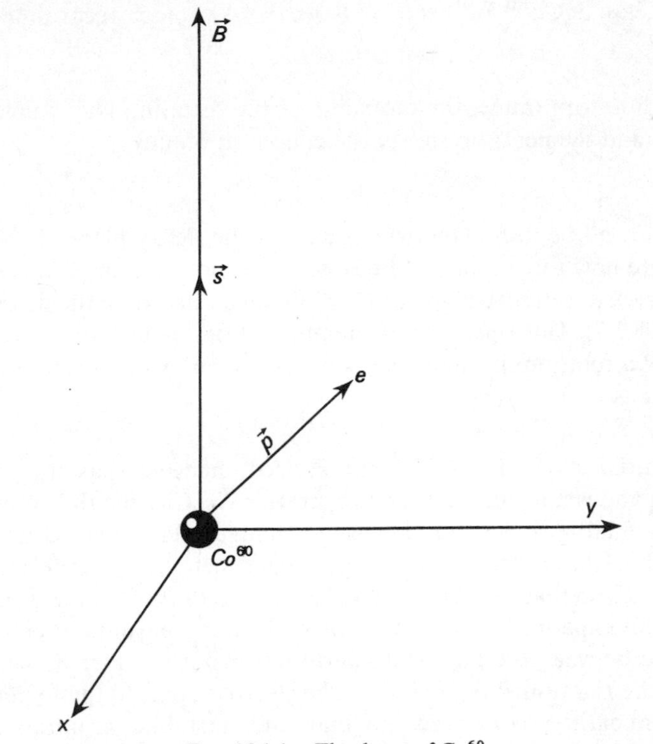

FIG. 15.1.1. The decay of Co^{60}.

the direction of the coordinate axes in space, then the momentum **p**, being a vector, changes sign, but the polarisation **s** does not, being an axial vector just like angular momentum. Thus, under such a transformation the probability becomes

$$A - B\mathbf{s} \cdot \mathbf{p}$$

hence it is not invariant.

(c) *Parity violation and polarisation*

Parity violation in β decay can be detected also by another method due to Frauenfelder, consisting in a measurement of the polarisation of electrons* emitted from nuclei which are not polarised themselves. To understand the results of the experiment we consider first what would be the consequences of parity conservation if it did apply in this situation. (The discussion is analogous to that of chapter 6, section 7.)

* The polarisation of electrons can be measured by investigating the angular distribution of the electromagnetic radiation (bremsstrahlung) which they emit when they are slowed down in matter. This effect, being an electromagnetic one, is well understood quantitatively.

In the case of Co^{60} for instance, β decay takes place according to

$$Co^{60} \rightarrow Ni^{60} + \bar{e} + \bar{\nu}_e \tag{1.5}$$

In the laboratory frame, the momenta of the recoiling Ni^{60} nucleus, the electron, and the neutrino satisfy the conservation law

$$\mathbf{p}_{Ni} + \mathbf{p}_e + \mathbf{p}_\nu = 0$$

Hence they all lie in a plane which we call the decay plane P. Next, we investigate how the initial and final states of the reaction (1.5) transform under a certain symmetry operation S_P defined relative to the decay plane (Figure 15.1.2). This operation is the product of a reflection in the origin, Π, and of a rotation through an angle π about an axis $\mathbf{n}$ normal to P, and is written as

$$S_P = \Pi R_{\mathbf{n}}(\pi) \tag{1.6}$$

The initial state of Co^{60} is unpolarised; hence it has no preferred direction and is not changed by the operation S_P. Clearly, the momenta of the three final particles are likewise unchanged. However, consider the behaviour of the electron spin. Being an axial vector, it is unaffected by the reflection Π, so that the action of S_P reduces to the rotation $R_n(\pi)$. Hence if, as in this experiment, only the electron spin is measured, then the only difference between the final state and its transform under S_P lies in the effect of the rotation $R_n(\pi)$ acting on the electron spin. In the hypothetical case where parity is conserved, the final state must, like the initial state, be unchanged by the symmetry operation S_P; therefore the spin state of the

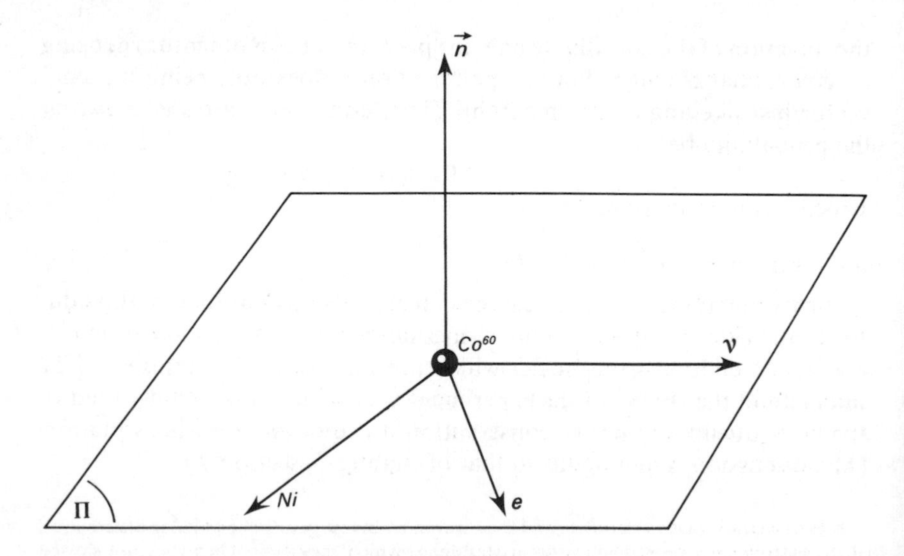

FIG. 15.1.2. Decay kinematics of Co^{60}.

electron must be invariant under the rotation $R_n(\pi)$, which means that it can be polarised only along the direction **n**. To sum up, parity conservation implies that the electron can be polarised only in the direction normal to the decay plane. In particular there can be no polarisation component along the electron momentum, i.e. no so-called longitudinal polarisation.

Here again experiment shows that parity is not conserved. The electron is found to have a longitudinal polarisation equal to $-v/c$ within experimental error, where v is the electron velocity. In the language of quantum mechanics this means that the state with helicity $-\frac{1}{2}$ is populated more than the state with helicity $+\frac{1}{2}$; denoting by N_λ the number of electrons detected with helicity λ, one has

$$\frac{N_{-1/2} - N_{+1/2}}{N_{-1/2} + N_{+1/2}} = \frac{v}{c}.$$

In similar experiments on β^+-emitting nuclei, one finds that the positrons are also polarised longitudinally, but this time with the opposite polarisation $+v/c$.

2. The neutrino

(a) *Neutrino mass*

The neutrino mass is extremely small (less than 1 keV). One can convince oneself of this by studying the shape of the electron energy spectrum in nuclear β decay. Perhaps the simplest way is to consider the form of the Dalitz plot representing, in triangular coordinates, the energies of the three particles emerging from reaction 1.1 or 1.5. Figure (15.2.1) shows the boundaries of the plot for zero and for nonzero neutrino mass.

We can see from the figure that a zero value of the neutrino mass results in a sharp corner at the end of the physical region.

It is easy to understand this corner if one recalls that in terms of the Mandelstam variables s, t, u one and the same curve serves as the boundary of the physical regions for all the reactions related by crossing to, say, $n + e^+ \to p + \bar{\nu}_e$. Now the decay is one of these crossed reactions. If the neutrino mass is zero, then its energy and momentum vanish simultaneously, and when they do it becomes kinematically impossible to distinguish between the two crossed reactions $n \to p + e + \bar{\nu}_e$ and $n + \nu_e \to p + e$. Therefore their boundary curves touch, and the double point involves a sharp angle; whence the Dalitz plot, being one of these physical regions, must indeed have such a corner.

Accordingly one can study the electron spectrum near maximum energy; its structure is completely different for zero and nonzero neutrino

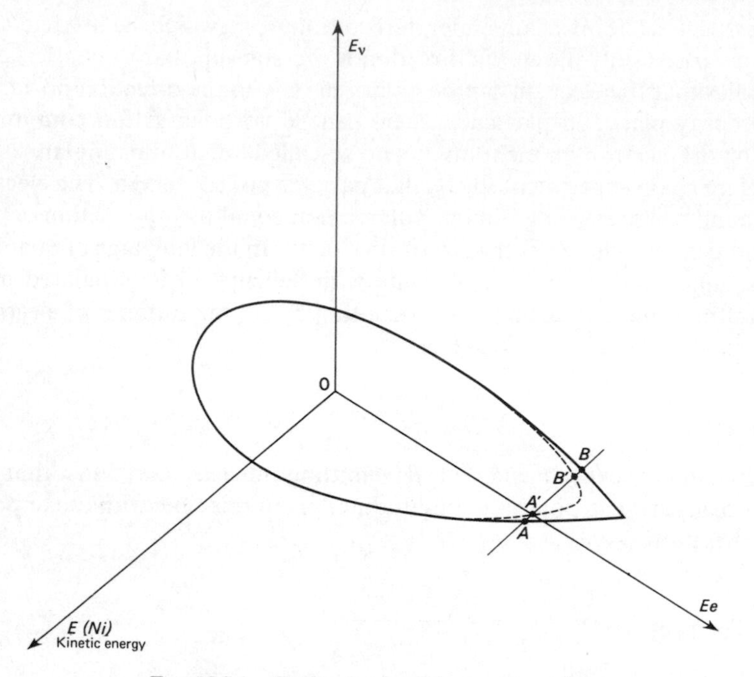

FIG. 15.2.1. Dalitz plot for β-decay of a nucleus.

mass, being proportional to phase space, i.e. to either AB or $A'B'$ as shown on figure 15.2.1. Careful measurements on the spectrum near its maximum can detect the neutrino mass. The spectrum will be linear if the mass is zero; otherwise the spectrum will be parabolic, with the curvature determining the neutrino mass. The most precise measurements have failed to indicate a mass, and we shall therefore adopt the simplest assumption, that the mass is exactly zero.

The same result with similar precision ($m_\nu < 1$ keV) can be obtained by comparing the maximum measured energy of the electron in nuclear β decay with the value predicted from the known energies of the initial and final nuclear states.

(b) *Neutrino spin*

Conservation of angular momentum in decays like

$$\pi \rightarrow \mu + \nu$$

requires the neutrino spin to be half-integer. We shall assume it to be $\frac{1}{2}$, which can be checked eventually in the course of the explicit construction of the weak-interaction Hamiltonian.

At this point it is useful to recall the results of chapter 4 about the spin states of zero-mass particles.

In view of the fact that neutrinos are produced exclusively by weak interactions which do not conserve parity, there is a priori no reason to attach any meaning to the action of the space-reflection operator on a single-neutrino state. In other words, it is reasonable to ask whether the helicity of the neutrino might not be a fixed quantity, remembering that for a zero mass particle an irreducible unitary representation of the Poincaré group (excluding parity) is uniquely determined by prescribing the momentum and the helicity. Thus, if we assume that the helicity of the neutrino is always $-\frac{1}{2}$, all neutrino states will be of the form

$$|\,\nu, \mathbf{p}_1\,, \lambda = -\tfrac{1}{2}\rangle$$

and one will never observe the state $|\nu, \mathbf{p}_1, \lambda = +\frac{1}{2}\rangle$.

Although it would be reasonable to call this assumption the one-component neutrino theory, it is in fact called the two-component neutrino theory because of the way it is formulated in field theory; we shall come to this presently.

The two-component neutrino theory has many consequences which one can test experimentally. Note that in a certain sense it implies that parity is violated maximally.

As one such consequence, consider π^+ decay

$$\pi^+ \rightarrow \mu^+ + \nu$$

and look at it in the pion rest frame, where the two daughter particles, μ^+ and ν, are emitted with opposite momenta. We explore the consequences of the fact that the neutrino helicity $-\frac{1}{2}$ is prescribed, by exploiting the conservation of angular momentum J_z along the neutrino direction. Since the neutrino and muon momenta are collinear, the orbital angular momentum cannot contribute to J_z, so that the final-state eigenvalue of J_z is just the difference between the neutrino and muon helicities. (To see that the difference and not the sum is appropriate, recall that the muon's helicity is the component of its spin along its momentum; and that its momentum is opposite in direction to the angular momentum quantization axis, which has been chosen parallel to the neutrino momentum.) Since the pion spin is zero, J_z is zero; this means that the μ^+ has the same helicity as the neutrino, whence it is *completely polarised longitudinally* in the pion rest frame, with helicity $-\frac{1}{2}$. We saw in chapter 11 how this polarisation is exploited to measure the muon gyromagnetic ratio. However, it is difficult to measure directly.

Much more can be learnt from muon decay, to which we now turn.

3. Muon decay*

(a) *Polarisation of muons at rest*

The two leptons μ^+ and μ^- decay according to

$$\begin{aligned} \mu^- &\to e + \nu_\mu + \bar{\nu}_e \\ \mu^+ &\to e^+ + \bar{\nu}_\mu + \nu_e \end{aligned} \tag{3.1}$$

Here we have indicated the difference between the electron-type neutrinos ν_e carrying electron number $+1$, and muon-type neutrinos carrying muon number $+1$. We explained in chapter 2 the experimental evidence for this distinction, and for the separate conservation of the two quantum numbers.

We now take the two-component neutrino theory as our working hypothesis for exploring these reactions. Thus we assume explicitly that neutrinos, ν_μ or ν_e, are always in states with helicity $-\frac{1}{2}$, while anti-neutrinos, $\bar{\nu}_\mu$ and $\bar{\nu}_e$, are always in states with helicity $+\frac{1}{2}$. We shall consider only muons from two-particle decays like

$$\begin{aligned} \pi^+ &\to \mu^+ + \nu_\mu \\ \pi^- &\to \mu^- + \bar{\nu}_\mu \\ K^+ &\to \mu^+ + \nu_\mu \\ K^- &\to \mu^- + \bar{\nu}_\mu \end{aligned} \tag{3.2}$$

As we have seen, at production they are completely polarised longitudinally. To avoid irrelevant technical details, we shall consider only experiments where the muons are not appreciably depolarised by passing through matter. Here it is important to realise that if, say, a μ^- produced in π^- decay is in a state with helicity $+\frac{1}{2}$, then it will be completely polarised also in its own rest frame. As we saw when studying helicity, a pure Lorentz transformation with relative velocity parallel to the μ momentum does not change the spin component in this direction. But the Lorentz transformation from the π to the μ rest frame does have just this property, so that in their own rest frame we can consider the μ's as completely polarised in a well-defined direction which we shall denote by Oz.

(b) *Polarisation of the electrons from muon decay*

Before describing the experimental results, we note that most electrons from muon decay are extreme-relativistic, because the muon's mass is more than two hundred times greater than the electron's. In the following

* Although 'β decay' denotes the process $\mu \to p + e + \bar{\nu}_e$, or related processes in nuclei, 'μ decay' denotes not a process where a muon is emitted, but where it decays: $\mu \to e + \bar{\nu}_e + \nu_\mu$.

we shall consider only such electrons, renouncing any information carried by the slow ones; we do this because it is much easier to discuss the spin states of an extreme-relativistic electron than those of a slow one. Indeed, as soon as the electron's energy much exceeds its mass, we can neglect the mass in first approximation and treat the electron as if it were a zero-mass particle.

The most striking experimental result is that the fast electrons emerging from μ^- decay are completely polarised longitudinally, in a direction opposite to their momentum. In other words, their helicity is always $-\frac{1}{2}$. By contrast, fast positrons emerging from μ^+ decay always have helicity $+\frac{1}{2}$.

This fact implies a remarkable property of the system consisting of the electron and its antineutrino, $(e, \bar{\nu}_e)$. Since we are treating the electron as a zero-mass particle, its helicity is practically invariant under Lorentz transformations, even under those that are not pure transformations with their velocity parallel to the electron momentum. Let us denote by (Σ) the centre of mass frame of the pair $(e, \bar{\nu}_e)$; then the electron helicity will remain equal to $-\frac{1}{2}$ even when we transform to the frame (Σ). By the two-component neutrino theory, the antineutrino has helicity $+\frac{1}{2}$. But in the frame (Σ) electron and antineutrino have equal and opposite momenta. Hence the total spin component of the pair along the electron momentum is always equal to -1; this is also the component of the total angular momentum, since the orbital angular momentum cannot contribute. From this it is natural to assume that the intrinsic angular momentum of the pair $(e, \bar{\nu}_e)$, which is a relativistic invariant, is always equal to 1. We shall see later how this leads to an explicit form of the weak-interaction Hamiltonian.

But before coming to that, we mention a test of the two-component neutrino theory.

(c) *A test of the two-component neutrino theory*

Let us confine ourselves to electrons with the maximum energy attainable in the muon rest frame. Such electrons can be emitted in directions Oz' making any arbitrary angle θ with the axis Oz along which the muons are polarised (Figure 15.3.1). But in order to maximise the electron energy, both neutrinos must have their momenta along Oz' and antiparallel to the electron; and both must have an absolute value equal to half the electron momentum.

From two-component neutrino theory, let us now calculate the component of the total angular momentum along Oz'. The helicity of the electron is $-\frac{1}{2}$; of the $\bar{\nu}_e$, $+\frac{1}{2}$; of the ν_μ, $-\frac{1}{2}$. Hence this component of the total angular momentum is $-\frac{1}{2}$. Consequently the entire final system is in

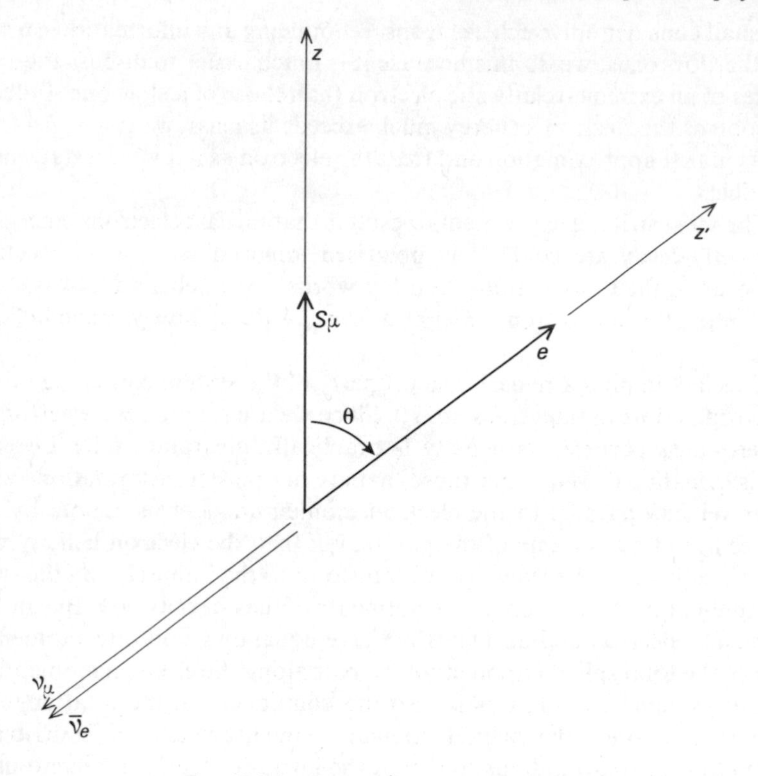

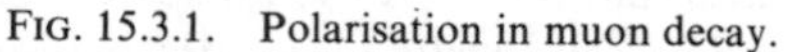

FIG. 15.3.1. Polarisation in muon decay.

the spin state

$$|\tfrac{1}{2}, -\tfrac{1}{2}\rangle_{z'},$$

where the first quantum number $\frac{1}{2}$ represents the total angular momentum, equal to the muon spin $\frac{1}{2}$. But we have seen that according to two-component neutrino theory the muon itself is in a state with spin $+\frac{1}{2}$ along the axis Oz, which we write as

$$|\tfrac{1}{2}, \tfrac{1}{2}\rangle_z .$$

Hence the amplitude for decay in the direction θ is proportional to the inner product

$${}_{z'}\langle\tfrac{1}{2}, -\tfrac{1}{2} \,|\, \tfrac{1}{2}\,\tfrac{1}{2}\rangle_z$$

which is simply a rotation matrix

$$d^{1/2}_{1/2-1/2}(\theta) = 1 - \cos\theta. \tag{3.3}$$

We see from this that the angular distribution of electrons with maximum energy is uniquely predicted by the two-component neutrino theory, if one takes into account also the experimental result about electron polarisation. The predicted distribution is indeed the one observed experimentally.

4. The weak-interaction Hamiltonian. Muon decay

(a) *Quantised fields*

Armed with the experimental confirmation of the two-component neutrino theory, we shall now try to establish the form of the weak-interaction Hamiltonian (also called the weak Hamiltonian for simplicity) responsible for μ decay; we shall rely on the assumption mentioned in the last section, that the intrinsic angular momentum of the $(e, \bar{\nu}_e)$ pair is always 1.

It seems advisable to use the formalism of field theory, in view of the fact that all the particles involved, $(\mu, e, \nu_e, \bar{\nu}_\mu)$, are created or destroyed in the reaction, and that we want to secure manifest relativistic invariance as far as possible. Hence we recall the quantised fields associated with the electron and the μ; they are expressed in terms of the corresponding creation and annihilation operators by the formulae

$$\psi_e(x) = \frac{1}{(2\pi)^{3/2}} \sum_r \int [u^{(r)}(\mathbf{q})\, a_e^{(r)}(\mathbf{q})\, e^{iqx} + v^{(r)}(\mathbf{q})\, b_e^{+(r)}(\mathbf{q})\, e^{-iqx}] \frac{d^3q}{\sqrt{q^0}}$$

$$\psi_\mu(x) = \frac{1}{(2\pi)^{3/2}} \sum_r \int [u^{(r)}(\mathbf{q})\, a_\mu^{(r)}(\mathbf{q})\, e^{iqx} + v^{(r)}(\mathbf{q})\, b_\mu^{+(r)}(\mathbf{q})\, e^{-iqx}] \frac{d^3q}{\sqrt{q^0}} \tag{4.1}$$

Here, $a_e^{(r)}(\mathbf{q})$ for instance is the annihilation operator for an electron with momentum $\mathbf{q}$ in a polarisation state (r); the latter is specified simply by the corresponding Dirac spinor $u^{(r)}(\mathbf{q})$. Similarly, $b_e^{+(r)}(\mathbf{q})$ is the creation operator for a positron with momentum $\mathbf{q}$ in a polarisation state (r). Recall that the spinor $v^{(r)}(\mathbf{q})$ is defined by

$$v_\alpha^{(r)}(\mathbf{q}) = C_{\alpha\beta}\bar{u}_\beta(\mathbf{q}) \tag{4.2}$$

where the matrix C is the one introduced in chapter 10.

In chapter 10 we saw that the wavefunction of a two-component neutrino can be expressed in the Dirac formalism by a four-component spinor satisfying the condition

$$(1 - \gamma_5)\, u(\mathbf{q}) = 0 \tag{4.3}$$

This amounts to considering an ordinary spinor $u(\mathbf{q})$ subject to the Dirac equation

$$(\gamma_\mu q_\mu)\, u(\mathbf{q}) = 0 \tag{4.4}$$

and never using any combination other than the quantity $(1 + \gamma_5)u(\mathbf{q})$; for a zero-mass particle this is equivalent to projecting onto the invariant state with helicity $-\frac{1}{2}$.

Let us now apply this two-component description of the neutrino to the wavefunction $u(\mathbf{q})$ occurring in the expression for the neutrino field analogous to (4.1); we see that the neutrino can be described by an ordinary field, i.e. by

$$\psi_\nu(x) = \frac{1}{(2\pi)^{3/2}} \sum_r \int [e^{iqx} u^{(r)}(\mathbf{q})\, a_\nu^{(r)}(\mathbf{q}) + e^{-iqx} v^{(r)}(\mathbf{q})\, b^{+(r)}(\mathbf{q})] \frac{d^3q}{\sqrt{q^0}} \tag{4.5}$$

where the wavefunction $u(\mathbf{q})$ obeys the zero-mass Dirac equation (4.4) and where $v(\mathbf{q})$ is defined by (4.2). To specify the polarisation states in the zero mass case, obviously one choses the wavefunctions $u^{(r)}(\mathbf{q})$, $(r = 1, 2)$, with helicities $\lambda = \pm\frac{1}{2}$. Then the operator $a_\nu^{(r)}(\mathbf{q})$ destroys a neutrino with momentum $\mathbf{q}$ and prescribed helicity. Next, by constructing the operator $(1 + \gamma_5)\psi_\nu(x)$ one selects automatically from (4.5) that part $(1 + \gamma_5)u(\mathbf{q})$ of the wavefunction which is nonzero only for helicity $-\frac{1}{2}$. We see that in this way the two-component neutrino is indeed described by the field $(1 + \gamma_5)\psi(x)$.

As regards the antineutrino, we see that its wavefunction is obtained from (4.2) by replacing v by $[(1 + \gamma_5)/2]v(\mathbf{q})$. In view of the explicit expressions for γ_5 and C given in chapter 10, this amounts to keeping for the antineutrino only the components with helicity $+\frac{1}{2}$. To sum up, the states both of a neutrino with helicity $-\frac{1}{2}$, and of an antineutrino with helicity $+\frac{1}{2}$, can be described simultaneously by the projection $(1 + \gamma_5)\psi(x)$ of a Dirac field.

(b) *The weak-interaction Hamiltonian*

Next, consider how to construct the weak Hamiltonian from the quantised fields. As our first clue we take the assumption, already encountered, that the $(e, \bar{\nu}_e)$ pair emitted in μ decay has intrinsic angular momentum 1. But we have seen how to construct, from two Dirac spinors, certain quantities transforming like various tensor components. In particular, from the electron and neutrino fields we can construct a vector

$$\bar{\psi}_e(x)\, \gamma_\lambda \psi_\nu(x) \tag{4.6}$$

and an axial vector

$$\bar{\psi}_e(x)\, \gamma_\lambda \gamma_5 \psi_\nu(x). \tag{4.7}$$

Acting on the vacuum, an operator like (4.6) creates, by virtue of the creation operators in the fields, a state consisting of an electron and an antineutrino. Since the vacuum state is a scalar and (4.6) is a vector, the state

$$\bar{\psi}_e(x)\, \gamma_\lambda \psi_\nu(x) \mid 0\rangle \tag{4.8}$$

is itself a four-vector. This must be reflected by simple total-angular-momentum properties of the electron-antineutrino pair. By explicit calculation, or by group theory, one can indeed show that the $(e, \bar{\nu}_e)$ pair described by (4.8) is in a superposition of states with total angular momenta $J = 1$ and $J = 0$; and that only the part with $J = 1$ survives for electrons with high energy. Hence it is tempting to assume that the weak Hamiltonian involves the electron and neutrino fields only in the combinations (4.6) and (4.7). But in order that neutrinos should have only one helicity component, the neutrino field must appear only in the combination $(1 + \gamma_5)\psi_\nu(x)$; it follows that the Hamiltonian must depend only on a well-defined combination of (4.6) and (4.7) given by

$$\bar{\psi}_e(x)\,\gamma_\lambda(1 + \gamma_5)\,\psi_{\nu_e}(x). \tag{4.9}$$

We note at once that this expression implies automatically the experimentally observed fact that high-energy electrons are created in a state of helicity $-\frac{1}{2}$. Since γ_5 and γ_μ anticommute, we see that (4.9) involves only the component $\bar{\psi}_e(1 - \gamma_5) = [(1 + \gamma_5)\psi_e(x)]^+\gamma_0$; hence the electron field, like the neutrino field, enters only in the combination $(1 + \gamma_5)\psi_e(x)$. This ensures that electrons, like neutrinos, appear only with helicity $-\frac{1}{2}$ in the high-velocity limit, where their mass can be dropped from the Dirac equation.

In order to complete the construction of the weak Hamiltonian, we must introduce the fields of the μ and of its antineutrino. Here we must use once again the combination $(1 + \gamma_5)\psi_\nu(x)$ for the neutrino, and from the fields construct a vector to compensate the vector character of (4.9). The simplest expression satisfying these conditions is

$$\begin{aligned} \mathrm{H} = \frac{g_\mu}{\sqrt{2}} \int d^3x[\bar{\psi}_e(x)\,\gamma_\lambda(1 + \gamma_5)\,\psi_{\nu_e}(x)][\bar{\psi}_{\nu_\mu}(x)\,\gamma_\lambda(1 + \gamma_5)\,\psi_\mu(x)] \\ + \text{Hermitean conjugate} \end{aligned} \tag{4.10}$$

We have indicated that one must add the Hermitean conjugate quantity (describing μ^+ decay) in order to make H Hermitean. The coefficient $1/\sqrt{2}$ is present simply by convention.

To see whether the suggested Hamiltonian (4.10) is the right one, we need merely calculate the μ decay probability by first order perturbation theory; since the matrix elements of the creation and annihilation operators are known, this is just a matter of routine. The result is good agreement between theory and experiment in every detail, and one finds the following value for g:

$$g_\mu = (1{,}434 \pm 0{,}001) \times 10^{-49} \text{ c.g.s.} \tag{4.11}$$

or

$$g_\mu \simeq 10^{-5}/(\text{proton mass})^2.$$

5. Currents

The Hamiltonian for μ decay appears to be constructed from simpler objects like

$$\bar{\psi}_e\gamma_\lambda(1+\gamma_5)\,\psi_{\nu_e} \qquad (5.1)$$
$$\bar{\psi}_\mu\gamma_\lambda(1+\gamma_5)\,\psi_{\mu_\nu}$$

which are strongly reminiscent of that part of the electromagnetic interaction Hamiltonian which multiplies the electromagnetic field, namely of

$$j_\lambda(x) = e\bar{\psi}_e(x)\,\gamma_\lambda\psi_e(x) \qquad (5.2)$$

As we learn more about the weak Hamiltonian, such quantities will become increasingly important; hence we shall investigate the operators (5.1) more closely and, to do this, we start with the electromagnetic current (5.2).

The electromagnetic current $j_\lambda(x)$ is a local operator, in the sense that it depends only on a single space-time point x, and that the commutation relations of the quantised electron field entail the causality condition

$$[j_\lambda(x), j_\mu(y)] = 0 \qquad (5.3)$$

when the space-time points x and y are spacelike separated. The consequences of (5.3) have been studied in detail by Bohr and Rosenfeld; they showed that when a system of elementary charges is in a state α under conditions where it can be treated macroscopically in a classical approximation, the matrix element

$$\langle\alpha\,|\,j_\lambda(x)\,|\,\alpha\rangle, \qquad (5.4)$$

is simply the classical current density. This is why $j_\lambda(x)$ is called the electromagnetic current operator. Further, the electromagnetic interaction Hamiltonian has the form

$$\mathrm{H}_1 = \int j_\mu(x)\,A_\mu(x)\,d^3x \qquad (5.5)$$

which is exactly the form encountered in classical electrodynamics, where j_μ is interpreted as the (classical) current density.

If (5.2) really is the current density, then the total charge contained in a well-defined state α must be given by the integral of the charge density over all space:

$$Q_\alpha(t) = \int \langle\alpha\,|\,j_0(\mathbf{x}, t)\,|\,\alpha\rangle\,d^3x. \qquad (5.6)$$

The constancy in time of this total charge is ensured by the classical condition of current conservation

$$\partial_\lambda\,j_\lambda(x) = 0. \qquad (5.7)$$

Indeed, assume that the condition (5.7) is satisfied and that the state α is roughly localised, in the sense that the current density (5.4) vanishes fast enough as **x** tends to infinity. Then one has

$$\frac{\partial Q_\alpha}{dt} = \int \langle\alpha \mid \partial_0 j_0(\mathbf{x}, t) \mid \alpha\rangle\, d^3x = -\int \operatorname{div}\langle\alpha \mid \mathbf{j}(\mathbf{x}, t) \mid a\rangle\, d^3x = 0$$

where the first step results from (5.7) and the second from the rapid decrease of (5.2) at infinity, combined with Stokes' theorem.

If one wants to take into account all the charges that can appear in electrodynamics, then it is obvious that protons and other hadrons will enter as well as electrons. Because of the analogy between protons and electrons one might hope to take the former into account simply by adding to the electromagnetic Hamiltonian a term

$$-e \int \bar{\psi}_P(x)\, \gamma_\lambda \psi_P(x)\, A_\lambda(x)\, d^3x \tag{5.8}$$

where $\psi_P(x)$ is the quantised proton field. The Feynman rules would reflect such an interaction through the possibility of direct couplings between protons and photons, described by vertices like those in figure (15.5.1).

But the situation cannot be so simple. The proton has strong interactions and can emit a π^+ which in turn interacts with a photon according to the diagram (15.5.2). Because the strong interactions are strong, one would need to take into account an infinite variety of strong-interaction diagrams; it is better to avert one's eyes by letting figure (15.5.3) symbolise the sum of all diagrams whose only external lines are one photon and two protons. Algebraically, this amounts to replacing the explicit interaction

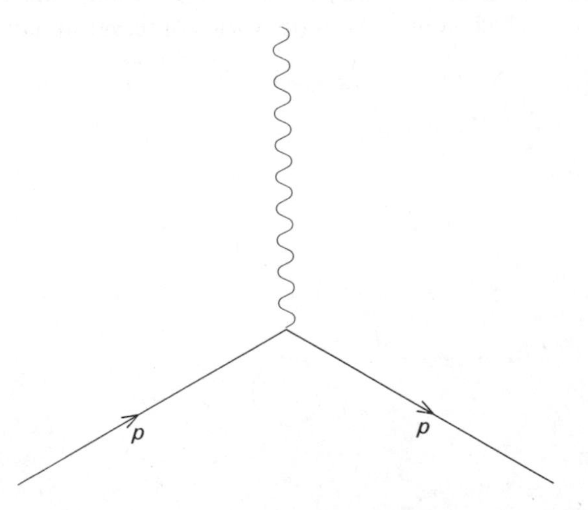

FIG. 15.5.1. The elementary vertex for proton-photon interaction.

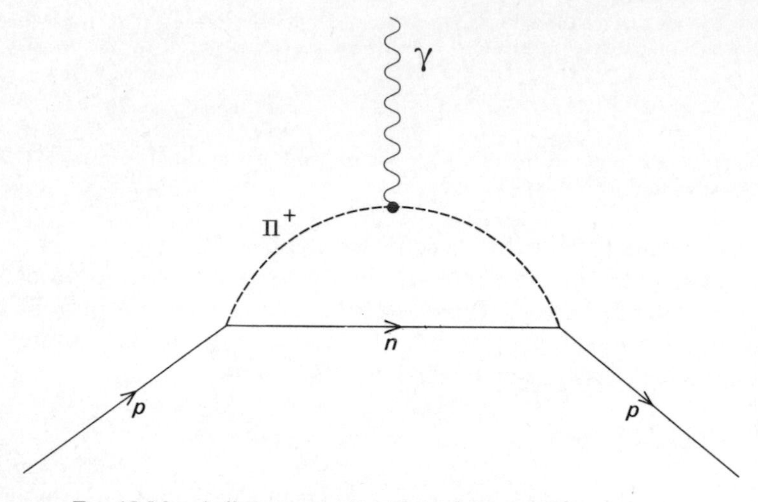

FIG. 15.5.2. A diagram contributing to the proton-photon vertex.

Hamiltonian (5.8) by

$$H_1 = \int j_\lambda{}^H(x)\, A_\lambda(x)\, d^3x \tag{5.9}$$

where $j_\lambda{}^H(x)$ is the current operator for all hadrons. We know little about it beyond the condition

$$\partial_\lambda\, j_\lambda{}^H(x) = 0 \tag{5.10}$$

imposed by charge conservation. Finally one can add the hadron current $j_\lambda{}^H(x)$ to the electron and muon currents to obtain the total current operator $j_\lambda(x)$, which continues to obey the conservation law

$$\partial_\lambda\, j_\lambda(x) = 0 \tag{5.11}$$

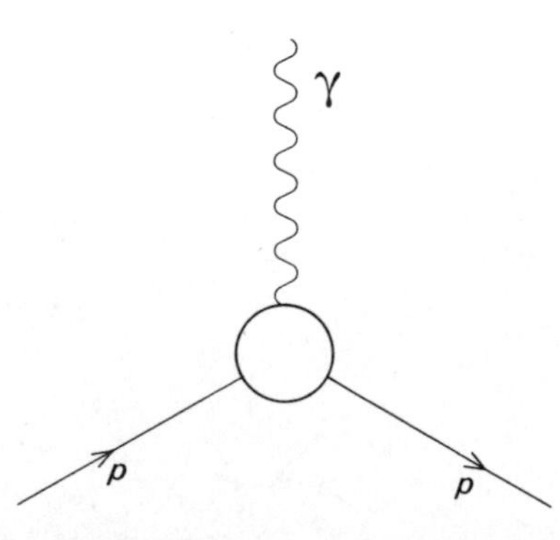

FIG. 15.5.3. Symbolic representation of the complete proton-photon vertex.

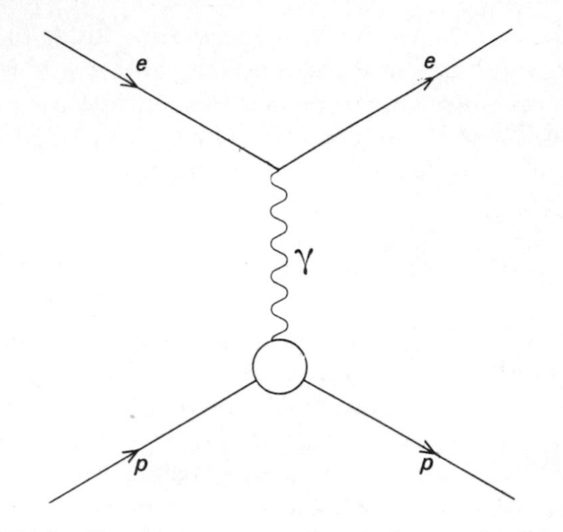

FIG. 15.5.4. Electron-proton scattering: the lowest-order diagram.

This condition has the extremely important consequence of ensuring the strict equality between charges that is needed for overall charge conservation. Thus the validity of (5.11) in the presence of weak interactions ensures charge conservation in the decays, ensuring thereby the equality (up to sign) between the proton and electron charges, which are linked via neutron decay. More generally the conservation law (5.11) entails, via charge conservation, that the charges of all elementary particles are multiples of a common universal unit, since particles appear in reactions necessarily in finite numbers. In other words, (5.11) expresses the *universality of electric charge*.

The hadron current operator can be explored in practice by investigating the scattering of electrons by protons and neutrons. To first order in the fine structure constant α this is described by the diagram (15.5.4); hence the matrix element of the current between single-proton states,

$$\langle p \mid j_\mu{}^H(x) \mid p\rangle \tag{5.12}$$

can be deduced from the experimental electron-proton scattering cross-section. If the proton had no spin, and if we denote its initial and final momenta by q_1 and q_2, then relativistic invariance would entail that this matrix element can be written in the form

$$(q_1 + q_2)_\mu \times e^{i(q_1 - q_2)J} F(t) \tag{5.13}$$

where $t = (q_1 - q_2)^2$. The function $F(t)$ is called the form factor. Basically, it determines the spatial distribution of charge inside the proton. The proton spin complicates formula (5.13) somewhat, making it necessary to introduce not just one but two form factors, corresponding essentially to the distributions of charge and of magnetisation inside the proton. The development of our present theory of strong interactions has been much influenced by measurements of the form factors and by their theoretical interpretation. But at the moment we are more concerned with the weak interactions, and it is time we returned to them.

6. The weak-interaction Hamiltonian

Let us return to β decay. Since the weak interactions can be treated by first order perturbation theory, it is again comparatively easy to deduce the Hamiltonian from the measured decay rates, angular correlations, and electron polarisations. Such measurements have been made on neutron decay, and for the β^+ and β^- decays of a good many nuclei. Note from the outset that in such decays the nucleons never accelerate to high velocity, but remain nonrelativistic; this makes it a delicate matter to determine the exact relativistic form of the Hamiltonian, and requires very high precision in some of the measurements.

Since the β processes involve the creation and annihilation of particles, it will again be convenient to describe them by means of the quantised fields associated with the electron, the neutrino, the proton, and the neutron. At this stage one must define precisely what one means by these fields. We saw in chapter 12 that particles entering and leaving a reaction could be described by free fields, which we denoted by $\Phi^{\text{in}}(x)$ and $\Phi^{\text{out}}(x)$. We can always use these fields since their introduction amounts to nothing more than a description of initial and final states by the formalism of creation and annihilation operators. By contrast, the interpolating field $\Phi(x)$ which links $\Phi^{\text{in}}(x)$ and $\Phi^{\text{out}}(x)$ is not known in practice until we can calculate strong interactions in full. But the difference between $\Phi(x)$ and $\Phi^{\text{in}}(x)$, say, can matter only when the structure of the particles is important, or, if we prefer, only when the decay process involves a high momentum transfer between an initial and a final nucleon. Indeed, taking strong interactions into account amounts roughly to taking into account the emission and reabsorption of virtual pions by the nucleus. But we saw in our analysis of nuclear forces that such pions cannot travel further from the nucleus than to a distance of the order of their Compton wavelength $\hbar/m_\pi c$, roughly one fermi (10^{-13} cm), which gives the order of magnitude of the nucleon radius. If, as in β decay, the momentum transfer between nucleons is much less than the pion mass m_π, then the de Broglie wavelengths of the nucleons are much longer than their radii, and therefore their structure will play no role. This means that in such reactions the nucleons behave like point particles without strong interactions, and that as far as they are concerned one need not distinguish between incoming, outgoing, and interpolating fields. Now that we have elucidated this semantic point, we can write down the β decay Hamiltonian deduced from experiment:

$$H_\beta = \frac{g}{\sqrt{2}} \int d^3x[\bar{\psi}_p(x)\,\gamma_\lambda\psi_n(x) + G_A\bar{\psi}_p(x)\,\gamma_\lambda\gamma_5\psi_n(x)][\bar{\psi}_e\gamma_\lambda(1+\gamma_5)\,\psi_{\nu_e}(x)]$$
$$+ \text{Hermitian conjugate} \tag{6.1}$$

where we have put

$$g_\beta = (1{,}418 \pm 0{,}004) \times 10^{-49}\ \text{erg-cm}^3 \tag{6.2}$$
$$G_A = 1{,}18 \pm 0{,}03. \tag{6.3}$$

We note a striking analogy between the forms of the Hamiltonians for μ decay and for β decay. Indeed, the coupling constants g_β and g_μ differ by less than 2%, though the difference is significant and outside the limits of error. If one equates g_β and g_μ nevertheless, the Hamiltonians (4.10) and (6.1) are seen to differ only insofar as G_A differs from 1.

So far, we have not tried to describe those parts of the weak Hamiltonian which are responsible for leptonic hyperon decays like

$$\begin{aligned} &\Lambda \to p + e + \bar{\nu}_e \\ &\Sigma^+ \to n + e^+ + \nu_e \\ &\Sigma^- \to n + e + \bar{\nu}_e \end{aligned} \tag{6.4}$$

for leptonic meson decays:

$$\begin{aligned} &\pi^+ \to \mu^+ + \nu_\mu \\ &\pi^+ \to e^+ + \nu_e \\ &K^- \to \mu^- + \bar{\nu}_\mu \quad \text{etc...} \end{aligned}$$

and for nonleptonic decays

$$\Lambda \to p + \pi \qquad K \to \pi + \pi \qquad K \to \pi + \pi + \pi \quad \text{etc...}$$

We shall now try to see whether it is possible to describe all these decays simultaneously by drawing for inspiration on the structure of those Hamiltonians which are already known. This will cause us to consider once again the remarkable equality between g_β and g_μ, and to revert to the interesting constant G_A.

7. Structure of the weak-interaction Hamiltonian

The β-decay Hamiltonian might allow one to describe also other processes like, say, charged pion decay. Since, so far, we have considered only β decay, where the only leptons to take part are electrons and their corresponding neutrinos, we shall concentrate first on the decay of $\pi^{\pm}$ into electron and neutrino:

$$\begin{aligned} \pi^- &\to e^- + \bar{\nu}_e \\ \pi^+ &\to e^+ + \nu_e \,. \end{aligned} \tag{7.1}$$

The strong interactions imply that these processes are not a priori independent of β decay. One could conceive that through the strong interactions the π^+ is converted into a virtual proton-antineutron pair, that the proton decays by β^+ decay into a positron, a neutrino, and a neutron, which then annihilates with the antineutron. The corresponding diagram is shown in figure 15.7.1.

Evidently this is not the only process whereby one can connect β decay and pion decay, and perhaps it would not be very realistic to accept it automatically as the dominant mechanism. The important point is the suggestion arising from this diagram, that the Hamiltonian H_β which we already know might be responsible for π^+ decay. Note first that in a diagram like (7.1) there is a large momentum transfer between proton and neutron. Let q_1 and q_2 be the four-momenta of the nucleons. Then the momentum transfer is $t = (q_1 - q_2)^2$; on the other hand, $-q_2$ can also be considered as the antineutron momentum in the virtual reaction $\pi^+ \to p + \bar{n}$, whence t is actually equal to the squared pion mass. Therefore, to analyse pion decay, we must know the β decay Hamiltonian in a form which is more general than (6.1) and which is applicable for large momentum transfers between the nucleons.

Once again we can rely on an analogy with electrodynamics. We saw that the operator

$$e\bar{\psi}_p(x)\,\gamma_\mu\psi_p(x) \tag{7.2}$$

has a well-defined meaning when one considers only small momentum transfers, but that it becomes ambiguous in cases where the momentum transfer is higher and where one must distinguish between incoming,

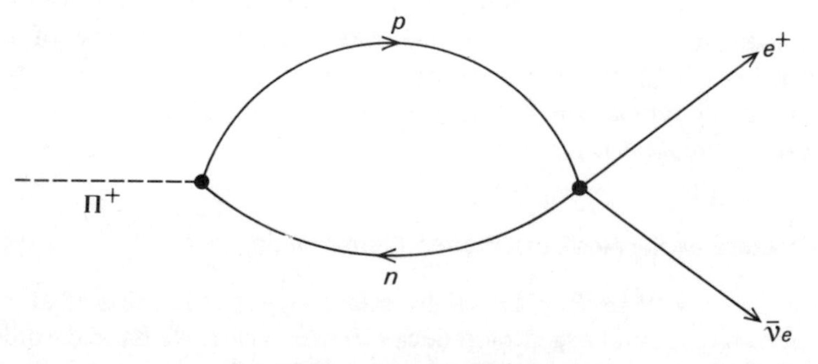

FIG. 15.7.1. A model of π^+ decay.

outgoing, and interpolating fields. Nevertheless there does exist a well-defined current operator $j_\lambda(x)$ which constitutes a well-defined generalisation of the operator (7.2), and which is the one to be used in electrodynamics. In π^- decay the strong interactions will enter only through matrix elements of the kind

$$\langle 0 \mid \frac{g}{\sqrt{2}} [\bar{\psi}_p(x)\, \gamma_\mu \psi_n(x) + G_A \bar{\psi}_p(x)\, \gamma_\mu \gamma_5 \psi_n(x)] \mid \pi^- \rangle \tag{7.3}$$

which it is our task to define correctly. The first term of (7.3) is a vector operator, and by virtue of the Wigner–Eckart theorem it cannot have a non-zero matrix element between the vacuum and the pseudoscalar pion state. Hence in this case only the second term needs to be kept and defined.

We now propose that this term should be understood as an axial vector current $J_\mu{}^A(x)$, which can be written explicitly in the form

$$\frac{g}{\sqrt{2}} G_A \bar{\psi}_p(x)\, \gamma_\mu \gamma_5 \psi_n(x)$$

when taking its matrix element between nucleon states at small momentum transfers. (In this notation the superfix A means 'axial'.) The pion decay amplitude is proportional to the matrix element

$$\langle 0 \mid J_\mu{}^A(x) \mid \pi^- \rangle \tag{7.4}$$

where $J_\mu{}^A$ is the operator which in the context of β decay was written out more explicitly as the second term of (6.1).

Can one test this assumption about the structure of weak interactions in terms of currents? To see how, let q be the four-momentum of the pion, and note to begin with that by virtue of relativistic invariance the matrix element (7.4) can be written in the form

$$\langle 0 \mid J_\mu{}^A(x) \mid \pi \rangle = f q_\mu\, e^{iqx}. \tag{7.5}$$

The quantity f can be determined experimentally by measuring the decay rate of the π into $e + v$. Our assumption is tested by noting that the structure of the current $J_\mu(x)$ depends only on the strong interactions; the weak interactions simply fix its normalisation, which for zero momentum transfer is proportional to $gG_A/\sqrt{2}$. One can then use dispersion relations to establish a connection, called the Goldberger–Treiman formula, between the quantities f, g, G_A, the nucleon mass m, and the pion-nucleon coupling constant $g_{NN\pi}$ (whose value is given by $g_{NN\pi}^2/4\pi = 14.6$). It reads

$$2m \frac{g}{\sqrt{2}} G_A = -f g_{NN\pi}. \tag{7.6}$$

This formula depends, as well as on other assumptions, also on the structure of the weak interactions in terms of currents; it is well satisfied experimentally, the discrepancy being of the order of 10% which is satisfactory theoretically.

The current-current structure

We are now in a position to put forward a form of the weak Hamiltonian. With each kind of lepton we associate a vector and an axial vector current:

$$j_\alpha^{(e)}(x) = \bar{\psi}_e(x)\,\gamma_\alpha \psi_{\nu_e}(x) + \text{H.c.} \tag{7.7}$$

$$j_\alpha^{(\mu)}(x) = \bar{\psi}_\mu(x)\,\gamma_\alpha \psi_{\nu_\mu}(x) + \text{H.c.} \tag{7.8}$$

$$\bar{j}_\alpha^{(e)}(x) = \bar{\psi}_e(x)\,\gamma_\alpha \gamma_5 \psi_{\nu_e}(x) + \text{H.c.} \tag{7.9}$$

$$\bar{j}_\alpha^{(\mu)}(x) = \bar{\psi}_\mu(x)\,\gamma_\alpha \gamma_5 \psi_{\nu_e}(x) + \text{H.c.} \tag{7.10}$$

Then the Hamiltonian for μ decay can be written in the condensed form

$$\frac{g_\mu}{\sqrt{2}} \int (j_\alpha^{(\mu)}(x) + \bar{j}_\alpha^{(\mu)}(x))(j_\alpha^{(e)}(x) + \bar{j}_\alpha^{(e)}(x))\, d^3x. \tag{7.11}$$

To generalise this expression we introduce currents for the set of all leptons:

$$j_\alpha^{(\ell)}(x) = j_\alpha^{(e)}(x) + j_\alpha^{(\mu)}(x) \tag{7.12}$$

$$\bar{j}_\alpha^{(\ell)}(x) = \bar{j}_\alpha^{(e)}(x) + \bar{j}_\alpha^{(\mu)}(x) \tag{7.13}$$

we introduce also a vector current $j_\mu^{(H)}(x)$ and an axial vector current $\bar{j}_\mu^{(H)}(x)$ for the hadrons. For small momentum transfers between hadrons, these currents reduce, respectively, to

$$j_\alpha^{(H)} \to \bar{\psi}_n(x)\,\gamma_\alpha \psi_p(x) + \text{H.c.} \tag{7.14}$$

$$\bar{j}_\alpha^{(H)} \to G_A \bar{\psi}_n(x)\,\gamma_\alpha \gamma_5 \psi_p(x) + \text{H.c.} \tag{7.15}$$

Finally we introduce total vector and axial vector currents, including both leptons and hadrons:

$$J_\alpha = j_\alpha^{(\ell)} + j_\alpha^{(H)} \tag{7.16}$$

$$\bar{J}_\alpha = \bar{j}_\alpha^{(\ell)} + \bar{j}_\alpha^{(H)} \tag{7.17}$$

and write the weak Hamiltonian as

$$\frac{G}{\sqrt{2}} \int (J_\alpha(x) - \bar{J}_\alpha(x))(J_\alpha - \bar{J}_\alpha(x))\, d^3x. \tag{7.18}$$

Note that this Hamiltonian includes simultaneously those for μ decay as well as for β decay. In constructing it we have for the moment ignored the fact that the β decay coupling constant is slightly different from that for μ decay, and have equated both to the same constant called G. Clearly, the Hamiltonian (7.18) has matrix elements for the decay of a hadron into leptons, like

$$\pi \to \mu + \nu \tag{7.19}$$

through the agency of terms like $j_\alpha^{(H)} j_\alpha^{(\ell)}$. Similarly it has matrix elements for non-leptonic decays involving only hadrons in initial and final states, through the agency of terms like $j_\alpha^{(H)} j_\alpha^{(H)}$. In the remainder of this chapter we shall investigate the consequences of this assumption.

Note that all the currents, hadronic as well as leptonic, have been constructed from the fields of one charged and one neutral particle. It follows that, in contrast to the electromagnetic current which has matrix elements only between states with the same charge, the currents we have introduced here have non-zero matrix elements only between states which differ in charge by one unit.

8. Weak magnetism

(a) *The universality of lepton couplings*

We have assumed a weak Hamiltonian which involves leptonic currents written directly as sums of electronic and muonic currents with equal coefficients. One says that the leptonic weak interactions have a universal coupling, and this property is called the universality of weak interactions. It can be tested by a direct comparison between μ and electron couplings in parallel processes like, say, the two charged-pion decay modes:

$$\pi \to \mu + \nu \tag{8.1}$$

$$\pi \to e + \nu. \tag{8.2}$$

We introduce Dirac spinors to describe the kinematic state of the leptons; by using the above form of the weak Hamiltonian, it is then easy to see that the probability amplitudes for these two decays are of the form

$$\frac{G}{\sqrt{2}} \langle 0 \mid \bar{j}_\alpha^{(H)}(0) \mid \pi \rangle \, \bar{u}^{(\mu, e)} \gamma_\alpha (1 + \gamma_5) \, u^{(\nu_\mu, \nu_e)} \tag{8.3}$$

Strong interactions can affect only the matrix element of the hadronic axial vector current between the vacuum and single-pion states. This is just a number f_π:

$$\langle 0 \mid \bar{j}_\alpha^{(H)}(0) \mid \pi \rangle = f_\pi q_\alpha^{(\pi)}, \tag{8.4}$$

the same for both processes (8.1) and (8.2). To sum up, by comparing the decay rates of the pion into $e + \nu$ and $\mu + \nu$, we can determine the ratio between the coupling constants of the weak electronic and muonic currents to the axial vector hadronic current; thus their equality can be verified experimentally.

(b) *Universality of the hadronic vector coupling*

Let us pursue further the near-equality between g_β and g_μ, which we have already merged into the same constant G. Clearly we are faced with a certain difference between the vector and axial vector couplings of the hadrons. The vector couplings take effect with a strength g_β which differs only little from g_μ. But while the axial vector coupling equals the vector coupling insofar as leptons are concerned, (this is expressed simply by the factor $(1 + \gamma_5)$ in the Dirac description of two-component neutrinos), the two couplings differ by the factor $g_A = 1.18$ in the case of non-relativistic nucleons.

Actually we should not be surprised by such a difference between hadrons and leptons, if we keep in mind their different properties with respect to quantum field theory. As we have often seen already, the strong interactions endow the hadrons with structure. Such structure has a marked effect on their electromagnetic properties, e.g. on the magnetic moments of proton and neutron, and we must expect an equally strong influence on the weak currents. This is often expressed by saying that a priori the strong interactions should appreciably renormalise the weak coupling constant. From this point of view, one is surprised not by the difference between the axial vector couplings of hadrons and leptons, but rather by the near-equality of the vector couplings; the latter suggests some regularity which calls for an explanation.

In fact there exists another and even more striking equality between lepton and hadron couplings; namely the equality of their respective electromagnetic couplings, or in other words, between the magnitudes of electron and proton electric charges. Here, too, one is dealing with identical couplings of two vector currents, namely of the electronic and the hadronic components of the electromagnetic current.

The fact that electron and proton charges are equal, or at least that they are commensurable (that their ratio is a rational number), is a necessary condition for charge conservation. Thus it reflects a conservation law, which leads us to look more closely at the correspondence between conservation laws and currents in general.

(c) *Currents and conservation laws*

One knows that in classical field theory there is a direct connection between invariance properties of the dynamics and the existence of currents related to conserved quantities. This connection is formulated most precisely by Noether's theorem; for a *classical* dynamical system it can be expressed in terms of fields $\Phi_r(x)$, where the index r enumerates the independent degrees of freedom of the field. In practice, r is a spinor or tensor index indicative of the relativistic transformation properties of the field. Assume that the dynamical system may be described by a Lagrangian density depending only on the fields Φ_r and on their derivatives:

$$\Phi_{r,\mu}(x) = \partial_\mu \Phi_r(x) = \frac{\partial \Phi_r(x)}{\partial x_\mu}. \tag{8.5}$$

Denote this Lagrangian density by

$$L(\Phi_r(x), \Phi_{r,\mu}(x)) \tag{8.6}$$

Noether's theorem applies under the assumption that the Lagrangian density (8.6) is invariant under a simultaneous transformation of the fields $\Phi_r(x)$ and of the coordinates x_μ under the operations of a group G:

$$\Phi_r(x) \to \Phi_r'(x') = \Phi_r(x) + \delta\Phi_r(x) \tag{8.7}$$

$$\delta\Phi_r(x) = M_r^{(\alpha)}\, \delta\omega^\alpha$$

$$x_\mu \to x_\mu' = x_\mu + \delta x_\mu \tag{8.8}$$

$$\delta x_\mu = \Lambda_\mu^{(\alpha)}\, \delta\omega^\alpha$$

where the ω^α are the infinitesimal parameters of the group.

One obvious example of such a group G are the Lorentz transformations.

Noether's theorem concludes that under these conditions there exists a conserved current, or rather as many conserved currents as there are independent parameters of the group G; they are defined explicitly by the formula

$$j_\mu{}^\alpha(x) = \frac{\partial L}{\partial \Phi_{r,\mu}} [\Lambda_\mu^{(\alpha)}\, \partial_\mu \Phi_r - M_r^{(\alpha)}] - L\Lambda_\mu^{(\alpha)} \tag{8.9}$$

The conservation of these currents is expressed, just like that of the electromagnetic current, by the condition

$$\partial_\mu\, j_\mu{}^\alpha = 0. \tag{8.10}$$

The conservation laws can be displayed equally well in integral form rather than in the differential form (8.10); one simply introduces the quantities

$$Q_{(t)}^{(\alpha)} = \int_{t=\text{const}} j_0^{(\alpha)}(\mathbf{x}, t)\, d^3x \tag{8.11}$$

Then Noether's theorem states that the $Q^{(\alpha)}(t)$ are constant in time. When $j_\mu^{(\alpha)}$ is the electromagnetic current, $Q^{(\alpha)}(t)$ is obviously just the total electric charge of the system.

It is worth noting another classical example, where G is group of translations

$$x_\mu \to x_\mu + a_\mu$$

the group parameters being the component of the vector a_μ. To every translation along a given coordinate axis α, the formula (8.9) assigns a current $j_\mu^{(\alpha)}(x)$; in this particular case the indices μ and (α) are physically similar, and we exploit their similarity by writing the current as $\theta_{\mu\alpha}(x)$. Indeed, since α is a vector index, $\theta_{\mu\alpha}(x)$ is a second rank tensor density, and the conservation laws (8.10) assume the form

$$\partial_\mu \theta_{\mu\alpha}(x) = 0. \tag{8.12}$$

$\theta_{\mu\nu}$ is called the energy-momentum tensor, because the conserved quantities (like 8.11) associated with it are simply the total four-momentum components of the system:

$$P_\alpha = \int_{t=\text{const.}} \theta_{0\alpha}(\mathbf{x}, t)\, d^3x \tag{8.13}$$

The last example points to an especially interesting property of the conserved quantities (8.11). When we studied the translation group in chapter 4, we saw that in quantum mechanics the operators P_μ for the energy and momentum of the system are simply the infinitesimal generators of translations. It can be shown that this is a general result: in quantum mechanics the Q^α are Hermitean operators, representing the infinitesimal generators of the group G in the Hilbert space of state vectors.

(d) *The hypothesis of weak magnetism*

From the foregoing there emerges immediately one possible formulation for the equality between hadronic and leptonic vector couplings. Let us assume that the hadronic vector current involved in the weak interactions is the current associated with a quantity conserved by the strong interactions; then the coupling constant will be unaffected by renormalisations due to the latter, in the same way as the proton charge remains equal to the electron charge in spite of strong-interaction renormalisations. In view of the fact that the current has nonzero matrix elements only between states differing by one unit of charge, the corresponding conserved quantity could be simply the following combination $I^{(+)}$ of the isospin generators:

$$I^{(+)} = I_1 + iI_2 \tag{8.14}$$

In other words, the Hermitean operators I_k are the infinitesimal generators of the group of isospin transformations acting on the physical states. To each of these generators there corresponds a current $j_\alpha^{(k)}(x)$, and the weak hadronic vector current is

$$j_\alpha(x) = j_\alpha^{(1)}(x) + ij_\alpha^{(2)}(x) \tag{8.15}$$

This is called the 'weak magnetism' hypothesis, because it establishes a parallel between weak and electromagnetic currents. The cleanest test of the hypothesis is a comparison between the matrix elements of the weak and electromagnetic currents governing certain nuclear decays

$$A \to B + \gamma$$

$$A' \to B' + e + \nu$$

where the members of each pair of nuclei (A and A') and (B and B') belong to the same isotopic multiplet. Then the weak magnetism hypothesis predicts that these particular elements of the weak and the electromagnetic currents are connected by known isotopic Clebsch–Gordan coefficients (in cases where the isotopic spins of A and B differ by one unit, so that only the part I_3 of the charge $Q = I_3 + \frac{1}{2}$ is effective). It is clear that these matrix elements differ only by the values of the third component of isotopic spin of the nuclear states and of the currents.

9. Current algebra

In contrast to the vector current, the hadronic axial vector current is coupled differently from the leptonic axial vector current by the weak Hamiltonian. This is reflected by the constant $g_A = 1.18$ representing the ratio of axial vector to vector couplings. We shall now see how one can develop a theory for the relation between the currents, from which to predict this number correctly. The theory is called *current algebra* and was put forward by Gell–Mann.

We saw that the hadronic vector current is associated with one of the generators of the isospin group. It was shown that the three generators of isotopic spin transformations, I_1, I_2, and I_3, can be expressed in terms of the hadronic vector currents $j_\alpha^{\ 1}(x)$, $j_\alpha^{\ 2}(x)$, and $j_\alpha^{\ 3}(x)$, by the relation (8.11):

$$I_k(t) = \int j_0^{(k)}(\mathbf{x}, t)\, d^3x \tag{9.1}$$

The operator $I_k(t)$ is actually independent of t because of isospin conservation, or equivalently, because of the fact that the isospin currents obey the conservation law

$$\partial_\mu j_\mu^{(k)}(x) = 0 \tag{9.2}$$

The infinitesimal isospin generators obviously satisfy the commutation rule

$$[I_1\, , I_2] = iI_3 \tag{9.3}$$

and the rules arising from it by cyclic permutation of the indices 1, 2, 3.

In terms of the hadronic axial vector current $\bar{j}_\alpha^{(H)}(x)$ one can define a quantity $\bar{I}(t)$ by analogy with (9.1):

$$\bar{I}(t) = \int \bar{j}_0^{(H)}(\mathbf{x}, t)\, d^3x \tag{9.4}$$

There is no reason whatever to believe that this quantity is independent of time. But at least we can note that, just like the vector current, the axial vector current was inserted into the β decay Hamiltonian, phenomenologically, in such a way that its matrix elements connect only pairs of states differing by one unit of charge. Hence it seems natural to assume that the axial vector as well as the vector current will have simple isotopic transformation properties. In other words we shall take the hadronic axial vector current $\bar{j}_\alpha^{(H)}(x)$ as a combination of two Hermitean axial vector currents $\bar{j}_\alpha^{(1)}(x)$ and $\bar{j}_\alpha^{(2)}(x)$:

$$\bar{j}_\alpha^{(H)}(x) = \bar{j}_\alpha^{(1)}(x) + i\bar{j}_\alpha^{(2)}(x) \tag{9.5}$$

Moreover, we shall assume that the two new currents are members of a triplet of axial vector currents, $(\bar{j}^{(1)}, \bar{j}^{(2)}, \bar{j}^{(3)})$, which behaves like a vector under isotopic spin transformations. The last-mentioned property can be expressed through commutation relations with the generators of such transformations

$$[I_1\, , \bar{j}_\alpha^{(2)}(x)] = i\bar{j}_\alpha^{(3)}(x) \tag{9.6}$$

$$[I_1\, , \bar{j}_\alpha^{(1)}(x)] = 0$$

By analogy with (9.4), let us define the 'axial charges'

$$\bar{I}_k(t) = \int \bar{j}_0^{(k)}(\mathbf{x}, t)\, d^3x \tag{9.7}$$

Then the commutation rules (9.6) imply

$$\begin{aligned} [I_1, \bar{I}_2(t)] &= i\bar{I}_3(t) \\ [I_1, \bar{I}_1(t)] &= 0. \end{aligned} \tag{9.8}$$

The basic principle of current algebra is to assume that the commutators of the axial charges with each other are given by the isospin generators, or in other words that one has

$$[\bar{I}_1(t), \bar{I}_2(t)] = iI_3 \tag{9.9}$$

This assumption originated from the quark model, though obviously the assumption could apply independently of the model. By exploiting (9.9) and by interpreting the Goldberger–Treiman relation (7.6), it becomes possible to express the constant g_A in terms of pion-nucleon scattering cross-sections; the result is called the Adler–Weissberger formula, and is in satisfactory agreement with experiment.

Current algebra represents a completely new kind of structuring of the strong interactions. It makes several predictions which, taken as a whole, agree fairly well with experiment. Just like $SU(3)$, it is a very bold hypothesis, with respectable experimental support, but whose foundations are still, to say the least, poorly understood.

10. Weak interactions and $SU(3)$

(a) *Hyperon β decays*

All the weak interactions we have considered so far involved nothing but leptons, and hadrons with the same strangeness; none of the hadronic currents that were introduced have matrix elements between states differing in strangeness. Evidently our Hamiltonian is still incomplete since it does not allow us to discuss processes as important as Λ and K decay:

$$\Lambda \to N + \pi$$

$$K \to \pi + \pi$$

$$K \to \pi + \pi + \pi$$

In order to discover how our weak Hamiltonian should be generalised, we shall consider, not these nonleptonic decays, but leptonic modes like

$$\Lambda \to p + e + \bar{\nu}_e \tag{10.1}$$

$$\Sigma^+ \to n + e^+ + \nu_e \tag{10.2}$$

$$\Sigma^- \to n + e^- + \bar{\nu}_e \tag{10.3}$$

which have the advantage of being more directly comparable with β decay.

The most striking characteristic of hyperon β-decays is their rarity. Indeed, the decay amplitudes deduced from the mean lives are generally found to be smaller by one order of magnitude than those which involve only nucleons. The reaction (10.2) is rarer still, since no candidate for it has yet been seen.

In view of these differences between weak interactions conserving and violating strangeness, and in view of the fact that strangeness is one of the generators of $SU(3)$, it is natural to enquire into the structure of the interactions responsible for reactions like (10.1–10.3), and also into the connection between weak interactions and $SU(3)$.

(b) *Structure of the hadronic currents*

It seems reasonable to try to retain the structure of the weak Hamiltonian as a product of currents even when extending it to strangeness-nonconserving reactions. If one adopts this procedure, then the problem is to discover the precise structure of the hadronic currents. We have seen that as far as strangeness-conserving reactions are concerned, the hadronic vector current is a combination of the currents associated with the isotopic generators; if current algebra is to be trusted, then the axial vector current has very similar properties at least with respect to isotopic transformations.

This structure is generalised to include the reactions (10.1–10.3) by assuming that the hadronic vector current is a combination of currents which on integration yield generators of $SU(3)$. The generators of $SU(3)$ belong to an octet representation of $SU(3)$, and constitute a set of tensor operators with the same quantum numbers as any other such representation, that for the mesons for instance. Since the leptonic current changes the charge carried by the leptons by one unit, the hadronic current must do likewise. But in the meson octet there are only two independent particles with one and the same nonzero value of the charge, for instance K^+ and π^+; similarly, there are only two currents which increase by one unit the charge of the states on which they act. Under $SU(3)$, one of them transforms like the π^+ and therefore does not change strangeness; this is the hadronic current we have been dealing with hitherto, and which we shall now write as $j^{\Delta S=0}$. The other transforms like K^+ and therefore increases strangeness by one unit; we shall write it as $j^{\Delta S=1}$. The most general combination of the two is a current of the form

$$j_\alpha^{(H)}(x) = \cos\theta_c j_\alpha^{\Delta S=0}(x) + \sin\theta_c j_\alpha^{\Delta S=1}(x) \qquad (10.4)$$

This expression was first proposed by N. Cabibbo. An analogous representation is assumed for the hadronic axial vector current.

The expression (10.4) contains a single parameter θ_c; it can be determined by calculating the β-decay amplitude for any one of the processes (10.1), (10.3), or

$$K \to \mu\nu$$

and comparing them to the amplitudes for neutron β-decay and for

$$\pi \to \mu\nu.$$

Once θ_c has been found from one of these reactions, the probabilities of all the others are predicted by (10.4), and provide a test of the theory. There is good agreement, with θ_c of the order

$$\theta_c \sim 0{,}25 \text{ radians.} \tag{10.5}$$

(c) *Selection rules for decays*

Note that one consequence of (10.4) follows from the fact that the representation $\{8\}$ contains no meson with positive charge and negative strangeness. Similarly, the current (10.4) and its Hermitean conjugate have matrix elements only between pairs of states α and β whose differences in strangeness, $\Delta S = S_\beta - S_\alpha$, and in charge, $\Delta Q = Q_\beta - Q_\alpha$, satisfy the relation

$$\frac{\Delta S}{\Delta Q} = +1. \tag{10.6}$$

Thus, in $\Lambda \to p + e + \nu$, the initial state has $S = -1, Q = 0$, while the hadrons in the final state (just the proton in this case) have $S = 0, Q = 1$; hence

$$\Delta S = 1 \qquad \Delta Q = 1 \qquad \frac{\Delta S}{\Delta Q} = +1.$$

But in the reaction (10.2), the initial Σ^+ has $S = -1, Q = +1$, while the final neutron has $S = 0, Q = 0$, whence

$$\Delta S = 1 \qquad \Delta Q = -1 \qquad \frac{\Delta S}{\Delta Q} = -1.$$

Consequently the current (10.4) has zero matrix element between Σ^+ and neutron, which implies that according to Cabibbo's theory the reaction (10.2) is forbidden; this agrees with the fact that it has not been observed experimentally.

The form (10.4) of the hadronic current has another consequence which deserves mention. Consider for instance the K decay channel

$$K \to \pi + \pi + e + \bar{\nu}_e \tag{10.7}$$

The amplitude for this process must be proportional to the matrix element of the hadronic current

$$\langle \pi\pi \mid j^{\Delta S=1} \mid K \rangle \tag{10.8}$$

But under isotopic spin transformations, $j^{\Delta S=1}$ transforms like a K^+; hence it has isotopic spin $I = \frac{1}{2}$. Therefore the isospin of the initial K in (10.7), and of the final $\pi\pi$ system, must obey the rule

$$|\mathbf{I}_K - \mathbf{I}_{\pi\pi}| = \tfrac{1}{2} \tag{10.9}$$

In other words, since the K isospin is $\frac{1}{2}$, one must have $I_{\pi\pi} = 0$ or 1. The selection rule (10.9), as applied to the leptonic decay of charged particles in general, is expressed in the condensed notation

$$|\Delta I| = \tfrac{1}{2} \tag{10.10}$$

(d) *Universality*

We can now return to the slight difference between the weak coupling constant g_β for β-decay, and the coupling for μ decay. With the value of the Cabibbo angle θ_c quoted earlier, one finds easily that

$$g_\beta = g_\mu \cos\theta_c \tag{10.11}$$

Hence, putting $g_\mu = g$, the weak Hamiltonian can be written in the form

$$\mathrm{H}_W = \frac{g}{\sqrt{2}}[j_\alpha^H + j_\alpha^\ell + \bar{j}_\alpha^H + \bar{j}_\alpha^\ell][j_\alpha^H + j_\alpha^\ell + \bar{j}_\alpha^\ell + \bar{j}_\alpha^\ell] \tag{10.12}$$

where j_α^H is given by (10.4). In this way Cabibbo's theory clarifies the meaning of the universality of weak interactions.

(e) *Current algebra*

Current algebra can be extended immediately to a set of eight vector and eight axial vector currents transforming like members of two octet representation of $SU(3)$. If one defines the 'charges' I_k, $(k = 1, 2, 3, \ldots, 8)$ and the 'axial charges' $\bar{I}_k$ as space integrals of $j_0^{(k)}$ and $\bar{j}_0^{(k)}$, then current algebra implies

$$\begin{aligned}[][I_j, I_k] &= if_{jkl}I_l \\ [I_j, \bar{I}_k] &= if_{jkl}\bar{I}_l \\ [\bar{I}_j, \bar{I}_k] &= if_{jkl}I_l\end{aligned} \tag{10.13}$$

where the coefficients f_{jk} are the structure constants of the $SU(3)$ group as defined in chapter 13.

(f) *Nonleptonic weak interactions*

The designation 'nonleptonic weak interactions' is applied to weak reactions which involve nothing but hadrons, like, for instance,

$$\Lambda \to N + \pi \tag{10.14}$$

$$\Sigma \to N + \pi \tag{10.15}$$

$$K \to \pi + \pi \tag{10.16}$$

$$K \to \pi + \pi + \pi \tag{10.17}$$

The part of the weak Hamiltonian responsible for such processes is

$$\frac{G}{\sqrt{2}}(j_\alpha{}^H + \bar{j}_\alpha{}^H)(j_\alpha{}^H + \bar{j}_\alpha{}^H) \tag{10.18}$$

No-one as yet has found a rigorous method for dealing with this Hamiltonian and for deducing its consequences. The most striking and relevant experimental result is the *validity of the* $\Delta I = \frac{1}{2}$ *rule* to a good approximation.

For instance, in the two-pion decay of the K^+, the $\Delta I = \frac{1}{2}$ rule determines the quantum numbers of the two-pion system. First of all, angular momentum conservation entails that in their centre of mass frame the angular momentum of the pions equals that of the K, and therefore vanishes. In other words they are in an S state, which is even under reflection and therefore even under interchange of the pion coordinates. In view of the Bose statistics obeyed by the pions, their isotopic wavefunction must therefore be even, so that their isotopic spin must be 0 or 2. Since the system has charge, the isotopic spin must be 2. But the $\Delta I = \frac{1}{2}$ rule forbids the change from the isospin $\frac{1}{2}$ of the K to the isospin 2 of the pions, and thereby forbids the reaction. And indeed the two-pion decay probability of the K^+ is much smaller than that of the K^0; this is ascribed to the fact that the former is forbidden by the $\Delta I = \frac{1}{2}$ rule.

The decay (10.15) of the Σ possibly provides a clearer illustration. The Σ decays into a pion-nucleon system which must have isospin $I = \frac{1}{2}$ or $\frac{3}{2}$. If the weak Hamiltonian is an isospin $\frac{1}{2}$ tensor operator in the sense of the Wigner–Eckart theorem, then there are only two independent matrix elements

$$\langle \Sigma, I = 1 \| \mathrm{H}_{\mathrm{weak}}^{i=1/2} \| \pi N, I = \tfrac{1}{2} \text{ or } \tfrac{3}{2} \rangle \tag{10.19}$$

But there are three such decay modes,

$$\begin{aligned} \Sigma^+ &\to n + \pi^+ \\ \Sigma^+ &\to p + \pi^0 \\ \Sigma^- &\to n + \pi^- \end{aligned} \tag{10.20}$$

and experiment indeed confirms that there exists a linear relation between their amplitudes.

So far, no one has found a sound argument deducing the $\Delta I = \frac{1}{2}$ rule for nonleptonic decays from known properties of the currents.

11. K^0 decay

(a) $K_1{}^0$ *and* $K_2{}^0$

As long as one takes into account only the Hamiltonian H of the strong and electromagnetic interactions, which conserve strangeness, the particles K^0 and $\bar{K}^0$ are two different though degenerate eigenstates of H.

But the weak Hamiltonian H_W does not conserve strangeness, and lifts the degeneracy between K^0 and $\bar{K}^0$. However, in view of the fact that the K's become unstable with the introduction of H_W, one cannot really speak of them as eigenstates of the Hamiltonian $H + H_W$. By using the S matrix one could easily define new states forming a natural basis for discussing weak interactions. But to avoid introducing a new formalism, we shall assume that in first approximation the weak interactions conserve the product ΠC of parity and charge conjugation, which is certainly the case for the Hamiltonian we have considered hitherto. Under these conditions the total Hamiltonian commutes with ΠC, and it is convenient to define two new states,

$$\begin{aligned} | K_1^0\rangle &= -\frac{i}{\sqrt{2}}(| K^0\rangle - | \bar{K}^0\rangle) \\ | K_2^0\rangle &= \frac{1}{\sqrt{2}}(| K^0\rangle + | K^0\rangle) \end{aligned} \tag{11.1}$$

From the fact that K^0 is pseudoscalar and from

$$C\,| K^0\rangle = | K^0\rangle \tag{11.2}$$

one sees at once that the states $|K_1{}^0\rangle$ and $|K_2{}^0\rangle$ are eigenstates of ΠC:

$$\begin{aligned} \Pi C\,| K_1^0\rangle &= | K_1^0\rangle \\ \Pi C\,| K_2^0\rangle &= -\,| K_2^0\rangle \end{aligned} \tag{11.3}$$

(b) *Decay into pions*

We known that the K^0 can decay into several pions:

$$\begin{aligned} K^0 &\rightarrow \pi + \pi \\ K^0 &\rightarrow \pi + \pi + \pi \end{aligned} \tag{11.4}$$

and it is interesting to consider the connection between these modes and the states $K_1{}^0$ and $K_2{}^0$.

A two-pion state arising from K^0 decay must have angular momentum 0, and is therefore orbitally symmetric. Further, for both $\pi^+\pi^-$ and $\pi^0\pi^0$ final states, charge conjugation amounts to an intercharge of the two pions, whence the system is in an eigenstate of C belonging to eigenvalue $+1$. On the other hand, two pions in an S state have parity $+1$. Therefore the two pions are in an eigenstate of ΠC belonging to eigenvalue $+1$. It follows that only the $K_1{}^0$ can decay into two pions.

Since phase space favours the two-pion mode much more than the three-pion mode, one expects that the $K_1{}^0$ will have a shorter mean life than the $K_2{}^0$. Their experimental mean lives are

$$\begin{aligned} \tau(K_1^0) &= (0{,}866 \pm 0{,}014) \times 10^{-10}\ \text{sec} \\ \tau(K_2^0) &= (5{,}62 \pm 0{,}68) \times 10^{-8}\ \text{sec.} \end{aligned} \tag{11.5}$$

Consider next the predictions for this case of the $\Delta I = \frac{1}{2}$ rule. Two pions in an S state have isotopic spin 0 or 2. The $\Delta I = \frac{1}{2}$ rule allows only states with $I = 0$ or 1 to be reached from the initial ispin $\frac{1}{2}$ K state, so the final state necessarily has isotopic spin 0. In view of

$$\frac{1}{\sqrt{2}} | \pi^+\pi^- + \pi^-\pi^+\rangle = \sqrt{\tfrac{1}{3}} \, | I = 2\rangle + \sqrt{\tfrac{2}{3}} \, | I = 0\rangle$$
$$| \pi^0\pi^0\rangle = \sqrt{\tfrac{2}{3}} \, | I = 2\rangle - \sqrt{\tfrac{1}{3}} \, | I = 0\rangle \tag{11.6}$$

the $\Delta I = \frac{1}{2}$ rule predicts the $K_1{}^0$ branching ratio

$$\frac{\Gamma(K_1{}^0 \to \pi^+\pi^-)}{\Gamma(K_1{}^0 \to \pi^0\pi^0)} = 2. \tag{11.7}$$

The measured value is 2.2 ± 0.1.

(c) ΠC *violation*

It has been observed recently that the long-lived K^0 can decay into two pions. This means that, contrary to our assumptions, the weak Hamiltonian does not conserve ΠC, and consequently that our understanding of the weak interactions is far from complete.

In the foregoing argument, one must now distinguish between the $K_1{}^0$, which is an eigenstate of ΠC, and that superposition of $K_1{}^0$ and $K_2{}^0$ which has a short mean life, and which we denote by K_S. Similarly, one must distinguish between $K_2{}^0$ and the long-lived combination K_L. Though the states K_S and K_L differ but little from $K_1{}^0$ and $K_2{}^0$ respectively, they are certainly distinct from the latter. ΠC violation is reflected by the nonzero value of the branching ratio

$$\frac{\Gamma(K_L{}^0 \to \pi^+\pi^-)}{\Gamma(K_L{}^0 \to \text{all modes})} \cong 2 \times 10^{-3}. \tag{11.8}$$

Bibliography

There is a certain number of books devoted to a study of elementary particles in general; the following two are particularly recommended, and should make easy reading after the introduction contained in the present volume:

1. G. Källén, Introduction to elementary particle physics. Addison-Wesley.
2. S. Gasiorowicz, Elementary particle physics. Wiley and Sons.

The book by Källen excels in its discussion of experimental facts, and that by Gasirowicz in its theoretical formulation.

Because of its originality, we mention also

3. R. P. Feynman, The theory of fundamental processes. Benjamin.

We recommend that before tackling original papers, one should acquire some familiarity with the above books, where techniques are developed further than they are here. Hence we have refrained from giving direct references to original research work, except in the chronology; and in the following we refer only to books. They include monographs, textbooks, and lecture notes; for easier classification they have been arranged in the order in which their subject matter is discussed in the chapters of the present volume.

Chapter 1

A clear and concise account of accelerators is given in

4. M. S. Livington, High energy accelerators. Interscience tracts on physics and astronomy.

A more complete discussion, from both the physical and the engineering points of view, is

5. M. S. Livington and J. P. Blewett, Particle accelerators. McGraw-Hill.

As an introduction to the problems involved in exploiting experimental data, see for instance

6. A. H. Rosenfield and W. E. Hunfray, Analysis of bubble chamber data. In Annual review of nuclear science (1963).

Chapter 3

There are several texts on group theory applied to physics, amongst them the classic by

7. E. P. Wigner, Group theory and its application to atomic spectra.

The following book is easier to read and includes several results, many due to Wigner himself, which are not in reference (7):

8. M. Hammermesh, Group theory.

The next reference is an introduction to continuous groups which is well adapted to physics:

9. G. Racah, Lectures on Lie groups. In 'Group theoretical concepts and methods in elementary particle physics', Istanbul summer school of theoretical physics (1962). Gordon and Breach.

The connections between the foundations of quantum mechanics and the mathematical aspects of group theory are given in (7), (8), and

10. E. P. Wigner. The role of invariance principles in natural philosophy. Varenna summer school (1964). Academic Press.
11. A. S. Wightman, L'invariance dans la mechanique quantique relativiste, Les Houches summer school (1966). Hermann.

Chapter 4

Representations of the Poincaré group are discussed in reference (11) and in
12. E. P. Wigner, Unitary representations of the inhomogeneous Lorentz group including reflections, Istanbul summer school, loc. cit.
A clear and detailed discussion is given in
13. R. Hagedorn, Relativistic kinematics of scattering with spin. Benjamin.
13a. F. R. Halpern, Special relativity and quantum mechanics. Prentice-Hall (1968).

Chapter 7

There exist two very complete treatises on scattering:
14. M. L. Goldberger and E. M. Watson, Collision theory. Wiley.
15. R. G. Newton, Scattering theory of waves and particles. McGraw-Hill.
The second is confined to the non-relativistic case.
It is also profitable to consult
16. R. Haag, Quantum theory of collision processes. In 'Lectures in theoretical physics', 1960, Boulder, Colorado. University of Colorado Press.
17. R. Haag, Particles and cross-sections. In 'Proceedings of the Pacific international summer school (1965).' Gordon and Breach.
Reference (17) investigates closely the connection between experiments and the mathematical abstractions from them. See also reference (13).
The problems arising from spin are discussed in detail in (14) and in
18. J. D. Jackson, Helicity. In Les Houches (1965). Gordon and Breach.

Chapter 8

There is a magisterial treatment of the analyticity properties of non-relativistic scattering amplitudes in
19. V. de Alfaro and T. Regge, Potential scattering. North Holland.
Reference (15) is also highly recommended.

Chapter 9

The phenomenology of particles is analysed every year at many summer schools and conferences, the most recent being evidently the most useful. Reference (1) is an excellent general introduction. See also the contribution of M. Jacob to
20. M. Jacob and G. F. Chew, Strong interaction physics. Benjamin.
On one particular topic, see the small elementary book
21. R. Wilson, Nucleon-nucleon interaction, Interscience tracts on physics and astrophysics.
The Lawrence Radiation Laboratory of the University of California at Berkeley publishes every year tables of particle and resonance properties, known as the Rosenfeld tables.

Chapter 10

The problems of relativistic kinematics are discussed in references (11) and (12), and in
22. J. D. Bjorken and S. D. Drell, Relativistic quantum mechanics, McGraw-Hill.
Consult also

23. D. W. Williams, Dirac algebra for any spin. In 'Lectures in theoretical physics', Boulder, Colorado (1964). University of Colorado Press.
24. W. C. Parke and H. Fehle, Covariant formulation of relativistic wave equations under the homogeneous Lorentz group. Same reference as in (23).

For the problems attending zero-mass particles consult

25. S. Weinberg, Quantum theory of massless particles. Brandeis University summer school (1964). Gordon and Breach.

For treatments of the helicity formalism, recall reference (18) and the very clear article

26. M. Jacob, Le formalisme de l'hélicité. Services de publications du Commissariat à l'Energie Atomique de Saclay.

Chapter 11

The most illuminating introduction to electrodynamics is the small book

27. R. P. Feynman, Quantum electrodynamics. Benjamin.

A more conventional work, very well set out, is

28. J. D. Bjorken and S. D. Drell, Relativistic quantum fields. McGraw-Hill.

The following is very thoroughgoing but harder to read:

29. G. Källén, Quanten-Elektrodynamik. In Handbuch der Physik.

We list also some older books:

30. J. M. Jauch and F. Rohrlich, The theory of photons and electrons. Addison-Wesley.
31. Akhiezer and Berestetsky, Quantum Electrodynamics. (In Russian, or translated into English.) Gasteditcke, Moscow.

This last book contains many useful explicit formulae.

The theoretical problems are treated at great depth by

32. N. N. Bogoliubov and V. Shirkov, Introduction to the theory of quantised fields. (In Russian, English, or French.)

To some of these rather weighty tomes one might prefer a less complete discussion such as the very clear one by G. Källén:

33. G. Källén, Topics in quantum electrodynamics. Brandeis University summer school (1963). Gordon and Breach. On the calculation of the Lamb shift and on infrared divergences, we recommend
34. D. R. Yennie, Topics in quantum electrodynamics. Same reference as in (33).

Chapter 12

The most reliable part of what is discussed in this chapter are the dispersion relations. How they are used is explained in reference (14) and in the lectures by

35. M. L. Goldberger, Introduction to the theory and application of dispersion relations. Les Houches summer school (1960). Hermann.

Dispersion relations have been applied to pion-nucleon scattering, to photo-production, to nucleon-nucleon scattering, and to form factors. See references (14), (35), (20), (24), and

36. D. Y. Wong, Dispersion relations and their applications. Proceedings of the 1964 Varenna summer school. Academic Press.
37. C. Lovelace: Rapporteur's talk in the 'Proceedings of the High Energy Physics Conference, Heidelberg 1967'.

There is a clear explanation of the techniques useful for this kind of analysis in

38. G. Barton, Introduction to dispersion techniques in field theory. Benjamin.

S-matrix theory is discussed in several books (see also reference (20)):

39. R. J. Eden, P. V. Landshoff, D. I. Olive and J. C. Polkinghorne, The analytic *S* matrix. Cambridge 1966.
40. G. F. Chew, *S*-matrix theory of strong interactions. Benjamin.

On this topic, consult also

41. F. E. Low, Report to the international conference on high energy physics, Sienna (1963).

There exist also other models of strong interactions. There is an illuminating discussion of the optical model in

42. R. J. Glauber, High energy collision theory. In 'Lectures in theoretical physics, Boulder (1958)'. The University of Colorado Press.

The peripheral model, with and without absorption corrections, is discussed in

43. J. D. Jackson, The peripheral model. Les Houches summer school (1965). Gordon and Breach.

An explanation of what is sometimes called the 'bootstrap philosophy' is given in

44. F. Zachariasen, Lectures on bootstrap. In 'Recent developments in particle physics'. Pacific international summer school (1965). Gordon and Breach.

Interesting applications to the proton-neutron mass difference and to $SU(3)$ symmetry breaking in the same volume:

45. S. Frautschi, The bootstrap theory of symmetry breaking.
45a. A. C. Hearn and S. D. Drell, Peripheral processes, in High energy physics, ed. E. H. S. Burhop, Academic Press. Volume 2, 1968.

Part II

By now there exists a vast literature on axiomatic field theory. Let us mention

46. R. Jost, Introduction to axiomatic field theory.
47. R. F. Streater and A. S. Wightman, PCT, spin and statistics, and all that. Benjamin.

The Lehmann–Symanzik–Zimmerman formalism and the reduction formulae are introduced very clearly at a mathematically less ambitious level in the small book

48. G. Barton, Introduction to advanced field theory. Interscience tracts in physics and astronomy.

In the same spirit one can consult

49. R. Hagedorn, Introduction to field theory and dispersion relations. Pergamon Press.

Proofs of dispersion relations are given in (32), (49), and in

50. R. Omnès, Démonstration des relations de dispersion. Les Houches (1960). Hermann.
51. M. Froissart, The proof of dispersion relations. Varenna (1964). Academic Press.

Axiomatic field theory is applied to proving the ΠCT theorem and the connection between spin and statistics in (46) and (47). A clear discussion of the representation theory of the canonical commutation relations between operators is given in

52. R. Haag, Canonical commutation relations in field theory and functional integration. Boulder (1966). University of Colorado Press.

Chapter 13

Amongst the books on $SU(3)$ we mention

53. M. Gell-Mann and Y. Ne'eman, The eightfold way. Benjamin. This is a pioneering book but difficult to read.
54. P. Carruthers, Unitary symmetry, Interscience. Clear and elementary.
55. M. Gourdin, Unitary symmetries, North Holland. Clear and formal, at a higher mathematical level than (54).
55a. P. T. Matthews, Unitary Symmetry, in High energy physics, ed. E. H. S. Burhop, Academic Press (1967); Volume 1.
55b. B. Sakita, Higher Symmetries of Hadrons, in Advances in particle physics, Interscience; Volume 1 (1968).

The quark model is discussed in the lectures by

56. B. W. Lee, Brandeis summer school (1965). Gordon and Breach.

A more critical discussion is

58. F. Gürsey, Les Houches summer school (1965). Gordon and Breach.

Chapter 14

Amongst books devoted to Regge poles, we mention reference (15) and

59. R. Omnès and M. Froissart, Mandelstam theory and Regge poles. Benjamin.

60. E. J. Squires, Complex angular momenta and particle physics. Benjamin.
61. S. C. Frautschi, Regge poles and S-matrix theory. Benjamin.
62. R. G. Newton, The complex j-plane. Benjamin.
Reference (61) is especially useful.
Applications to high energy physics are to be found in the following reports:
63. L. Van Hove, Report to the International conference on high energy physics, Berkeley (1966).
64. R. J. Eden, High energy collisions of elementary particles. Cambridge University Press (1967).
64a. P. D. B. Collins and E. J. Squires, Regge poles in particle physics. Springer (1968).

Chapter 15

We quote some lecture courses and books on weak interactions and recall that references (1) and (2) are excellent on this subject.

65. V. Telegdi, in 'Weak interactions, Bergen summer school (1963)'. Benjamin.
This studies weak magnetism and compares it with experiment.
66. V. L. Fitch, same reference. ΠC violation.
67. C. S. Wu, Beta decay. Varenna (1966). Academic Press.
68. L. M. Lederman, same reference. Experimental muon physics.
69. S. Treiman, in references (65) and (50), studies the structure of weak interactions.
70. N. Cabibbo, Varenna (1966). Academic Press. Discusses the connection between weak interactions and $SU(3)$.
71. H. J. Lipkin, Beta decay for pedestrians. North Holland. Very elementary.
72. T. D. Lee and C. S. Wu, Annual Reviews of Nuclear Science (1965 and 1966).
A very illuminating discussion.
73. H. Primakoff, Varenna (1966). Academic Press. Theory of μ decay.
74. J. S. Bell, Les Houches (1965). Gordon and Breach. Clear and elegant; includes ΠC violation.
75. N. Cabibbo, Brandeis (1965). Equally lucid.
75a. A very complete reference is: R. E. Marshak, Riazuddin, and C. P. Ryan, Theory of weak interactions in particle physics. Wiley-Interscience (1969). On current algebra, consult the lectures by
76. M. Gell-Mann and N. Cabibbo, 'Ettore Majorana' summer school (1966). Academic Press.
Alternatively,
77. S. L. Adler and R. F. Dashen, Current algebra and applications to particle physics. Benjamin. A clear discussion accompanied by reprints.
77a. B. Renner, Current algebras and their application. Pergamon (1968).
On high energy neutrino physics consult the lectures by
78. T. D. Lee, M. Schwartz, and L. M. Ledermann, Varenna (1966). Academic Press.

Recent developments:

For periodically bringing oneself up to date the best sources of references are the Proceedings of the International Conferences on High Energy Physics.

Chronology

We quote some dates in order to provide a better overall view of the evolution of elementary particle physics. It is to some extent arbitrary which events one chooses to mark turning points; and we have concentrated on discoveries, to the exclusion of the studies in depth which have also played a large part in the progress of physics. The following table must not be considered as a complete list of major researches, and is intended simply as a modest aid to orientation.

1911: Development of the ionisation chamber, (the so-called Wilson Cloud Chamber) by C. T. R. Wilson. Discovery of cosmic rays by V. F. Hess. For a long time to come these rays will provide the only high energy particles obtainable.

1919: Invention of the first cyclotron by Lawrence. The first machine is 10 cm across.

1922: A. Compton observes the Compton effect (scattering of X rays by electrons).

1927: Dirac quantises the electromagnetic field, considered as a superposition of oscillators. In this way he explains the emission and absorption of light by atoms(1).

1928: Quantisation of the electron field by Jordan and Wigner(2).

1930: Fermi shows the difficulty in quantising the Coulomb field(3). Oppenheimer proves that quantum electrodynamics implies an infinite shift of spectral lines. This is the first appearance of infinities in electrodynamics(4).

1932: Chadwick discovers the neutron, which marks the beginnings of modern nuclear physics(5).
Dirac, Fock, and Podolsky quantise the electromagnetic field including the Coulomb field(6). This shows that the Coulomb force is due to the exchange of photons between charged particles. The Wilson cloud chamber is much improved (automation, photographic recording) by P. M. S. Blackett. This apparatus will lead to the discovery of the positron, and to the study of cosmic ray showers produced by the collision of primary cosmic ray protons with nuclei in the upper atmosphere.

1933: Discovery of the positron in cosmic rays by C. D. Anderson(7).

1934: Heisenberg investigates the fluctuations occurring when one measures electric or magnetic field in vacuo(8).
Dirac, investigating the existence of fluctuations when measuring the electron field, introduces the idea of virtual pairs and of vacuum polarisation(9).

1935: By analogy with electrodynamics, H. Yukawa suggests that the nuclear binding forces are due to the exchange of an unknown particle, the meson(10). The meson must exist in two charge states, corresponding to $\pi^{\pm}$.

1936: Weisskopf proves that vacuum polarisation is infinite(11). The first results in nuclear physics lead Breit, Gordon, and Present to assume that nuclear forces are charge-independent(12). Cassen and Condon immediately translate this assumption into the isotopic spin formalism just for nucleons(13). The formalism had already been used by Heisenberg without reference to charge independence.

1938: By analogy with classical electron theory, Kramers proposes the programme for renormalising the electron mass(14). The charge-independence of nuclear forces and Yukawa's theory lead Froelich, Heitler, and Kemmer to predict the existence of π^0. This is already the complete isospin formalism(15).

1939: Wigner introduces time reversal[16].
Weisskopf notes the divergence of wave renormalisation in electrodynamics[17].
Between 1939 and 1945 Powell develops the technique for detecting particles by means of photographic emulsions. This will allow a more far-reaching investigation of cosmic rays than the Wilson cloud chamber, and will distinguish the muon from the pion in showers.

1940: Wigner studies the unitary representations of the Poincaré group, and thereby lays the foundations of quantum kinematics[18].
Pauli proves the connection between spin and statistics in field theory[19].

1941: Lawrence's new cyclotron at Berkeley begins to produce particles.

1943: Heisenberg defines the S-matrix[20].

1944: Leprince–Ringuet and Lhéritier find indications in cosmic rays of the existence of particles with mass $>1000m_e$[21].

1946: Tomonaga proposes a formalism for the covariant quantisation of electrodynamics[22].

1947: The techniques of high-frequency radio waves developed during the war allow the utilisation of resonance methods. By this means Lamb and Retherford measure the energy difference between the $2s_{1/2}$ and $2p_{1/2}$ levels in hydrogen[23].
Bethe shows that the Lamb shift results from quantum electrodynamics and correctly calculates its order of magnitude[24]. To go further one needs a covariant formalism which can accommodate the renormalisation programme as a practical proposition. The discussion of the new methods proposed by Feynman and by Schwinger continues in the following years.
Lattes, Occhialini, and Powel discover the pion in cosmic rays[25].
Rochester and Butler discover 'V'-shaped tracks in their emulsions, the first indication of new particles which, unlike the pion, were not expected[26]. They are dubbed 'curious' or 'strange'.

1948: Foley and Kusch measure the anomalous magnetic moment of the electron[27]. It is calculated by Schwinger, who also publishes the principles of his covariant method for electrodynamics[28].

1949: Feynman publishes his theory of electrodynamics which leads to Feynman diagrams[29]. Dyson shows the equivalence of Feynman's and Schwinger's methods[30]. He proves, further, that in principle these methods enable one to implement the renormalisation programme, and to isolate the divergences to all orders of perturbation theory[31]. The Lamb shift is calculated by Kroll and Lamb[32]. Wigner stresses the idea of baryon conservation[33]. Discovery of the τ meson (three-pion decay mode of the K) in cosmic rays[34].

1951: First studies of pion-nucleon scattering[35] and of $\pi^{\pm}$-d reactions[36]. Discovery of the Λ^0, and θ^0 (two-pion decay mode of the K)[37]. As time goes on the difference between the measured θ^0 and τ masses becomes smaller and smaller, and will lead to the proposal that parity is not conserved.

1952: Detailed calculation of the Lamb shift by Karplus and Klein[38].
Marshak applies perturbation theory to the strong interactions[39]. Comparison with experiment reveals the problems inherent in strong interactions.

1953: Determination of the π^+ spin using a principle due to Marshak and Chaston[40]. First laboratory production of strong particles at the Brookhaven Cosmotron. Discovery of the $\Sigma^{\pm}$ [41], and of the $K^{\pm}$ [42] in the θ mode.
Hofstadter investigates electron scattering from protons and measures the proton form factors, i.e. the charge distribution.

1954: K decay mode $K \to \mu + \nu$[44].
Lüders gives the first proof of the PCT theorem, soon improved by Pauli[45].

1955: The π^- parity is established through the low rate of the reaction $\pi^- d \to n + n + \pi^0$ [46]. The underlying principle was suggested in 1946 by Ferretti.
Lehmann, Symanzik, and Zimmermann propose the first version of axiomatic field theory[47].
Conservation of lepton number[48].
Discovery and observation of the antiproton at the Berkeley Bevatron which can accelerate protons to over 7 GeV[49].

First applications of dispersion relations, mainly by Goldberger and Gell-Mann. Attempt to link them with causality.

In the hands of Chew and Low, the static model for the pion-nucleon interaction reveals that it is possible to build a dynamical theory on analyticity, unitarity, and crossing(50).

1956: Dispersion relations are applied to pion-nucleon scattering and to photo-production, and explain many observed effects(51).

Bogoliubov gives the first proof of a dispersion relation for arbitrary momentum transfer, based on axiomatic field theory(51). (A proof for $\Delta^2 = 0$ had already been suggested by Symanzik(52)).

Lee and Yang suggest that the equality of the θ and τ masses is explained by the parity non-conserving decay of one and the same particle, the K^0. They propose an experiment to test this hypothesis(53).

The neutrino is detected experimentally for the first time, exploiting the intense neutrino flux from a nuclear reactor(54).

1957: Following the suggestion of Lee and Yang, Mrs. Wu investigates the decay of Co^{60}, and finds evidence for parity non-conservation in weak interactions(55). It is discovered simultaneously in μ decay by Garwin, Lederman, and Weinrich(56).

Exploiting an old remark by Herman Weyl, Salam, Landau, and Lee and Yang independently propose the two-component theory of the neutrino(57).

Jost proves the *PCT* theorem from axiomatic field theory(58).

Nambu suggests the existence of boson resonances to explain the structure of nucleon form-factors observed by Hofstadter(59).

1958: Mandelstam puts forward the two-variable dispersion relations from which analytic *S*-matrix theory is to develop(60).

The form of the weak-interaction Hamiltonian is established in more detail. Its structure as a product of currents becomes clear. Feynman and Gell-Mann put forward the hypothesis of 'weak magnetism', that the weak-interaction vector current belongs to the same isotopic multiplet as the electromagnetic current(61).

The bubble chamber, whose principles had been put forward by Glaser, becomes operational.

1959: The π^0 parity is confirmed(62).

Parity of the K(63).

Regge discovers the poles in angular momentum of the non-relativistic scattering amplitude(64).

1960: Frazer and Fulco apply dispersion-relation methods to the form factors, and once again predict boson resonances(65).

Fadeev solves the three-body scattering problem in quantum mechanics(66).

1961: The idea of Regge poles is applied to the classification of particles and to high energy scattering(67,68).

Discovery of the ρ, ω, η, K^*, Y^{1*} (69).

Measurement of the K^0 mass difference in an experiment whose basic principle had been suggested by Pais and Piccioni(70).

Gell-Mann and Ne'eman suggest that strong interactions are partially invariant under the group $SU(3)$(71).

The first spark chambers become operational.

1962: Measurement of the gyromagnetic ratio of the μ(72).

First analysis of the spin and parity of a Y^* resonance(73).

Discovery of the Ξ^* (74), f^0 (75), and of Y^*'s(76).

Evidence for two neutrinos: separate conservation of electron and muon numbers(77).

1963: Discovery of the φ(78) and the B(79).

Application of $SU(3)$ to the leptonic weak interactions(80).

1964: Gell-Mann puts forward the algebra of currents, connecting the vector and axial vector currents of weak interactions(81).

Discovery of *PC* violation by two-pion decay of the long-lived K_0(82).

Discovery of the Ω^- whose existence and mass had been predicted from $SU(3)$(83).

The quark model is formulated by Gell-Mann and Zweig(84).

1965: Substantial extension of the proof of dispersion relations by Martin[85].

Adler and Weisberger calculate the axial vector coupling constant from current algebra[86].

1966: Indications of the existence of Regge trajectories, in the pion-nucleon system, rising to high spins.

Applications of current algebra to weak and to strong interactions involving the emission of slow pions.

The number of boson and baryon resonances becomes very considerable.

1968: Measurement of the reaction $e^+e^- \rightarrow \rho$.

References

(1) P. A. M. Dirac, Proc. Roy. Soc. *A114*, 243 (1927).
(2) P. Jordan and E. Wigner, Zeit. für Physik *47*, 631 (1928).
(3) E. Fermi, Atti della Reale Accademia Nazionale dei Lincei *12*, 431 (1930).
(4) J. R. Oppenheimer, Phys. Rev. *35*, 461 (1930).
(5) J. Chadwick, Proc. Roy. Soc. *A136*, 692 (1932).
(6) P. A. M. Dirac, V. A. Fock and B. Podolsky, Physikalische Zeitschrift der Sovjetunion 8, Band 2, Heft 6 (1932).
(7) C. D. Anderson, Phys. Rev. *43*, 491 (1933).
(8) W. Heisenberg, Sachsische Akademie der Wissenschaften *86*, 317 (1934).
(9) P. A. M. Dirac, Rapport du 7ème Conseil Solvay de Physique, Structure et Propriétés des Noyaux Atomiques, page 203 (1934).
(10) H. Yukawa, Proc. Phys. Math. Soc. of Japan *17*, 48 (1935).
(11) V. S. Weisskopf, Kondelige Dan. Viden. Selsk, Mat-Fys. Med XIV, N^0 6 (1936).
(12) G. Breit, E. V. Condon and R. D. Present, Phys. Rev. *50*, 825 (1936).
(13) B. Cassen, E. V. Condon, Phys. Rev. *50*, 846 (1936).
(14) H. A. Kramers, Quantentheorie der Elektron und der Strahlung, Leipzig 1938.
(15) H. Fröhlich, W. Heitler and N. Kemmer, Proc. Roy. Soc. *A166*, 154 (1938).
(16) E. P. Wigner, Ann. Math. *40*, 149 (1939).
(17) V. S. Weisskopf, Phys. Rev. *56*, 72 (1939).
(18) E. P. Wigner, Annals of math. *40*, 149 (1939).
(19) W. Pauli, Phys. Rev. *58*, 716 (1940).
(20) W. Heisenberg, Z. Physik, 120, 513, 673 (1943).
(21) L. Leprince-Ringuet and M. Lhéritier, Comptes Rendus Acad. Sciences *219*, 618 (1944).
(22) S. Tomonaga, Progr. Theor. Phys. *1*, 27 (1946).
(23) W. E. Lamb Jr. and R. C. Retherford, Phys. Rev. *72*, 241 (1947).
(24) H. A. Bethe, Phys. Rev. *72*, 339 (1947).
(25) C. M. G. Lattes, G. P. S. Occhialini and C. F. Powell, Nature *160*, 453 (1947).
(26) G. D. Rochester and C. L. Butler, Nature *160*, 855 (1947).
(27) H. M. Foley and P. Kusch, Phys. Rev. *73*, 412 (1948).
(28) J. Schwinger, Phys. Rev. *73*, 416 (1968), Phys. Rev. *74*, 1439 (1948).
(29) R. P. Feynman, Phys. Rev. *76*, 749 (1949).
(30) F. J. Dyson, Phys. Rev. *75*, 486 (1949).
(31) F. J. Dyson, Phys. Rev. *75*, 1436 (1949).
(32) N. M. Kroll and W. E. Lamb Jr. Phys. Rev. *75*, 388 (1949).
(33) E. P. Wigner, Proc. Amer. Phil. Soc. *93*, 521 (1949).
(34) R. H. Brown, U. Camerini, P. H. Fowler, H. Muirhead, C. F. Powell, D. M. Ritson, Nature *163*, 82 (1949).
(35) C. Chedester, P. Isaacs, A. Sachs, J. Steinberger, Phys. Rev. *82*, 958 (1951), H. L. Anderson, E. Fermi, E. A. Long, R. Martin, D. E. Nagle, Phys. Rev. *85*, 934 (1952).
(36) R. Durbin, H. Loar, J. Steinberger, Phys. Rev. *83*, 646 (1951) $\pi^+ d \to pp$,
D. L. Clark, A. Roberts, R. Wilson, Phys. Rev. *83*, 649 (1951) $\pi^+ d \to pp$.
W. K. H. Panofsky, R. L. Aamodt, J. Hadley, Phys. Rev. *81*, 565 (1951) $\pi^+ d \to 2n$, $2n + \gamma$.

(37) R. Armenteros, K. H. Barker, C. C. Butler, A. Cachan, A. H. Chapman, Nature *167*, 501 (1951).
(38) R. Karplus and A. Klein, Phys. Rev. *87*, 848 (1952).
(39) R. E. Marshak, *Meson Physics*, McGraw Hill, New York 1952.
(40) W. F. Cartwright, C. Richman, M. N. Whitehead, H. A. Wilcox, Phys. Rev. *91*, 677 (1958) $pp \to \pi$–d.
(41) A. Bonetti, R. Levi-Setti, M. Panetti, Nuovo Cimento *10*, 1736 (1953).
(42) R. W. Thomson, A. V. Duskirk, L. R. Etter, C. J. Karzmark, R. H. Rediker, Phys. Rev. *91*, 445A (1953).
(43) M. Gell-Mann, Phys. Rev. *92*, 833 (1953).
T. Nakato, K. Nishijima, Progr. Theor. Phys. *10*, 581 (1953).
K. Nishijima, Progr. Theor. Phys. *12*, 107 (1954), *13*, 285 (1955).
(44) B. P. Gregory, A. Lagarrigue, L. Leprince-Ringuet, F. Müller, C. Peyrou, Nuovo Cimento *11*, 292 (1954).
(45) G. Lüders, Mat. Fys. Medd. Kongl. Dan. Vid. Selsk. *28*, N^0 5 (1954).
W. Pauli in '*Niels Bohr and the development of Physics*' Pergamon (1955).
(46) W. Chinowsky and J. Steinberger, Phys. Rev. *100*, 1476 (1955).
(47) H. Lehmann, K. Symanzik, W. Zimmermann, Nuovo Cimento *1*, 205 (1955).
(48) R. Davis, Phys. Rev. *97*, 766 (1955).
(49) O. Chamberlain, E. G. Segré, E. C. Wiegand, T. Ypsilantis, Phys. Rev. *100*, 947 (1955).
(50) G. F. Chew, F. E. Low, Phys. Rev. *101*, 1570 (1956).
(51) G. F. Chew, A. L. Goldberger, F. E. Low and Y. Nambu, Phys. Rev.
(52) K. Symanzik. Seattle Conference (1956).
(53) T. D. Lee and C. N. Yang, Phys. Rev. *104*, 254 (1956).
(54) C. L. Cowan, F. Reines, F. B. Harrison, H. W. Kruse, A. D. McGuire, Science *124*, 103 (1956).
(55) C. S. Wu, E. Ambler, R. W. Hayward, D. D. Hoppes, R. P. Hudson, Phys. Rev. *105*, 1413 (1957).
(56) R. L. Garwin, L. M. Lederman, M. Weinrich, Phys. Rev. *105*, 1415 (1957).
(57) A. Salam, Nuovo Cimento *5*, 299 (1957).
L. O. Landau, Nucl. Phys. *3*, 127 (1957).
T. D. Lee and C. N. Yang, Phys. Rev. *105*, 167 (1957).
(58) R. Jost, Helv. Phys. Acta *30*, 409 (1957).
(59) Y. Nambu, Phys. Rev. *106*, 1366 (1957).
(60) S. Mandelstam, Phys. Rev. *112*, 1344 (1958).
(61) R. P. Feynman, M. Gell-Mann, Phys. Rev. *109*, 193 (1958).
J. J. Sakurai, Nuovo Cimento *7*, 649 (1958).
E. C. G. Sudarshan, R. E. Marshak, Phys. Rev. *109*, 1860 (1953).
(62) R. Plano, A. Prodell, N. Samios, M. Schwartz, J. Steinberger, Phys. Rev. Letters, *3*, 525 (1959).
C. N. Yang, Phys. Rev. *77*, 242 (1950).
N. Kroll, W. Wada, Phys. Rev. 1355 (1955).
(63) M. M. Block, E. D. Brückner, I. S. Hughes, T. Kikuchi, C. Meltzer, F. Anderson, A. Pevsner, E. M. Harth, J. Leitner, H. O. Cohn, Phys. Rev. Letters *3*, 291 (1959).
(64) T. Regge, Nuovo Cimento, *14*, 951 (1959).
(65) W. R. Frazer and J. R. Fulco, Phys. Rev. *117*, 1609 (1960).
(66) L. D. Faddeev, Zh. Eksptl, i. Teoret. Fiz. *39*, 1659 (1960).
(67) G. F. Chew and S. C. Frautschi, Phys. Rev. Letters *7*, 399 (1961).
(68) S. C. Frautschi, M. Gell-Mann and F. Zachariasen, Phys. Rev. *126*, 2, 204 (1962).
(69) A. R. Erwin, R. March, W. D. Walker, E. West, Phys. Rev. Letters *6*, 628 (1961).
E. Pickup, D. K. Robinson, E. D. Salant, Phys. Rev. Letters *7*, 192 (1961).
B. C. Maglic, L. W. Alvarez, A. H. Rosenfeld, M. L. Stevenson, Phys. Rev. Letters *7*, 178 (1961).
N. H. Xuong, G. R. Lynch, Phys. Rev. Letters *7*, 327 (1961).
A. Pevsner, R. Kraemer, M. Nussbaum, C. Richardson, P. Schlein, R. Strand, P. Toohig, M. Block, A. Engler, R. Gessaroli, C. Meltzer, Phys. Rev. Letters *7*, 421 (1961).
M. H. Alston, L. W. Alvarez, P. Eberhard, M. L. Good, W. Graziano, H. K. Ticho, S. G. Wojcicki, Phys. Rev. Letters *6*, 300 (1961) (Y_1*).

(70) A. Païs, O. Piccioni, Phys. Rev. *100*, 1487 (1955).
R. H. Good, R. P. Matsen, F. Müller, O. Piccioni, W. M. Powde, H. S. White, W. B. Fowler, R. W. Birge, Phys. Rev. *124*, 1223 (1961).
(71) M. Gell-Mann, Cal. Tech. Report, CTSL 20 (1961) (unpublished).
Y. Ne'eman, Nucl. Phys. *26*, 222 (1961).
(72) G. Charpak, F. J. M. Farley, R. L. Garwin, T. Muller, J. C. Sens, A. Zichichi, Phys. Letters *1*, 16 (1962).
(73) M. Ferro-Luzzi, R. D. Tripp, M. B. Watson, Phys. Rev. Letters *8*, 28 (1962).
(74) G. M. Pjerrou, D. J. Prowse, P. Schlein, W. E. Slater, D. H. Stork, H. K. Ticho, Phys. Rev. Letters *9*, 114 (1962).
L. Bertanza, V. Brisson, P. L. Connolly, E. L. Hart, I. S. Mitra, G. S. Moneti, R. R. Rau, N. P. Samios, I. O. Skillicorn, S. S. Yamamoto, M. Goldberg, L. Gray, J. Leitner, S. Lichtman, J. Westgard, Phys. Rev. Letters *9*, 180 (1962).
(75) W. Selove, V. Hagopian, H. Brody, A. Baker, E. Leboy, Phys. Rev. Letters *9*, 272 (1962).
(76) M. H. Alston, L. W. Alvarez, M. Ferro-Luzzi, A. H. Rosenfeld, H. K. Ticho, S. G. Wojcicki, Proc. 1962, Ann. Int. Conf. High-Energy Physics at CERN p. 311, CERN Scientific Information Service (1962).
(77) G. Danby, J. M. Gaillard, K. Goulianos, L. M. Lederman, N. Mistry, M. Schwartz, J. Steinberger, Phys. Rev. Letters *9*, 36 (1962).
(78) P. Schlein, W. E. Slater, L. T. Smith, D. H. Stork, H. K. Ticho, Phys. Rev. Letters *10*, 368 (1963).
(79) M. Abolins, R. L. Lander, W. A. W. Mehlhop, N. Xuong, P. M. Yager, Phys. Rev. Letters (1963).
(80) M. Gell-Mann, Physics *1*, 63 (1964).
N. Cabbibo, Phys. Rev. Letters *10*, 531 (1963).
(81) M. Gell-Mann, Physics *1*, 63 (1964).
(82) J. H. Christenson, J. W. Cronin, V. L. Fitch, R. Turlay, Phys. Rev. Letters *13*, 138 (1964).
(83) V. E. Barnes, P. L. Connolly, D. J. Crennell, B. B. Culwick, W. C. Delanay, W. B. Fowler, P. E. Hagerty, E. L. Hart, N. Horwitz, P. V. C. Hough, J. E. Jensen, J. K. Kopp, K. W. Lai, J. Leitner, J. L. Loyd, G. W. London, T. W. Morris, Y. Oren, R. B. Palmer, A. G. Prodell, D. Radojicic, D. C. Rahm, C. R. Richardson, N. P. Samios, J. R. Sanford, R. P. Schutt, J. R. Smith, D. L. Stonehill, R. C. Strand, A. M. Thorndike, M. S. Webster, W. J. Willis, S. S. Yamamoto, Phys. Rev. Letters *12*, 204 (1964).
(84) M. Gell-Mann, Phys. Letters *8*, 214 (1964).
G. Zweig, CERN, preprint Th. 412 (1964).
(85) A. Martin, Nuovo Cimento *42*, 930 (1966).
(86) Adler, Phys. Rev. Letters *14*, 1043 (1965).
Weisberger, Phys. Rev. Letters *14*, 1047 (1965).

INDEX